Barron's Review Course Series

Let's Review:

Sequential Mathematics, Course I

Second Edition

Lawrence S. Leff
Assistant Principal, Mathematics Supervision
Franklin D. Roosevelt High School
Brooklyn, New York

BARRON'S

To Rhona
For the understanding,
for the love,
and with love.

All inquiries should be addressed to:
Barron's Educational Series, Inc.
250 Wireless Boulevard
Hauppauge, New York 11788

Library of Congress Catalog Card No. 95-2341

International Standard Book No. 0-8120-9036-5

Library of Congress Cataloging-in-Publication Data

Leff, Lawrence S.
    Let's review : sequential mathematics, course I / Lawrence S.
Leff. — 2nd ed.
        p.   cm. — (Barron's review course series)
    Includes index.
    ISBN 0-8120-9036-5
    1. Mathematics. 2. Mathematics—Study and teaching (Secondary)
—New York (State)   I. Title. II. Series.
QA39.2.L42   1995
510—dc20   95-2341
                                                CIP

PRINTED IN THE UNITED STATES OF AMERICA

67   8800   987654

TABLE OF CONTENTS

PREFACE

For which course can the book be used?

This book offers complete topic coverage and Regents Exam preparation for Course I of the New York State Three-Year Sequence in High School Mathematics. The book is suitable for use in *any* Course I class as either the primary textbook or as a supplementary study aid. The straightforward approach, the numerous guided and independent practice exercises, and the built-in Regents Examination preparation make this book an ideal choice as the primary textbook in evening and summer school Course I classes.

What special features does this book have?

- *A Compact Format Designed for Self-Study and Rapid Learning*
  For easy reference, the major topics of the book are grouped by their branch of mathematics. The clear writing style quickly identifies essential ideas while avoiding unnecessary details. Helpful diagrams, convenient summaries, and numerous step-by-step demonstration examples facilitate learning. These features will be greatly appreciated by students who will want to use this book to help prevent or resolve difficulties that may arise as they move from topic to topic in their Course I classrooms or in their Regents preparation.

- *Key Ideas*
  Each section of each chapter begins with a KEY IDEAS box that highlights and motivates the material that follows. The KEY IDEAS, together with the annotated demonstration examples that follow, try to anticipate and answer students' "why" questions.

- *Regents Examination Preparation*
  The guided and independent practice exercises include Regents Exam types of questions. Each chapter closes with a set of REGENTS TUNE-UP exercises selected from actual past Course I Regents Examinations. These Regents problems provide a comprehensive review of the material covered in the chapter while previewing the types and levels of difficulty of problems found on actual Regents examinations. As a culminating activity, several full-length Course I Regents Examinations with answers are included at the end of the book.

- *Answers to Odd-Numbered Practice Exercises*
  Each set of practice exercises includes questions at different levels of difficulty that are designed to build understanding, skill, and confidence. The answers to odd-numbered practice and Regents Tune-Up exercises are provided at the end of each chapter. This will give students valuable feedback that will lead to greater mastery of the material.

Who should use this book?

Students who wish to improve their classroom performance and test grades will benefit greatly from the straightforward explanations, helpful examples, numerous practice exercises, and past Regents problems organized by topic. Students studying for classroom exams or preparing for the Course I Regents Examination will find this book a handy resource whenever they need further explanation or additional guided and independent practice on a troublesome topic.

Teachers who desire an additional instructional resource that is convenient to use will want this book in their personal and school libraries. Teachers will find this book to be an ideal companion to any of the existing Course I textbooks, providing a valuable lesson-planning aid as well as many practice and test problems.

LAWRENCE S. LEFF
January, 1995

NUMBERS, VARIABLES, AND LOGIC

CHAPTER 1

NUMBERS AND VARIABLES— BASIC CONCEPTS

1.1 UNDERSTANDING SETS, VARIABLES, AND NUMBER SYSTEMS

===== KEY IDEAS =====

Algebra is generalized arithmetic in which letters, called **variables**, are used as "general numbers." When letters are used in place of numbers, special operation symbols for multiplication are needed. Sets of numbers and their properties are the foundation of algebra.

Set Notation

A **set** is a collection of objects, called **elements**, that is listed within braces, { }. The set that contains the first three letters of the English alphabet is the set $\{a, b, c\}$. Since the order of the elements within the braces does not matter, $\{b, a, c\}$ and $\{a, b, c\}$ are the same set.

If each element of a set is also an element of a second set, then the first set is said to be a **subset** of the second set. For example, if set B is $\{2, 4\}$ and set A is $\{1, 2, 3, 4, 5\}$, then set B is a *subset* of set A.

A set may have no elements, in which case it is called the **empty set**. For example, the set of elephants that can fly is the empty set, { }. The empty set is sometimes called the **null set**, which is denoted by $\varnothing$.

Comparison Symbols

In the statement $1 + 1 = 2$, the symbol = is read as "is equal to" and indicates that whatever appears on the left side of the symbol is equal in value to whatever appears on the right side of the symbol. A statement of this type is an **equation**.

A number statement that two quantities are not equal is called an **inequality**, as in $1 + 1 \neq 3$. The symbol $\neq$ is read as "is not equal to." If two numbers are not equal, then there are various possibilities, which are summarized in

Table 1.1. Keep in mind that statements such as $8 > 7$ ("8 is greater than 7") and $7 < 8$ ("7 is less than 8") are equivalent.

TABLE 1.1 COMPARISON SYMBOLS

Example	Symbol	Read as . . .
$8 > 7$	$>$	8 *is greater than* 7.
$5 \geq 2$	$\geq$	5 *is greater than* or *equal to* 2.
$1 < 4$	$<$	1 *is less than* 4.
$2 \leq 3$	$\leq$	2 *is less than* or *equal to* 3.

Variables and Constants

Suppose you pick any number from the set $\{1, 2, 3, 4, 5, 6, 7, 8\}$ and then add 9 to that number. The sum that results will depend on the number you pick. If you pick 1, the sum is $1 + 9$ or 10; if you pick 2, the sum is $2 + 9$ or 11; if you pick 3, the sum is $3 + 9$ or 12, and so forth. If the number you will pick is not known, the sum can be represented by $x + 9$, where the letter x serves as a placeholder for the number that you will pick from the set. In the sum $x + 9$, x is called a *variable* and 9 is called a *constant*. In general:

- A **variable** is a symbol that refers to some member of a given set. Letters of the alphabet are frequently used as variables.
- A **constant**, unlike a variable, has a fixed value that does not change.

Representing Multiplication

In arithmetic, we usually use the symbol $\times$ to indicate multiplication. Thus, $2 \times 3 = 6$. Multiplication can be represented in other ways.

- A raised dot between numbers or variables indicates multiplication. For example,

  $2 \cdot 3$ means 2 times 3 or 6,
  $2 \cdot y$ means 2 times y,
  $a \cdot b$ means a times b.

- Parentheses around numbers or variables with no other operation symbol indicates multiplication. For example,

  $2(3)$ means 2 times 3 or 6,
  $(2)(3)$ means 2 times 3 or 6,
  $2(x + y)$ means 2 times the sum of x and y,
  $(a)(b)$ means a times b.

- A number and a variable, or two variables, written next to each other indicates their product. For example,

$$2y \text{ means } 2 \text{ times } y,$$
$$ab \text{ means } a \text{ times } b.$$

When a number and a variable are written consecutively, as in $3n$, the number is called the **coefficient** of the variable. Thus, $3n$ means 3 times n and 3 is the coefficient of n. If $n = 4$, then $3n = 3(4) = 12$. The expression $2ab$ means 2 times a times b. If $a = 5$ and $b = 7$, then $2ab = 2(5)(7) = 10(7) = 70$.

Signed Numbers

In the winter months temperatures may be below or above zero. If a temperature of 3 degrees *above* 0 is represented by +3 (read "positive 3"), then ⁻3 (read "negative 3") represents a temperature of 3 degrees *below* 0. The positive sign in +3 is the plus sign, +. The negative sign in ⁻3 is a raised dash, ⁻.

If a nonzero number has no sign in front of it, then the number is assumed to be positive. Thus, you can interpret 7 as +7. Zero has no sign since it is neither positive nor negative.

Number Opposites

Positive 3 and negative 3 are *opposites* in the same sense that winning 3 games (+3) and losing 3 games (⁻3) are opposite situations. The *opposite* of a number may be indicated by writing an ordinary (centered) subtraction sign in front of the number. Thus, –3 is read as "the opposite of 3." Since number pairs like –3 (the opposite of winning 3 games) and ⁻3 (losing 3 games) refer to the same number concept, we will simplify the notation by using the centered subtraction sign (–) to represent both types of numbers. For example, negative 3 will be written as –3, rather than ⁻3.

Keep in mind that:

- The opposite of a positive number is a negative number. Since –(+4) means take the opposite of +4,

$$-(+4) = -4.$$

- The opposite of a negative number is a positive number. Since –(–4) means take the opposite of –4,

$$-(-4) = +4 \text{ or } 4.$$

- The opposite of 0 is 0.

Number Line

The size relationship between negative and positive numbers can be represented pictorially by a number line. A **number line** is a nonending, ruler-like
horizontal line on which any convenient point, called the **origin**, is labeled 0.
Positive numbers lie in increasing size order to the right of the origin (above 0),
negative numbers appear in decreasing size order to the left of the origin
(below 0). As shown in Figure 1.1, each positive number and its opposite are
located at points on either side of the origin that are the same distance from
the origin.

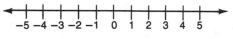

Figure 1.1   A Number Line

The number associated with each point on a number line is called its
coordinate. The set of *all* coordinates on a number line corresponds to the set
of **real numbers**.

As we move along the number line from left to right, the coordinates
increase in value. For example, –2 > –3, –1 > –2, 0 > –1, 1 > 0, and so forth.

MATH FACTS

ORDER PROPERTY OF NUMBERS

Any number on the number line is greater than any
number to its left.

The Set of Integers

The numbers 0, 1, 2, 3, 4 . . . and so on are called **whole numbers**. The set of
integers includes negative whole numbers, 0, and positive whole numbers.
Thus,

$$\{\text{Integers}\} = \{\ldots, -3, -2, -1, 0, 1, 2, 3, \ldots\}.$$

The three trailing dots indicate that the pattern continues without ever
ending. The **positive integers** are the numbers 1, 2, 3, . . . and the numbers –1,
–2, –3, . . . are the **negative integers**. The set of **nonnegative integers** includes
0 and the positive integers.

Rational Numbers

Numbers like 3, –8, $\frac{4}{9}$, and 0.5 are rational numbers. A **rational number** is any number that can be written in fractional form as the quotient of two integers, provided the denominator is not 0. Integers are rational numbers since each integer can be put into this form by writing it as a fraction with a denominator of 1. For example, 3 and –8 are rational numbers since $3 = \frac{3}{1}$ and $-8 = \frac{-8}{1}$.

Decimal numbers that can be expressed as the quotient of two integers also represent rational numbers. For example, 0.5 [$= \frac{1}{2}$] and 0.33333 . . . [$= \frac{1}{3}$] are examples of rational numbers.

The sets of whole numbers and integers are subsets of the set of rational numbers. **Real numbers** include rational numbers as well as numbers like π (approximately 3.14) that are not rational.

Absolute Value

The **absolute value** of a number x is written as $|x|$ and may be interpreted as the *distance* between the point on the number line whose coordinate is x and the origin. Since distance cannot be negative, $|x|$ is always nonnegative and can be evaluated by writing the number *without* a sign. For example, $|+5| = 5$, $|0| = 0$, and $|-3| = 3$. Here are two more examples:

$$\textbf{1.}\ |-5| + |-4| = 5 + 4 = 9$$
$$\textbf{2.}\ 3|-2| = 3(2) = 6$$

Example

1. Which number of the set {–2, 5, –8} has the largest absolute value?

Solution: Although **–8** is the smallest number, it has the largest absolute value since $|-2| = 2$, $|5| = 5$, and $|-8| = 8$.

Exercise Set 1.1

1. Which set is a subset of the set of integers?
(1) Real numbers   (3) Rational numbers
(2) Fractions   (4) Whole numbers

2. Which of the following sets of numbers is *not* a subset of any of the other sets?
(1) {Rationals}   (2) {Reals}   (3) {Integers}   (4) {Decimals}

3–8. Evaluate each of the following:

3. $|-2| + |1|$

4. $|-5| |-3|$

5. $\dfrac{|-10|}{10}$

6. $|-1.8| + |0.7|$

7. $|-1\frac{3}{4}| |-2|$

8. $\dfrac{|-12|}{20}$

9–17. Replace □ with a comparison symbol (<, >, or =) that makes the resulting statement true.

9. -5 □ -1

10. $|-3|$ □ 0

11. 2 □ $|-4|$

12. $|-5|$ □ 5

13. $|-8|$ □ $4(|-2|)$

14. $|-10|$ □ -2

15. $\left|\dfrac{-1}{3}\right|$ □ 0.43

16. $\dfrac{2}{5}$ □ $\left|\dfrac{-1}{2}\right|$

17. $(1.5)(|-3.2|)$ □ 5

1.2   REVIEWING SOME NUMBER FACTS

KEY IDEAS

The numbers 0 and 1 have special properties. Adding 0 to a number, or multiplying a number by 1, always results in that number. The sum of a number and its opposite is always 0, while the product of a nonzero number and its reciprocal is always 1.

An integer is *divisible* by a nonzero integer if their quotient has a remainder of 0.

Identity and Inverses for Addition

If 0 is added to any real number, the sum is that number. For example, $4 + 0 = 4$ and $0 + 4 = 4$. The number 0 is called the **additive identity**.

MATH FACTS

ADDITION PROPERTY OF ZERO

If *a* represents any real number, then
$a + 0 = a$ and $0 + a = a$.

If the sum of two numbers is 0, then either number is called the *additive inverse* of the other number. Thus, the **additive inverse** of any real number is the opposite of that number. For example, -5 is the additive inverse (opposite) of 5 since $5 + (-5) = 0$.

MATH FACTS

ADDITIVE INVERSE PROPERTY

If *a* represents any real number, then
$a + (-a) = 0$ and $(-a) + a = 0$.

Identity and Inverses for Multiplication

The number 1 has the same properties for multiplication that 0 has for addition. If any real number is multiplied by 1, the product is that number. For example, $4 \times 1 = 4$ and $1 \times 4 = 4$. The number 1 is called the **multiplicative identity**.

MATH FACTS

MULTIPLICATION PROPERTY OF ONE

If *a* represents any real number, then
$a \times 1 = a$ and $1 \times a = a$.

If the product of two numbers is 1, then either number is called the *multiplicative inverse* of the other number. Thus, the **multiplicative inverse** of any nonzero real number is the reciprocal of that number. For example, the multiplicative inverse (reciprocal) of 5 is $\frac{1}{5}$ since $5 \times \frac{1}{5} = 1$. The multiplicative inverse of $\frac{2}{3}$ is $\frac{3}{2}$ since $\frac{2}{3} \times \frac{3}{2} = 1$.

MATH FACTS

MULTIPLICATIVE INVERSE PROPERTY

If *a* represents any real number except 0, then

$$a \times \frac{1}{a} = 1 \text{ and } \frac{1}{a} \times a = 1.$$

Example

1. What value of x makes each of the following true?

(a) $x + 8 = 0$ **(b)** $\left(\frac{3}{4}\right)x = 1$

Solutions: **(a)** The value of x is -8 since the sum of a number (8) and its opposite (-8) is 0.
 (b) The value of x is $\frac{4}{3}$ since the product of a number $\left(\frac{3}{4}\right)$ and its reciprocal $\left(\frac{4}{3}\right)$ is 1.

Divisibility

The number 12 is *divisible* by 4 (or 4 *divides* 12 *evenly*) since the quotient is 3 with a *remainder* of 0. The number 12 is *not* divisible by 5 since, when 12 is divided by 5, there is a remainder of 2. An **even number** is any integer that is divisible by 2, while an **odd number** is any integer that is *not* divisible by 2:

$$\{Even \text{ numbers}\} = \{\dots, -6, -4, -2, 0, 2, 4, 6, \dots\},$$
$$\{Odd \text{ numbers}\} = \{\dots, -5, -3, -1, 1, 3, 5, \dots\}.$$

The number 0 may be written in different forms as a fraction. For example, $\frac{0}{1} = 0$, $\frac{0}{2} = 0$, and so forth. In other words, 0 divided by any nonzero number is always 0. *Dividing by 0 is not allowed.* An expression such as $\frac{3}{0}$ is said to be "meaningless" or "not defined."

The number 1 may also be represented in different ways as a fraction. For example, $\frac{3}{3} = 1$ and $\frac{x}{x} = 1$ (provided that $x \neq 0$). In general, any nonzero number divided by itself is 1.

Examples

2. For what value of x is $\dfrac{1}{x-4}$ not defined?

Solution: If $x = 4$, then the denominator of the fraction will be equal to 0 since $4 - 4 = 0$. Since division by 0 is not allowed, x cannot have the value of 4.

3. Express $\dfrac{2}{5}$ as an equivalent fraction having 20 as its denominator.

Solution: Multiply $\dfrac{2}{5}$ by 1 in the form of $\dfrac{4}{4}$:

$$\frac{2}{5} \times \frac{4}{4} = \frac{8}{20}.$$

A **prime number** is a whole number greater than 1 that is divisible only by itself and 1. The numbers 3, 5, 7, 11, 13, 17, 19, and 23 are examples of prime numbers. The number 2 is the only even number that is also a prime number.

A **composite number** is any natural number greater than 1 that is *not* a prime number.

The numbers 5, 10, 15, 20, 25, 30, 35, ... are *multiples* of 5 since each number is obtained by multiplying 5 by a different positive integer. This means that each multiple of 5 must be divisible by 5. In general, a number m is a **multiple** of a number n if m is divisible by n.

The **least common multiple (LCM)** of two or more numbers is the smallest number that is a multiple of the given numbers. For example, the least common multiple of 8 and 12 is 24 since 24 is the smallest natural number that is divisible by 8 *and* divisible by 12.

The **least common denominator (LCD)** of two or more fractions is the least common multiple of their denominators. The LCD of $\frac{1}{6}$ and $\frac{5}{9}$ is 18 since 18 is the smallest positive integer that is evenly divisible by both 6 and 9.

Example

4. Which of the following is an example of an odd number that is not a prime number?

(1) 2   (2) 3   (3) 9   (4) 11

Solution: The number 9 is an odd number that is divisible by 3. The correct choice is **(3)**.

Exercise Set 1.2

1. How many prime numbers are greater than or equal to 30 and less than or equal to 43?

2. Which of the following numbers is a prime number *and*, when divided by 8, has a remainder of 3?
(1) 27   (3) 24
(2) 37   (4) 19

3. Change $\frac{2}{3}$ to an equivalent fraction whose denominator is 12.

4–7. Find the **(a)** *additive inverse and* **(b)** *multiplicative inverse for each of the following numbers:*

4. 3

5. –8

6. $\frac{1}{6}$

7. 0.2

8. Which statement is true for the set of whole numbers?
(1) The multiplicative identity is 0.
(2) The additive identity is 1.
(3) All elements of the set are rational numbers.
(4) Each whole number has an additive inverse that is also a whole number.

9. Which number is equal to its multiplicative inverse?
(1) 0   (3) $\frac{1}{2}$
(2) 1   (4) –2

10. If $x + y = x$, then what is the value of xy?

11–14. Find the value of x *that makes each of the following statements true:*

11. $3 + x = 0$

12. $\frac{1}{6}x = 1$

13. $x - 5 = 0$

14. $\frac{7}{3}x = 1$

15. Determine the LCD of the fractions whose denominators are:
(a) 8 and 28   (b) 3 and 7   (c) 5 and 40

16–20. For what value of x *is the reciprocal of each of the following not defined?*

16. $x - 7$

17. $\dfrac{3 + x}{2 - x}$

18. $4x$

19. $\dfrac{x}{3}$

20. $2x - 1$

TABLE 1.2 RULES FOR WORKING WITH SIGNED NUMBERS

Operation	Sign of Numbers	Procedure
Multiplication and division	SAME $(+)(+)\ \ = +$ $(-)(-)\ \ = +$ $(+) \div (+) = +$ $(-) \div (-) = +$	Multiply (or divide) numbers while ignoring their signs. Make the sign of the answer *positive.* (a) $(+5)(+8) =$ **+40** (b) $(-5)(-8) =$ **+40** (c) $(+40) \div (+8) =$ **+5** (d) $(-40) \div (-8) =$ **+5**
	DIFFERENT $(+)(-)\ \ = -$ $(-)(+)\ \ = -$ $(+) \div (-) = -$ $(-) \div (+) = -$	Multiply (or divide) numbers while ignoring their signs. Make the sign of the answer *negative.* (a) $(+5)(-8) =$ **−40** (b) $(-5)(+8) =$ **−40** (c) $(+40) \div (-8) =$ **−5** (d) $(-40) \div (+8) =$ **−5**
Addition	SAME $(+) + (+) = +$ $(-) + (-) = -$	Add numbers while ignoring their signs. Write the sum using their *common* sign. (a) $(+5) + (+8) =$ **+13** (b) $(-5) + (-8) =$ **−13**
	DIFFERENT	Subtract numbers while ignoring their signs. The answer has the same sign as the number having the *larger absolute value.* (a) $(+5) + (-8) =$ **−3** (b) $(-5) + (+8) =$ **+3**
Subtraction	SAME	(a) $(+5) - (+8) = (+5) + (-8) =$ **−3** ↓ Take the **opposite** and *add.*
	DIFFERENT	(b) $(+5) - (-8) = (+5) + (+8) =$ **+13** ↓ Take the **opposite** and *add.*

1.3 WORKING WITH SIGNED NUMBERS

$$\bigwedge \text{ Key Ideas}$$

The rules for multiplying and dividing signed numbers are similar to those for unsigned numbers. The rules for adding signed numbers depend on whether the signs of the numbers are the same or different. Subtracting signed numbers involves changing the example into an equivalent addition example.

Parentheses

Parentheses may be used to make an addition or subtraction problem easier to read.

- The sum of +7 and –2 is written as (+7) + (–2), rather than as +7 + –2.
- The difference between +5 and –3 is written as (+5) – (–3), rather than as + 5 – –3.

Parentheses may also be used in multiplication and division operations.

- The product of +3 and –4 may be written in several different ways, including: (+3)(–4) and 3(–4).
- The quotient of –6 and +7 may be written in several different ways, including: (–6) ÷ (+7) and $\frac{-6}{+7}$. Also note that the following expressions are equivalent:

$$-\frac{6}{7}, \quad \frac{-6}{7}, \quad \frac{6}{-7}, \quad -\left(\frac{6}{7}\right).$$

Operations with Signed Numbers

Review Table 1.2 on page 11, which summarizes the rules for arithmetic operations with signed numbers.

Examples

1. Find the value of $\left|-2\right| + \left|7\right| - \left|-13\right|$.

Solution: $\left|-2\right| + \left|7\right| - \left|-13\right| = 2 + 7 - 13$
$$= 9 - 13$$
$$= 9 + (-13)$$
$$= -4$$

2. Subtract – 3 from 10.

Solution: $10 - (-3) = 10 + 3 = 13$

The subtraction may also be performed by writing the number that is being subtracted under the other number. This subtraction example is changed into an equivalent addition example by changing the subtraction sign to an addition sign *and* replacing the number that is being subtracted with its opposite.

Original Subtraction Example

$$-\quad\frac{10}{\underline{-3}}$$

Equivalent Addition Example

$$+\quad\frac{10}{\underline{+3}}$$
$$13$$

3. Simplify: $\dfrac{15-50}{-5}$.

Solution: $\dfrac{15-50}{-5} = \dfrac{-35}{-5} = \mathbf{7}$

4. Multiply: $(-3)(-2)(-4)$.

Solution: $(-3)(-2)(-4) = (+6)(-4) = \mathbf{-24}$

5. Divide: $\dfrac{4}{21} \div \left(-\dfrac{2}{7}\right)$.

Solution: Change into an equivalent multiplication example by inverting the second fraction.

$$\frac{4}{21} \div \left(-\frac{2}{7}\right) = \frac{4}{21} \times \left(-\frac{7}{2}\right)$$
$$= \frac{4}{21} \times \left(-\frac{7}{2}\right)$$
$$= \frac{2(-1)}{3(1)} = -\frac{2}{3}$$

6. Subtract: $-\dfrac{2}{3} - \left(-\dfrac{1}{4}\right)$.

Solution: First change from a subtraction example into an equivalent addition example.

$$-\frac{2}{3} - \left(-\frac{1}{4}\right) = -\frac{2}{3} + \frac{1}{4}$$

The LCD of the two fractions is 12 since 12 is the smallest positive integer that is divisible by both 3 and 4. Multiply each fraction by a form of 1 so that an equivalent fraction results that has the LCD as its denominator.

$$-\frac{2}{3}+\frac{1}{4} = \frac{4}{4}\left(-\frac{2}{3}\right)+\frac{3}{3}\left(\frac{1}{4}\right)$$

$$=-\frac{8}{12}+\frac{3}{12}$$

$$=\frac{-8+3}{12}$$

$$=-\frac{5}{12}$$

Exercise Set 1.3

1–21. Perform the indicated operation.

1. $(-7)(-3)$

2. $(-18) \div (+2)$

3. $(-5)(+3)(-2)$

4. $(-3) + (-7)$

5. $(-5) + (+12)$

6. $(-9) + (+7)$

7. $-11 + 3$

8. $(-8) - (-5)$

9. $(-14) - (+5)$

10. $(-21) \div (-7)$

11. $3.2 \div (-0.8)$

12. $0.2(-4.9)$

13. $|-10 - (-7)|$

14. $-8 - (-2)$

15. $(-1.5)(-0.75)$

16. $\dfrac{-1.05}{-0.35}$

17. $\dfrac{56}{-8}$

18. $-8.1 - 1.9$

19. $-8.1 + 1.9$

20. $8.1 - 1.9$

21. $8.1 - (-1.9)$

22. Subtract 2 from -18.

23. Subtract -5 from $+3$.

24. From the sum of -4 and $+2$, subtract 6.

25. To the product of -2 and -5, add -3.

26. From the product of -3.2 and 4.1, subtract 1.38.

27. To the quotient of -9.8 and 4.9, add -0.7.

28–35. Perform the indicated operation with the signed fractions.

28. $\left(-\dfrac{3}{5}\right) + \left(-\dfrac{1}{5}\right)$

29. $\left(-\dfrac{8}{9}\right) + \left(\dfrac{7}{9}\right)$

30. $\left(-\dfrac{3}{7}\right)\left(-\dfrac{14}{33}\right)$

31. $\left(+\dfrac{1}{5}\right) \div \left(-\dfrac{3}{10}\right)$

32. $\left(-\dfrac{2}{3}\right) + \left(+\dfrac{1}{4}\right)$

33. $\left(-\dfrac{2}{5}\right) + \left(\dfrac{3}{10}\right)$

34. $\left(-\dfrac{3}{4}\right)\left(\dfrac{10}{21}\right)$

35. $-21 \div \left(-\dfrac{1}{3}\right)$

1.4 LEARNING MORE ABOUT REAL NUMBERS

△
Kᴇʏ Iᴅᴇᴀs
△

Real numbers behave in predictable ways. The order in which real numbers are added or multiplied does not matter. The **distributive property** links the operations of multiplication and addition by stating that the parentheses in an expression like $3(2 + 4)$ can be removed by writing $3(2 + 4) = 3 \cdot 2 + 3 \cdot 4$.

Commutative Properties

We know that $2 + 1 = 1 + 2$ and $3 \cdot 4 = 4 \cdot 3$. The commutative law states that this property holds for any pair of real numbers. The commutative law does not hold for subtraction. For example, $5 - 3 \neq 3 - 5$, since $5 - 3 = 2$ and $3 - 5 = -2$. Since $\frac{5}{3} \neq \frac{3}{5}$, there is no commutative property for division.

Mᴀᴛʜ Fᴀᴄᴛs

COMMUTATIVE LAWS

For all real numbers a and b,
$a + b = b + a$ and $a \cdot b = b \cdot a$.

Using Parentheses to Indicate Order

When parentheses enclose an operation between two numbers, then this operation is performed first. Thus,

$$2 + (1 + 3) = 2 + 4 = 6 \text{ and } 2(3 \cdot 4) = 2(12) = 24.$$

Associative Property

You can easily verify that $2 + (1 + 3) = (2 + 1) + 3$ and $2(3 \cdot 4) = (2 \cdot 3)4$. The associative law states that this property holds for any three real numbers. The associative law does not hold for subtraction. For example, $5 - (2 - 3) \neq (5 - 2) - 3$. Since $12 \div (6 \div 2) \neq (12 \div 6) \div 2$, there is no associative law for division.

ASSOCIATIVE LAWS

For all real numbers *a*, *b*, and *c*,
$$a + (b + c) = (a + b) + c \text{ and } a\,(b \cdot c) = (a \cdot b)\,c.$$

Distributive Property

The expression $3(2 + 4)$ can be evaluated by first performing the operation inside the parentheses. Thus,

$$3(2 + 4) = 3(6) = 18.$$

The distributive property states that the same result is obtained when each term inside the parentheses is multiplied by 3 and then the products are added. Thus,

$$3(2 + 4) = 3 \cdot 2 + 3 \cdot 4 = 6 + 12 = 18.$$

DISTRIBUTIVE PROPERTY

If *a*, *b*, and *c* are real numbers, then
$$a(b + c) = ab + ac \text{ and } (b + c)a = ba + ca.$$

The commutative, associative, and distributive laws can be applied to expressions that contain variables since these expressions also represent real numbers. For example, $x + 3$ and $3 + x$ are equivalent expressions.

Examples

1. Remove the parentheses by applying the distributive property:
    (a) $5(x + 2)$ (b) $6(2y - 3)$ (c) $-2(x - 4)$

Solution: In each case remove the parentheses by multiplying each of the numbers that are inside the parentheses by the number that is in front of the parentheses.

$$\begin{aligned}
\textbf{(a)} \ \ 5(x + 2) &= 5x + 5(2) &&= \mathbf{5x + 10} \\
\textbf{(b)} \ \ 6(2y - 3) &= 6(2y) + 6\,(-3) &&= \mathbf{12y - 18} \\
\textbf{(c)} \ \ {-2}(x - 4) &= -\,2x + (-2)(-4) &&= \mathbf{-2x + 8}
\end{aligned}$$

17

2. Find the value of $9 \cdot 37 + 9 \cdot 63$.

Solution: The given expression has the form $a \cdot b + a \cdot c$, where $a = 9$, $b = 37$, and $c = 63$. Applying the distributive property in reverse gives

$$a \cdot b + a \cdot c = a(b + c)$$

$$9 \cdot 37 + 9 \cdot 63 = 9(37 + 63)$$

$$= 9\,(100)$$

$$= \mathbf{900}$$

Algebraic Expressions

Here are some examples of *algebraic expressions:*

$$2x, \; x + 2, \; 5x - y, \; \frac{4}{x + 1}, \text{ and } x^2 - 3x + 7.$$

A sequence of operations that includes at least one variable is an **algebraic expression**. An algebraic expression is composed of **terms** that are separated by the operations of addition or subtraction. The terms of $x^2 - 3x + 7$ are x^2, $3x$, and 7.

Like Terms

Terms such as $4x$ and $5x$ are **like terms** since they have the same variable factor, x, and differ only in their coefficients, 4 and 5. The reverse of the distributive property allows like terms to be combined into a single term. For example,

$$4x + 5x = (4 + 5)x = 9x.$$

To simplify an algebraic expression, combine any like terms by adding or subtracting their numerical coefficients. Here are some more examples.

1. $3b + 2b = 5b$
2. $7y - y = 7y - 1y = 6y$
3. $2a + 3a + 4a = 9a$
4. $2x + 3y = $ *Cannot be combined since* 2x *and* 3y *are not like terms.*
5. $4(x + 1) - 9x \qquad\qquad = 4x + 4 - 9x$
Use the Commutative law: $\qquad = 4x - 9x + 4$
Combine like terms: $\qquad\quad\; = -5x + 4$

Notice in the last example that more than one law is needed in grouping the like terms together so that they can be combined.

Exercise Set 1.4

1–6. Name the law of arithmetic that is illustrated.

1. $x + (y + z) = x + (z + y)$

2. $(xy)z = x(yz)$

3. $(x + y) + z = x + (y + z)$

4. $x + (y + z) = (y + z) + x$

5. $x(yz) = x(zy)$

6. $ab + ac = a(b + c)$

7–10. Remove the parentheses by applying the distributive property.

7. $4(x - 2)$

8. $6(1 - 2y)$

9. $-3(x - 4)$

10. $4 - (2x + 9)$

11–16. Simplify by combining like terms, where possible.

11. $6y - y$

12. $3(x + 1) - 4x$

13. $2(n - 3) + 5$

14. $-5(x - 1) + 9x$

15. $7 - 3(x - 1)$

16. $3(2c - 2) - 7(5 - c)$

1.5 WORKING WITH EXPONENTS

The numbers 2 and 15 are factors of 30 since $2 \times 15 = 30$. Each of the numbers being multiplied to obtain a product is called a **factor** of the product. A number may be written as the product of more than two factors. For example, $2 \times 3 \times 5 = 30$. If the same number appears as a factor in a product more than once, it may be written using an *exponent:*

$$\text{exponent} \longrightarrow$$
$$\underbrace{2 \cdot 2 \cdot 2 \cdot 2}_{} = 2^4$$
$$\text{four factors} \longrightarrow \qquad \qquad \sqcup\!\!- \text{base}$$

Exponents

Repeated multiplication of the same number may be expressed in a more compact way by writing the number only once with an exponent. The exponent tells the number of times a number is repeated in the multiplication process. For example, $2 \cdot 2 \cdot 2 \cdot 2 = 2^4 = 16$. The number being multiplied (2) is called the base. The number of times the base is used as a factor (4) is called the **exponent**. The number with its exponent is called the **power** of the number.

If the exponent is 1, it is usually omitted. For example, $5^1 = 5$. When the exponent is 2, we sometimes refer to the base as being *squared*. For example, 5^2 is usually read as "5 squared" or "the square of 5." Also 5^3 is usually read as "5 cubed" or "the cube of 5."

Example

1. Evaluate:
   (a) $(-4)^3$ **(b)** $(-3)^4$

Solution: **(a)** $(-4)^3 = (-4)(-4)(-4) = (+16)(-4) = \mathbf{-64}$
   (b) $(-3)^4 = (-3)(-3)(-3)(-3) = (+9)(-3)(-3)$
   $= (-27)(-3) = \mathbf{81}$

Note: A negative number raised to an *odd* integer power will always have a negative value. A negative number raised to an *even* integer power (except 0) will always have a positive value.

Multiplying and Dividing Powers

Simple rules can be discovered for multiplying and dividing powers having the *same* base. The product of 3^6 and 3^2 can be obtained by writing each power in expanded form:

$$6 + 2 = 8$$

$$3^6 \cdot 3^2 = (3 \cdot 3 \cdot 3 \cdot 3 \cdot 3 \cdot 3)(3 \cdot 3) = 3^8.$$

eight factors of 3

This example suggests the following rule:

MATH FACTS

MULTIPLICATION RULE FOR EXPONENTS

To **multiply** powers with *like* bases, keep the base and *add* the exponents: $a^x a^y = a^{x+y}$.

A shortcut method for dividing powers having the same base may also be developed:

$$6 - 2 = 4$$

$$\frac{3^6}{3^2} = \frac{3 \cdot 3 \cdot 3 \cdot 3 \cdot 3 \cdot 3}{3 \cdot 3} = 3 \cdot 3 \cdot 3 \cdot 3 = 3^4.$$

four factors of 3

This example suggests the following rule:

MATH FACTS

QUOTIENT RULE FOR EXPONENTS

To **divide** powers with *like* bases, keep the base and *subtract* the exponents:

$$a^x \div a^y = a^{x-y}.$$

Examples

2. Multiply: $5^3 \cdot 5^8$.

Solution: $5^3 \cdot 5^8 = 5^{3+8} = \mathbf{5^{11}}$

3. Multiply: $n^4 \cdot n^2$.

Solution: $n^4 \cdot n^2 = n^{4+2} = \boldsymbol{n^6}$

4. Multiply: $a^2 \cdot b^4$.

Solution: Since the bases are *not* the same, the multiplication rule cannot be used.

5. Divide: $b^7 \div b^2$.

Solution: $b^7 \div b^2 = b^{7-2} = \boldsymbol{b^5}$

6. Divide: $\dfrac{c^2}{c}$.

Solution: $\dfrac{c^2}{c} = \dfrac{c^2}{c^1} = c^{2-1} = c^1$ or $\boldsymbol{c}$

Zero Exponent

Notice that there are two ways to evaluate $2^3 \div 2^3$:

1. Apply the quotient rule: $2^3 \div 2^3 = 2^{3-3} = 2^0$.

2. Substitute 8 for 2^3: $2^3 \div 2^3 = 8 \div 8 = 1$.

Since the left sides of the equations in methods 1 and 2 are the same, the right sides must be equivalent. Therefore $2^0 = 1$. In general, *any nonzero number or variable raised to the zero power is equal to 1.*

MATH FACTS

ZERO POWER RULE

If $a \neq 0$, then $a^0 = 1$.

Example

7. Simplify: **(a)** $2x^0$ **(b)** $(2x)^0$.

Solution: **(a)** $2x^0 = 2(x^0) = 2 \cdot 1 = \mathbf{2}$

(b) $(2x)^0 = \mathbf{1}$

Negative Exponents

Using the quotient rule gives us $3^2 \div 3^6 = 3^{2-6} = 3^{-4}$. By writing the quotient in fraction form and simplifying, we obtain

$$\frac{3^2}{3^6} = \frac{\overset{1}{\cancel{3}} \cdot \overset{1}{\cancel{3}}}{\cancel{3} \cdot \cancel{3} \cdot 3 \cdot 3 \cdot 3 \cdot 3} = \frac{1}{3 \cdot 3 \cdot 3 \cdot 3} = \frac{1}{3^4}$$

Since the answers must be the same, $3^{-4} = \frac{1}{3^4}$. Similarly, $\frac{1}{3^{-4}} = 3^4$.

This suggests that a nonzero number raised to a negative power is equivalent to its reciprocal with the sign of the exponent changed from negative to positive.

MATH FACTS

NEGATIVE EXPONENT RULE

If n is a positive integer and $a \neq 0$, then

$$a^{-n} = \frac{1}{a^n} \text{ and } \frac{1}{a^{-n}} = a^n.$$

Examples

8. Simplify: $\dfrac{3}{10^{-2}}$.

Solution: $\dfrac{3}{10^{-2}} = 3(10^2) = 3(100) = \mathbf{300}$

9. Write: $\dfrac{9}{2m^{-4}}$ with a positive exponent.

Solution: $\dfrac{9}{2m^{-4}} = \dfrac{9m^4}{2}$

10. Evaluate: $(2 + 3)^{-1}$.

Solution: $(2 + 3)^{-1} = 5^{-1} = \dfrac{\mathbf{1}}{\mathbf{5}}$ *or* **0.2**

Scientific Notation

In scientific notation 37,000 is written as 3.7×10^4 and 0.00000408 is written as 4.08×10^{-6}. A positive number is in **scientific notation** when it is written as a number between 1 and 10 times a power of 10.

- To write a number greater than 10 in scientific notation:

  Step 1. Move the decimal point k places to the left so that the number is now between 1 and 10.

  $$1\,2\,0\,0\,0\,0\,0.$$
  $$k = 6$$

  Step 2. Multiply the number obtained in step 1 by 10^k.

  $$1.2 \times 10^6$$

  Thus $1{,}200{,}000 = 1.2 \times 10^6$.

- To write a number between 0 and 1 in scientific notation:

  Step 1. Move the decimal point k places to the right so that the number is now between 1 and 10.

  $$0.0\,0\,0\,0\,3\,0\,8$$
  $$k = 5$$

  Step 2. Multiply the number obtained in step 1 by 10^{-k}.

  $$3.08 \times 10^{-5}$$

  Thus, $0.0000308 = 3.08 \times 10^{-5}$.

- To write a positive number between 1 and 10 in scientific notation, multiply the number by 10^0. In scientific notation, $6.14 = 6.14 \times 10^0$.

Exercise Set 1.5

1–3. Evaluate to a single number.

1. $(-2)^5 + (-5)^2$

2. $4^0 + 4^{-1}$

3. $(2^3)\,(3^2)$

4–7. Replace $\square$ with the symbol $<$, $>$, or $=$ so that the resulting statement is true.

4. $8^0 \,\square\, 9^5 \div 9^5$

5. $(2 + 3)^2 \,\square\, 2^2 + 3^2$

6. $2^{-3} \,\square\, 2^{-2}$

7. $3^{-2} + 3^{-1} \,\square\, \dfrac{5}{9}$

8–13. Find the product or quotient.

8. $(-1)^{99} \times (99)^{-1}$

9. $y \cdot y^3 \cdot y^4$

10. $t^5 \div t$

11. $\dfrac{n^3}{n^5}$

12. $x^{10} \div x^{10}$

13. $(12a^9) \div (4a^5)$

14–17. Find the value of n.

14. $63{,}000{,}000 = 6.3 \times 10^n$

15. $0.000000071 = 7.1 \times 10^n$

16. $(5 \times 10^5)(8 \times 10^3) = 4 \times 10^n$

17. $981{,}000{,}000{,}000 = 9.81 \times 10^n$

18–23. Express in scientific notation.

18. 0.001001

19. $\dfrac{3}{10{,}000}$

20. $108{,}000{,}000$

21. $(91{,}000) \times (1{,}000)$

22. $(800)^2$

23. $(0.0005)^3$

24–25. Express in standard decimal form.

24. 3.08×10^6

25. 5.37×10^{-5}

26–29. Simplify by combining like terms, where possible.

26. $x^2 + 3x^2$

27. $2w^2 - 3w^2$

28. $3y^2 + 2y^3$

29. $5(n^2 + 1) + 3(n^2 + 1)$

1.6 EVALUATING AND SIMPLIFYING EXPRESSIONS

⋀
KEY IDEAS
⟋_⟍

Evaluating an arithmetic expression like $16 + 2^3 \div 4$ means finding the number the expression equals. When evaluating an arithmetic expression, arithmetic operations are *not* necessarily performed in the order in which they are encountered. Instead, they are performed according to an agreed upon **order of operations**.

ORDER OF OPERATIONS

To evaluate $16 + 2^3 \div 4$, follow these steps:

1. Evaluate all exponents: $16 + 2^3 \div 4 = 16 + 8 \div 4$
2. Working from left to right,
   do all multiplications and divisions: $= 16 + 2$
3. Next, do all additions and subtractions,
   working from left to right: $= \mathbf{18}$

If an expression contains parentheses, then evaluate the expression inside the parentheses first. The first letter of each of the words of the easy-to-remember phrase

"*P*lease *E*xcuse *M*y *D*ear *A*unt *S*ally"

corresponds to the first letter of what you need to evaluate when following the complete order of operations: *P = Parentheses; E = Exponents; M = Multiplication; D = Division; A = Addition; S = Subtraction.*

Examples

1. Evaluate: $(16 + 2^3) \div 4$

Solution: $(16 + 2^3) \div 4 = (16 + 8) \div 4$
                   $= 24 \div 4$
                   $= \mathbf{6}$

2. Evaluate: **(a)** -5^2 **(b)** $(-5)^2$

Solution: **(a)** Since exponentiation is performed before any of the four basic arithmetic operations, $-5^2 = \mathbf{-25}$.
         (b) $(-5)^2 = (-5)(-5) = \mathbf{25}$

26

Evaluating Algebraic Expressions

Algebraic expressions represent real numbers. To find out the real number that an algebraic expression represents, we need to replace each variable with its assigned value. The resulting *arithmetic* expression can be evaluated following the order of operations.

Examples

3. Find the value of $-a^2 + b^3$ if $a = -3$ and $b = -2$.

Solution: $-a^2 + b^2 = -(-3)^2 + (-2)^3$
$= -(9) + (-8)$
$= \mathbf{-17}$

4. If $y = 3x^2 + 5x - 7$, find the value of y when $x = -2$.

Solution: Substitute -2 for x in $y = 3x^2 + 5x - 7$. Thus,
$y = 3(-2)^2 + 5(-2) - 7$
$= 3(4) - 10 - 7$
$= 12 - 17$
$y = \mathbf{-5}$

5. If $x = -3$ and $y = 4$, find the value of:

(a) $\dfrac{1}{2}x^2y$ **(b)** $\left(\dfrac{1}{2}xy\right)^2$ **(c)** $\dfrac{y + x}{y^2 - x^2}$

Solution:

(a) $\dfrac{1}{2}x^2\, y = \dfrac{1}{2}(-3)^2(4) = \dfrac{1}{2}(9)(4) = \dfrac{1}{2}(36) = \mathbf{18}$

(b) $\left(\dfrac{1}{2}xy\right)^2 = \left(\dfrac{1}{2}(-3)(4)\right)^2 = \left(\dfrac{1}{2}(-12)\right)^2 = (-6)^2 = \mathbf{36}$

(c) $\dfrac{y + x}{y^2 - x^2} = \dfrac{4 + (-3)}{4^2 - (-3)^2} = \dfrac{1}{16 - 9} = \dfrac{\mathbf{1}}{\mathbf{7}}$

Exercise Set 1.6

1–10. Find the numerical value of each arithmetic expression.

1. $-2^3 + (-3)^2$

2. $12 + 16 \div 4$

3. $(12 + 16) \div 4$

4. $-27 \div 3^2 + 1$

5. $1 - 6^2 \div (13 - 4 \times 5)$

6. $(-4)(-7) \div (2 - 9)$

7. $-0.6(-5.1 + 3.7) + 1.8$

8. $4(-3)^2$

9. $(6^2 + 8^2) \div (1^2 + 7^2)$

10. $40 \div (3 + 16 \div 2^3)$

11–16. If x $= -2$ *and* y $= 5$, *find the value of each algebraic expression.*

11. $(xy)^2$

12. $xy + y^2$

13. $y^2 - 3y + 7$

14. $(y - x)(y + x)$

15. $\dfrac{x^2 - y^2}{x + y}$

16. $\dfrac{1}{3}(y + 1)^2$

17. If $y = -x^2 + 3$, what is the largest possible value of y?

18. If $y = (4 - x)^2$, what is the smallest possible value of y?

19–22. If r $= 24$ *and* s $= -2$, *find the value of each algebraic expression.*

19. $r \div 3s^3$

20. $(r \div 3)s^3$

21. $\dfrac{1}{3}(r + 9s)^2$

22. $\left(\dfrac{r}{s + 5}\right)^2$

REGENTS TUNE-UP: CHAPTER 1

Each of the questions in this section has appeared on a previous Course I Regents Examination. Here is an opportunity for you to review Chapter 1 and, at the same time, prepare for the Course I Regents Examination.

1. Find the value of $a^2 - b$ if $a = 3$ and $b = -4$.

2. If $a = -2$ and $b = 3$, what is the value of $-3a^2b$?

3. If 340,000 is expressed in the form 3.4×10^n, what is the value of n?

4. If $m = 3$ and $v = -4$, find the value of $\frac{1}{2}mv^2$.

5. For which value of x is the expression $\frac{x-1}{x+2}$ undefined?

6. If $x = 5$ and $y = -2$, what is the value of $\frac{2x - y}{3}$?

7. Which expression represents the number 0.00017 written in scientific notation?
   (1) 1.7×10^{-4} (2) 1.7×10^4 (3) 1.7×10^{-3} (4) 1.7×10^3

8. The additive inverse of $a - b$ is
   (1) $a + b$ (2) $-a + b$ (3) $-a - b$ (4) $\frac{1}{a - b}$

9. The value of 5^{-2} is
   (1) $-\frac{1}{25}$ (2) $\frac{1}{25}$ (3) -10 (4) -25

10. Which is the additive inverse of $-\frac{a}{3}$?
    (1) $\frac{a}{3}$ (2) $\frac{3}{a}$ (3) $-\frac{3}{a}$ (4) 0

11. The product of $-\frac{1}{a}$, $a \neq 0$, and its reciprocal is
    (1) 1 (2) -1 (3) $-\frac{1}{a^2}$ (4) 0

12. Written in standard notation, 7.2×10^3 is equivalent to
    (1) 0.0072   (2) 0.00072   (3) 7,200   (4) 72,000

13. Which sentence illustrates the commutative property for addition?
    (1) $(a + b) + c = a + (b + c)$
    (2) $a(b + c) = ab + ac$
    (3) $a + 0 = a$
    (4) $a + b = b + a$

29

14. Expressed in decimal form, the number 1.23×10^{-3} is
    (1) 1230   (2) 0.000123   (3) 0.00123   (4) 123,000

15. If $x > 0$ and $y = 0$, which statement is true?
    (1) $x - y = 0$ (2) $xy = 0$ (3) $x + y < 0$ (4) $\dfrac{x + y}{x} = 0$

16. Find the numerical value of $4xy^2$ when $x = \dfrac{1}{2}$ and $y = -3$.

ANSWERS TO ODD-NUMBERED EXERCISES: CHAPTER 1

Section 1.1
1. (4)   **5.** 1   **9.** $<$ **13.** $=$ **17.** $<$
3. 3   **7.** 3.5   **11.** $<$ **15.** $<$

Section 1.2
1. (3)   **9.** (2)   **17.** -3
3. $\frac{8}{12}$ **11.** -3 **19.** 0
5. (a) 8 (b) $-\frac{1}{8}$ **13.** 5
7. (a) -0.2 (b) 5   **15.** (a) 56 (b) 21 (c) 40

Section 1.3
1. 21   **7.** -8 **13.** 3   **19.** -6.2 **25.** 7   **31.** $-\frac{2}{3}$
3. 30   **9.** -19 **15.** 1.125   **21.** 10   **27.** -2.7 **33.** $-\frac{1}{10}$
5. 7   **11.** -4 **17.** -7 **23.** 8   **29.** $-\frac{1}{9}$ **35.** 63

Section 1.4
1. commutative   **7.** $4x - 8$ **13.** $2n - 1$
3. associative   **9.** $-3x + 12$ **15.** $-3x + 10$
5. commutative   **11.** $5y$

Section 1.5
1. -7 **7.** $<$ **13.** $3a^5$ **19.** 3×10^{-4} **25.** 0.0000537
3. 72   **9.** y^8 **15.** -8 **21.** 9.1×10^7 **27.** $-w^2$
5. $>$ **11.** $\frac{1}{n^2}$ **17.** 11   **23.** 1.25×10^{-10} **29.** $8n^2 + 8$

Section 1.6
1. 1   **5.** 5   **9.** 2   **13.** 17   **17.** 3   **21.** 12
3. 7   **7.** 2.64   **11.** 100   **15.** -7 **19.** -1

Regents Tune-Up: Chapter 1
1. 13   **5.** -2 **9.** (2)   **13.** (4)
3. 5   **7.** (1)   **11.** (1)   **15.** (2)

CHAPTER 2
LOGIC AND TRUTH TABLES

2.1 FINDING TRUTH VALUES OF SENTENCES

$$\bigwedge$$ **KEY IDEAS**

A **statement** is an English or mathematical sentence that can be judged to be either true or false. When a mathematical sentence contains a variable, as in $x + 1 = 6$, it is called an **open sentence** since its truth or falsity cannot be determined until variable x is replaced by a particular number.

Statements and Their Truth Values

Consider whether each of the following sentences is true or false.

1. Daffodils are prettier than roses.
2. It is a flower with five petals.
3. A daffodil is a flower.

The first sentence reflects an opinion and cannot be judged to be true or false. The second sentence is an **open sentence** since its truth value cannot be determined until *it* is replaced by the name of an actual type of flower. The third sentence is based on a fact that can be verified. This type of sentence is called a **statement**. A statement has a truth value of either TRUE (T) or FALSE (F), but not both.

Replacement and Solution Sets

The truth value of the *open sentence* $x + 1 = 6$ depends on the value of x. If x is replaced by 2, then the open sentence becomes the *statement* $2 + 1 = 6$, whose truth value is false; if x is replaced by 5, then the truth value of the statement $5 + 1 = 6$ is true. In an open sentence that contains a variable:

- The collection of all possible substitutions for the variable is called the **replacement set** or the **domain** of the variable.
- The **solution set** consists of those values from the replacement set, if any, that make the open sentence a true statement.

For example, if the replacement set for y is $\{-3, -2, -1, 0, 1, 2, 3\}$, then the solution set of $y^2 - 4 = 0$ is $\{-2, 2\}$. The sentence $y^2 - 4 = 0$ is a true statement when y is replaced by -2 since $(-2)^2 - 4 = 4 - 4 = 0$. The number 2 is also a

31

member of the solution set since $(2)^2 - 4 = 4 - 4 = 0$. Table 2.1 illustrates that a solution set may contain no members or an infinite number of members.

TABLE 2.1 FINDING SOLUTION SETS

Replacement Set	Open Sentence	Solution Set
1. {Whole numbers}	$x < 1$	{0}
2. {−2, −1, 0, 1, 2}	$x < 1$	{−2, −1, 0}
3. {Integers}	$x < 1$	{. . . , −4, −3, −2, −1, 0}
4. {2, 4, 6}	$x < 1$	{ } or ∅

Exercise Set 2.1

1–7. Determine the solution set.

Replacement Set	Open Sentence	Solution Set
1. {Whole numbers}	n is between 2 and 7.	?
2. {Natural numbers}	n is between 3 and 5, inclusive.	?
3. {Whole numbers ≤ 13}	x is divisible by 3.	?
4. {Real numbers}	$x + 1 > 4$.	?
5. {Real numbers}	$x = x + 1$	?
6. {Integers}	n is a prime number between 50 and 70.	?
7. {1, 3, 5}	$x + 1 = 3$.	?

8–11. The replacement set for each of the following open sentences is {−2, −1, 0, 1, 2}. Write the solution set for each open sentence.

8. $|x| = 2$ **9.** $|x| = x$ **10.** $|x| + x = 0$ **11.** $x^2 = 1$

2.2 NEGATING STATEMENTS

KEY IDEAS

A new statement, called the **negation** of a statement, is formed by inserting the word *not* so that the resulting statement has the opposite truth value of the original statement.

Forming a Negation of a Statement

Here is an example of a statement and its negation:

    Original statement: Monday is the day after Sunday.   (True)
        Negation: Monday is *not* the day after Sunday.   (False)
                  or
        It is *not* true that Monday is the day after Sunday.

Notation

Sometimes it is convenient to use letters to represent statements. For example,

$$p: \text{Monday is the day after Sunday.}\quad\text{(True)}$$

is read as "Let *p* represent the statement 'Monday is the day after Sunday'." If *p* represents a statement, then ~*p* (read as "not *p*") represents the negation of statement *p*. Thus,

$$\sim p: \text{Monday is } not \text{ the day after Sunday.}\quad\text{(False)}$$

Truth Tables and Negations

If ~*p* represents the negation of statement *p*, then ~(~*p*) represents the negation of the negation of *p* (or "double" negation of *p*). Statements *p* and ~*p* have opposite truth values, whereas *p* and ~(~*p*) have the same truth value. For example,

    p: $2 + 3 = 5$.   (True)
    ~*p*: It is not true that $2 + 3 = 5$, *or, simply*, $2 + 3 \neq 5$.   (False)
    ~(~*p*): It is not true that $2 + 3 \neq 5$.   (True)

    The set of possible truth values of statements can be organized into **truth tables**. The truth table for statement *p* and its negations is given in Table 2.2. Notice that the first vertical column lists the possible truth values for statement *p* (T or F). The second column lists the corresponding truth values for ~*p*, while the last column negates (takes the opposite of) the truth values found in the second column.

TABLE 2.2 TRUTH TABLE FOR NEGATIONS OF STATEMENT *p*

p	~*p*	~(~*p*)
T	F	T
F	T	F

Negating Inequalities

Table 2.3 shows how to negate inequalities.

TABLE 2.3 NEGATING INEQUALITIES

Inequality	Negation
>	$\not>$ or $\leq$
$\geq$	$\not\geq$ or $<$
<	$\not<$ or $\geq$
$\leq$	$\not\leq$ or $>$

Example

Write the negation of p when:
(a) p: $7 < 10$ **(b)** p: 2 is not a prime number

Solution:
(a) p: $7 < 10$ (True)
       $\sim p$: $7 \geq 10$ (False)
(b) p: 2 is not a prime number.   (False)
        $\sim p$: It is not true that 2 is not a prime number   (True)
            or, simply, 2 is a prime number.   (True)

Exercise Set 2.2

1–7. Form the negation of each of the given statements. Determine the truth value of the original statement and its negation.

1. $3 + 4 = 4 + 3$.

2. $5 < 9$.

3. A week has 7 days.

4. $3 \geq 1$.

5. The sum of two even numbers is an odd number.

6. A triangle has three sides.

7. Parallel lines do not intersect.

8. If p is false, then the truth value of $\sim(\sim p)$ is _____.

9. If $\sim p$ is true, then the truth value of p is _____.

10. Let q represent the statement "$x > 3$." If the domain of x is the set of whole numbers, for what value(s) of x is $\sim q$ true?

11. If $x = 4$, then $\sim p$ is true for which of the following statements?
   (1) p: $x - 1 = 3$ (3) p: x is not a prime number.
   (2) p: $x > 4$ (4) p: $x - 1 < 5$

2.3 CONNECTING STATEMENTS WITH *AND* AND *OR*

KEY IDEAS

The words AND and OR are **logical connectives**. A **compound statement** is formed by combining two or more simple statements with a logical connective.

Much of our study of logic concerns forming compound statements, and then analyzing their truth values.

Conjunction

A new statement can be formed by taking two given statements and connecting them by placing the word AND between them. The resulting compound statement is called a **conjunction**. The conjunction of p and q is written symbolically as $p \wedge q$ (read as "p and q"). The conjunction $p \wedge q$ is true only when p and q are true at the same time. As an illustration, consider these statements:

$$p: \ x \text{ is divisible by 2.}$$
$$q: \ x \text{ is divisible by 3.}$$
$$p \wedge q: \ x \text{ is divisible by 2 and } x \text{ is divisible by 3.}$$

In this example, the truth value of $p \wedge q$ depends on the value of x. If $x = 6$, then p is true since 6 is divisible by 2, and q is true since 6 is divisible by 3. Hence $p \wedge q$ is true. If $x = 8$, however, p is true but q is false, so that $p \wedge q$ is false.

TABLE 2.4 TRUTH TABLE FOR CONJUNCTION $p \wedge q$

p	q	$p \wedge q$
T	T	T
T	F	F
F	T	F
F	F	F

The truth table given in Table 2.4 summarizes the different possible combinations of truth values of p and q. In the first row, p and q are both true so that the conjunction of these statements is true. In the remaining rows at least one of the statements, p or q that make up the conjunction is false, so that the conjunction is also false.

Example

1. Let p represent "A square has three sides" and q represent "2 is a prime number." Determine the truth of each of the following:

    (a) $p \wedge q$ **(b)** $\sim(p \wedge q)$ **(c)** $\sim p \wedge q$

    Solution:

(a) False (since p is false and q is true, $p \wedge q$ is false).

(b) True (since the conjunction is false, its negation is true).

(c) True (since $\sim p$ is true and q is true, their conjunction is true).

Disjunction

When two statements are joined with the word OR, a new statement that is called the **disjunction** of the original statements is formed. The disjunction of statements p and q is written as $p \vee q$ and read as "p or q." The disjunction $p \vee q$ is true when p is true or q is true or both p and q are true. As an example, consider these statements:

    p: x is an even number.
    q: x is a prime number.
    $p \vee q$: x is an even number or x is a prime number.

In this example the truth value of $p \vee q$ depends on the value of x. If $x = 8$, p is true, q is false, and $p \vee q$ is true. If $x = 9$, however, p and q are each false, which makes $p \vee q$ false.

Table 2.5 gives a truth table for the disjunction $p \vee q$. In the first three rows of this truth table, at least one of the component statements is true, so that the disjunction is true. In the last row of the table both p and q are false, so that the disjunction of p and q is false.

TABLE 2.5 TRUTH TABLE FOR DISJUNCTION $p \vee q$

p	q	$p \vee q$
T	T	T
T	F	T
F	T	T
F	F	F

Examples

2. If *p* represents "Math is fun," and *q* represents "Math is difficult," write in symbolic form, using *p* and *q*:

(**a**) Math is fun or math is difficult.
(**b**) Math is fun and math is not difficult.
(**c**) It is not true that math is not fun or math is difficult.

Solution: (**a**) $p \vee q$ (**b**) $p \wedge \sim q$ (**c**) $\sim(\sim p \vee q)$

3. If *p* represents "*x* is divisible by 3" and *q* represents "*x* is the LCM (least common multiple) of 4 and 6," then what is the truth value of each of the following statements when *x* = 24?

(**a**) $p \wedge q$ (**b**) $p \vee q$ (**c**) $\sim p \vee \sim q$ (**d**) $\sim(p \wedge q)$

Solution: The LCM of 4 and 6 is 12 since 12 is the smallest positive number that is divisible by both 4 and 6. If *x* = 24, then *p* is true (since 12 is divisible by 3) and *q* is false.

(**a**) $p \wedge q$ is **false** since *q* is false.
(**b**) $p \vee q$ is **true** since *p* is true.
(**c**) $\sim p \vee \sim q$ is **true** since $\sim q$ is true.
(**d**) $\sim(p \wedge q)$ is **true** since the conjunction is false, so that its negation is true.

4. Fill in the missing truth values.

(**a**)

p	*q*	*p* ∨ *q*	*p* ∧ *q*
?	F	T	?

(**c**)

p	*q*	*p* ∨ *q*	$\sim(p \vee q)$
F	?	?	T

(**b**)

p	*q*	$\sim p$	$\sim p \vee q$
?	F	?	T

(**d**)

p	*q*	$\sim q$	*p* ∧ $\sim q$
?	?	?	T

Solution:

(**a**) Since the disjunction is true and *q* is false, *p* must be **true**. The conjunction $p \wedge q$ is **false** since *q* is false.

(**b**) Since the disjunction is true and *q* is false, $\sim p$ must be **true**, so that *p* is **false**.

(**c**) In order for the statement in the last column to be true, the disjunction $p \vee q$ must be **false**. This implies that both *p* and *q* must be **false**.

(**d**) Since the conjunction is true, each of its members must also be true, so that *p* is **true** and $\sim q$ is **true**; therefore *q* is **false**.

Constructing a Truth Table for a Compound Statement

The truth value of a compound statement can be determined, for each possible combination of truth values of its component statements, by developing a truth table. To construct a truth table that analyzes the truth value of the compound statement

$$(q \wedge \sim p) \vee \sim q$$

proceed as follows:

Step 1. Label the first two columns of the truth table as p and q, and label the last column as $(q \wedge \sim p) \vee \sim q$. Then, working from left to right, label the columns in between with the component parts of the compound statement. Be sure to provide separate column headings for negations and statements involving a connective. For example,

(1)	(2)	(3)	(4)	(5)	(6)
p	q	$\sim p$	$q \wedge \sim p$	$\sim q$	$(q \wedge \sim p) \vee \sim q$

Step 2. Complete columns (1) and (2) by listing all possible combinations of truth values for p and q. It is common practice to list these combinations in the following order:

	(1)	(2)
	p	q
row 1	T	T
row 2	T	F
row 3	F	T
row 4	F	F

Step 3. In each row of column (3) enter the opposite (negation) of the corresponding truth value in column (1). (*Note: corresponding* means "in the same row.")

Step 4. In each row of column (4) enter the truth value obtained by taking the conjunction of the corresponding truth values in columns (2) and (3).

Step 5. In each row of column (5) enter the opposite (negation) of the corresponding truth value in column (2).

Step 6. In each row of column (6) enter the truth value obtained by taking the disjunction of the corresponding truth values in columns (4) and (5). Here is the completed truth table.

(1)	(2)	(3)	(4)	(5)	(6)
p	*q*	~*p*	*q* ∧ ~*p*	~*q*	(*q* ∧ ~*p*) ∨ ~*q*
T	T	F	F	F	F
T	F	F	F	T	T
F	T	T	T	F	T
F	F	T	F	T	T

Exercise Set 2.3

1–4. Let p *represent the statement "All rectangles are squares,"* q *represent the statement "The product of two odd numbers is an odd number," and* r *represent the statement "The product of two negative numbers is negative." Express each of the following compound statements as an English sentence and give its truth value:*

1. $p \land q$ **2.** $p \lor q$ **3.** $\sim r \lor q$ **4.** $\sim(p \land r)$

5–7. Fill in the blank with either true *or* false *so that the resulting statement is correct.*

5. If $p \lor \sim q$ is true and p is false, then q is _____.

6. $p \land q$ and $p \lor q$ have the same truth value when p is false and q is _____.

7. If $\sim(p \land q)$ is true and p is true, then q is _____.

8. Which statement is always true?
(1) $p \lor \sim p$ (2) $p \land q$ (3) $p \land \sim p$ (4) $(q \lor \sim p)$

9. Which statement is always false?
(1) $\sim q$ (2) $\sim p \lor \sim q$ (3) $q \land \sim q$ (4) $\sim(p \land q)$

10–11. Let p *represent "It is cold in January" and* q *represent "It snows in August."*

10. Which statement represents "It is cold in January or it does not snow in August."?
(1) $p \land \sim q$ (2) $\sim p \lor q$ (3) $p \lor \sim q$ (4) $\sim(p \land q)$

11. Which statement represents "It is not true that it is cold in January and it snows in August."?
(1) $\sim p \land \sim q$ (2) $\sim(p \land q)$ (3) $\sim p \land q$ (4) $\sim p \lor q$

12–13. Find the missing truth values.

12.

p	q	$\sim q$	$p \wedge \sim q$	$\sim(p \wedge \sim q)$
?	F	?	?	F

13.

p	q	$\sim p$	$\sim p \wedge q$	$\sim q$	$p \vee \sim q$
?	?	?	?	?	F

14. If p represents the statement "n is divisible by 4," and q represents the statement "n is divisible by 3," what is truth value of the compound statement $p \vee \sim q$ for each of the following values of n?
(a) $n = 6$ (b) $n = 12$ (c) $n = 7$ (d) $n = 8$

15. Let p represent the statement "The quotient of x and 4 has a remainder of 1," and q represent the statement "x is divisible only by itself and 1." Which statement is true when $x = 17$?
(1) $\sim(p \vee q)$ (2) $\sim(p \wedge q)$ (3) $p \wedge q$ (4) $\sim p \wedge q$

16–18. In each case construct a truth table for all possible combinations of truth values of p *and* q.

16. $\sim(p \wedge \sim q)$ **17.** $(\sim p \vee \sim q) \wedge p$ **18.** $(p \wedge q) \vee \sim q$

2.4   ANALYZING A CONDITIONAL STATEMENT

$$\overset{\wedge}{\underset{\diagup \quad \diagdown}{\text{Key Ideas}}}$$

A new statement called a **conditional** is formed from statements p and q by writing, "If p, then q." A conditional is false in the single instance when p is true and q is false.

Truth Value of a Conditional

The shorthand notation $p \rightarrow q$ (read as "p implies q") is used to represent the conditional, "If p, then q." The statement in the "If" clause is called **hypothesis** (or **antecedent**), while the statement in the "then" clause is called the **conclusion** (or **consequent**). For example, consider these statements:

p: I study.
q: Mr. Euclid will pass me.
$p \rightarrow q$: If I study, then Mr. Euclid will pass me.
        hypothesis   *conclusion*

A conditional may be thought of as being true provided that whatever "promise" is made in the conclusion part of the statement is not broken when the hypothesis is true. This is illustrated in the truth table given in Table 2.6. The truth table demostrates that a conditional is always true except in the single case in which the hypothesis p is true, and the conclusion q is false (see row 2 of the truth table).

TABLE 2.6 TRUTH TABLE FOR CONDITIONAL $p \to q$

	p	q	$p \to q$	Explanation
row 1	T	T	T	I study, and Mr. Euclid passes me.
row 2	T	F	(F)	Mr. Euclid fails to keep his promise.
row 3	F	T	T	No broken promise since I still pass.
row 4	F	F	T	No broken promise since I did not study.

Examples

1. Let p represent "x is divisible by 2," and q represent "The fraction $\dfrac{1}{x-9}$ is undefined." Which of the following statements is true when $x = 9$?

(1) $p \wedge q$ (2) $p \vee \sim q$ (3) $p \to \sim q$ (4) $q \to p$

Solution: When $x = 9$, p is false and q is true. In **choice (3)** the conditional is true since both the hypothesis and the conclusion are false.

2. Consider these statements:

        p: 9 is a prime number.
        q: 12 is divisible by 2.
        r: A square has four equal sides.

Write in words and determine the truth value of each of the following:
(a) $q \to p$ **(b)** $p \to \sim r$ **(c)** $\sim(p \to q)$

Solution:
  (a) $q \to p$: If 12 is divisible by 2, then 9 is a prime number. **False**, since the hypothesis (q) is true and the conclusion (p) is false.
  (b) $p \to \sim r$: If 9 is a prime number, then a square does not have four equal sides. **True**, since the hypothesis (p) is false and the conclusion ($\sim r$) is false.
  (c) $\sim(p \to q)$: It is not true that, if 9 is a prime number, then 12 is divisible by 2. **False**. The truth value of $p \to q$ is true since p is false and q is true. Therefore, the negation of $p \to q$ is false.

Constructing a Truth Table for a Conditional Statement

To construct a truth table for

$$[(p \wedge q) \vee \sim p] \to (p \vee q),$$

begin by labeling the first two columns as p and q. The statement enclosed by the brackets is the hypothesis, and the expression inside the parentheses, to the right of the arrow, is the conclusion. The next set of columns of the truth table breaks down the hypothesis into its component statements [see columns (3) through (5) in the accompanying truth table]. The truth value of the conclusion is given in column (6) of the accompanying truth table. The last column of the truth table is labeled with the original conditional. The columns of the truth table are completed as follows:

Columns (1) and (2): The truth values are entered, as usual, for p and q.

Column (3): The conjunction of statements in columns (1) and (2) is true only when both p and q are true. This happens in the first row.

Column (4): In each row the negation of the truth value found in the corresponding row of column (1) is entered.

Column (5): The disjunction is true except in the case in which corresponding rows of columns (3) and (4) are both false. This happens in the second row. This column represents the hypothesis of the given conditional.

Column (6): This disjunction is true except in the case in which corresponding rows of columns (1) and (2) are both false. This is the case in the last row. This column represents the conclusion of the given conditional.

Column (7): The conditional is false only when the hypothesis in column (5) is true and the conclusion in column (6) is false. This situation occurs in the last row of the truth table.

(1)	(2)	(3)	(4)	(5)	(6)	(7)
p	q	$p \wedge q$	$\sim p$	$[(p \wedge q) \vee \sim p]$	$p \vee q$	$[(p \wedge q) \vee \sim p] \to (p \wedge q)$
T	T	T	F	T	T	T
T	F	F	F	F	T	T
F	T	F	T	T	T	T
F	F	F	T	T	F	F

Forms of a Conditional

Implication is another name for a conditional. The "If p, then q" form of the conditional may be expressed in any one of several equivalent forms.

Equivalent Forms	Example
If p, then q.	If I am tired, then I will rest.
p implies q.	I am tired implies I will rest.
q if p.	I will rest if I am tired.
p only if q.	I am tired only if I will rest.

Exercise Set 2.4

1–5. Fill in each blank with either true *or* false *so that the resulting statement is correct.*

1. $p \rightarrow q$ is false when p is _____ and q is _____.

2. If $p \wedge q$ is true, then $p \rightarrow q$ is _____.

3. If $p \rightarrow p$ is false, then $p \vee q$ is _____.

4. $q \rightarrow \sim p$ is false when q is true and p is _____.

5. If $p \rightarrow q$ is true and q is false, then p is _____.

6. Let p represent "x is a prime number," and q represent "x is not divisible by 2." Write each of the following as an English sentence, and state its truth value when x is 23:
   (a) $p \rightarrow q$ (b) $q \rightarrow p$ (c) $\sim p \rightarrow \sim q$ (d) $\sim q \rightarrow \sim p$

7. Let p represent "x is a multiple of 3," and q represent "$x + 1 > 10$." Which of the following statement is true when $x = 9$?
   (1) $p \rightarrow q$ (2) $\sim p \rightarrow \sim q$ (3) $\sim(p \wedge q) \rightarrow q$ (4) $(p \vee q) \rightarrow (p \wedge q)$

8. Let p represent "A square is a rectangle," and q represent "A circle has three sides." Which of the following statements is true?
   (1) $p \rightarrow (p \vee q)$ (3) $(p \vee q) \rightarrow (p \wedge q)$
   (2) $\sim q \rightarrow q$ (4) $\sim q \rightarrow \sim(p \vee q)$

9. Which of the following statements is true for all possible truth values of p and q?
   (1) $p \rightarrow \sim q$ (3) $p \rightarrow (q \vee \sim q)$
   (2) $(p \vee \sim p) \rightarrow q$ (4) $\sim(p \wedge \sim p) \rightarrow q$

10. Let p represent the statement "x is divisible by a prime number that is less than x," and q represents the statement "x can be written as the product of two identical numbers." Which of the following statements are true when $x = 15$?
   (1) $p \rightarrow q$ (2) $p \rightarrow \sim q$ (3) $p \wedge q$ (4) $\sim q \rightarrow \sim p$

11–13. Construct a truth table for each of the following statements:

11. $(p \wedge q) \rightarrow \sim p$

12. $(p \wedge \sim q) \rightarrow (p \vee q)$

13. $[(p \rightarrow q) \wedge \sim q] \rightarrow \sim p$

2.5 FORMING THE CONVERSE, INVERSE, AND CONTRAPOSITIVE

KEY IDEAS

By interchanging or negating both parts of a conditional, or by doing both, three conditionals of special interest—the **converse**, the **inverse**, and the **contrapositive**—can be formed. The relationships between certain pairs of these derived conditionals illustrate that it is possible for pairs of statements to look different, yet always agree in their truth values.

Related Conditionals

Starting with the conditional $p \rightarrow q$, we can form three related conditionals.

Statement	Symbolic Form	To Obtain from Original...
Original	$p \rightarrow q$	
Converse	$q \rightarrow p$	Switch p with q.
Inverse	$\sim p \rightarrow \sim q$	Negate both p and q.
Contrapositive	$\sim q \rightarrow \sim p$	Switch and negate p and q.

As an example, let p represent "$x = 2$," and q represent "x is even." Then:

> *Original*: If $x = 2$, then x is even.   (True)
> *Converse*: If x is even, then $x = 2$.   (False)
> *Inverse*: If $x \neq 2$, then x is not even.   (False)
> *Contrapositive*: If x is not even, then $x \neq 2$.   (True)

As the preceding example indicates, the truth value of the converse *may* be different from the truth value of the original statement. The inverse and the converse will always have the same truth value. The contrapositive may be interpreted as the inverse of the converse or, equivalently, as the converse of the inverse. The original statement and its contrapositive will always have the same truth value.

Examples

1. What is the inverse of $\sim p \to q$?

Solution: Forming the inverse requires negating both parts of the conditional: $\sim(\sim p) \to \sim(q)$. The hypothesis of this conditional may be simplified by keeping in mind that the consecutive negations cancel out since one undoes the effect of the other. The correct answer is $\boldsymbol{p \to \sim q}$.

2. Let p represent "It is raining," and q represent "I have my umbrella." Express in words the contrapositive of $q \to p$.

Solution: The contrapositive of $q \to p$ is $\sim p \to \sim q$, so that:

$q \to p$: If <u>I have my umbrella</u>, then it is <u>raining</u>.

negation negation

$\sim p \to \sim q$: If <u>it is not raining</u>, then <u>I do not have my umbrella</u>.

Logical Equivalence

A pair of statements are *logically equivalent* if they always have the same truth value. Table 2.7 compares the truth values of the related forms of a conditional. Columns (1) and (5) show that a statement and its contrapositive will always have the same truth value. Therefore, a statement and its contrapositive are logically equivalent. Columns (2) and (4) show that the converse and the inverse also are logically equivalent since their truth values are always in agreement.

TABLE 2.7 TRUTH TABLE FOR RELATED CONDITIONALS

		(1) Original	(2) Converse	(3) Negations		(4) Inverse	(5) Contrapositive
p	q	$p \to q$	$q \to p$	$\sim p$	$\sim q$	$\sim p \to \sim q$	$\sim q \to \sim p$
T	T	T	T	F	F	T	T
T	F	F	T	F	T	T	F
F	T	T	F	T	F	F	T
F	F	T	T	T	T	T	T

logically equivalent statements

Examples

3. Which statement is logically equivalent to $r \rightarrow \sim s$?
   (1) $\sim s \rightarrow r$ (2) $r \rightarrow s$ (3) $\sim s \rightarrow \sim r$ (4) $s \rightarrow \sim r$

Solution: A conditional and its contrapositive are logically equivalent. Choice (1) represents the converse. Negating both parts of the converse leads to the statement in choice (4), which is the contrapositive. The correct choice is **(4)**.

4. Given the true statement "If a figure is a square, then the figure is a rectangle," which statement is also true?
   (1) If a figure is not a rectangle, then it is a square.
   (2) If a figure is a rectangle, then it is a square.
   (3) If a figure is not a rectangle, then it is not a square.
   (4) If a figure is not a square, then the figure is not a rectangle.

Solution: A conditional and its contrapositive always have the same truth value. Since the original statement is true, the contrapositive must also be true. The contrapositive of the given statement is given in choice (3). Hence, the correct choice is **(3)**.

5. Construct a truth table to show that the statements

$$q \rightarrow (p \wedge q) \text{ and } (p \vee q) \rightarrow p$$

are logically equivalent.

Solution:

(1)	(2)	(3)	(4)	(5)	(6)
p	q	$p \wedge q$	$q \rightarrow (p \wedge q)$	$p \vee q$	$(p \vee q) \rightarrow p$
T	T	T	T	T	T
T	F	F	T	T	T
F	T	F	F	T	F
F	F	F	T	F	T

The statement $q \rightarrow (p \wedge q)$ is logically equivalent to $(p \vee q) \rightarrow p$ since columns (4) and (6) have the same truth value for each row of the truth table.

Exercise Set 2.5

1–6. Fill in the blank in each of the following so that the resulting statement is true:

1. A conditional and its _____ are logically equivalent.

2. If the inverse of a statement is true, then the _____ must also be true.

3. If j represents the truth value of a conditional statement and k represents the truth value of the contrapositive, then the truth value of $j \wedge k$ is _____.

4. The inverse of "If $x > 0$, then x is positive" is the statement_____.

5. If the converse of a statement is $p \rightarrow \sim q$, then the original statement is _____.

6. The truth value of the converse of the statement "If the sum of two numbers is negative, then the two numbers are negative" is _____.

7–9. Form the converse, inverse, and contrapositive of each of the following:

7. If a triangle has two equal sides, then the triangle is isosceles.

8. If n is an odd integer, then $n + 1$ is an even integer.

9. $p \rightarrow \sim q$.

10. Which statement is logically equivalent to the statement "If it is sunny, then it is hot"?
(1) If it is hot, then it is sunny.
(2) If it is not hot, then it is not sunny.
(3) If it is not sunny, then it is not hot.
(4) If it is not hot, then it is sunny.

11. Which statement is logically equivalent to $\sim p \rightarrow q$?
(1) $p \rightarrow \sim q$ (2) $\sim q \rightarrow p$ (3) $q \rightarrow \sim p$ (4) $\sim q \rightarrow \sim p$

12. In which of the following pairs of statements are the statements logically equivalent?
(1) $k \rightarrow h$ and $\sim h \rightarrow \sim k$ (3) $k \rightarrow h$ and $\sim k \rightarrow \sim h$
(2) $h \rightarrow k$ and $\sim h \rightarrow \sim k$ (4) $h \rightarrow k$ and $k \rightarrow h$

13. Which statement is logically equivalent to the statement "If $n \leq 5$, then $n \leq 8$"?
(1) If $n > 8$, then $n > 5$.   (3) If $n > 5$, then $n > 8$.
(2) If $n \leq 5$, then $n \leq 8$.   (4) If $n \leq 8$, then $n \leq 5$.

14. Which statement is logically equivalent to the *inverse* of the statement "If I study hard, then I will pass."?
(1) If I do not study hard, then I will pass."
(2) If I do not pass, then I will not study hard.
(3) If I will pass, then I study hard.
(4) If I will pass, then I do not study hard.

15–18. Construct a truth table to determine whether, in each given pair of statements, the statements are logically equivalent.

15. ~($p \wedge q$) and (~$p \vee$ ~q)   **17.** ~$q \rightarrow$ ~p and $p \vee$ ~q

16. ($p \vee q$) $\rightarrow p$ and $p \vee$ ~q **18.** ($p \wedge q$)$\rightarrow$($p \vee q$) and ~$q \rightarrow$ ($p \vee q$)

2.6 PROVING THAT A STATEMENT IS A TAUTOLOGY

≡ KEY IDEAS ≡

Sometimes statements are combined in such a way that the resulting compound statement is always true or is always false.

Tautologies

What is the truth value of the compound statement $p \vee$ ~p when p is true? When p is false? The statement is true regardless of the truth value of p. A **tautology** is a compound statement that is always true regardless of the truth values of its component statements. The statement $p \vee$ ~p is an example of a tautology. Usually we are interested in determining whether more complicated statements are tautologies, in which case a truth table must be constructed.

Example

Construct a truth table to prove that [($p \rightarrow q$) $\wedge p$]$\rightarrow q$ is a tautology.

Solution:

p	q	$p \rightarrow q$	($p \rightarrow q$) $\wedge p$	[$p \rightarrow q \wedge p$] $\rightarrow q$
T	T	T	T	T
T	F	F	F	T
F	T	T	F	T
F	F	T	F	T

Since the last column of the truth table shows that [($p \rightarrow q$) $\wedge p$] $\rightarrow q$ is true for all combinations of truth values of p and q, this compound statement is a tautology.

Contradictions

A compound statement that is always false is called a **contradiction**. The statement $p \wedge {\sim}p$ is an example of a contradiction. It is always false since a statement and its negation cannot be true at the same time.

Exercise Set 2.6

1. Which of the following statements is a tautology?
(1) ${\sim}(p \wedge {\sim}p)$ (3) $p \vee {\sim}q$
(2) ${\sim}(p \vee q)$ (4) $q \wedge {\sim}q$

2. Which of the following statements is a tautology?
(1) ${\sim}(p \rightarrow {\sim}p)$ (3) $q \rightarrow {\sim}q$
(2) $p \rightarrow q$ (4) $(p \rightarrow q) \vee (q \rightarrow p)$

3–8. In each case, construct a truth table in order to determine whether the statement is a tautology.

3. $(p \wedge {\sim}q) \rightarrow ({\sim}q \rightarrow {\sim}p)$ **6.** $[(p \wedge q) \vee {\sim}(p \rightarrow q)] \rightarrow {\sim}p \vee q$

4. $(p \rightarrow {\sim}q) \rightarrow {\sim}(p \wedge q)$ **7.** $[p \wedge (p \rightarrow q)] \rightarrow q$

5. $[(p \vee q) \wedge ({\sim}p)] \rightarrow q$ **8.** $[p \wedge q) \rightarrow p] \wedge [q \rightarrow (p \vee q)]$

2.7 UNDERSTANDING THE BICONDITIONAL

Key Ideas

The conjunction of a conditional $p \rightarrow q$ and its converse $q \rightarrow$ p is called a **biconditional** and is written p $\leftrightarrow$ q. Since a conditional and its converse may have different truth values, a biconditional may or may not be true. When statements p and q are both true or are both false, $p \rightarrow q$ is true and, at the same time, $q \rightarrow$ p is also true so the biconditional $p \leftrightarrow q$ is true.

An Example of a True Biconditional

According to a fundamental law of mathematics, if the product of two real numbers is 0, then at least one of these numbers is 0. It follows that $5n = 0$ "if and only if" $n = 0$. We mean by this that the compound statement

"If $5n = 0$, then $n = 0$." *AND* "If $n = 0$, then $5n = 0$."

is a true statement. This compound statement is a *biconditional*. When $5n = 0$ and $n = 0$ are both true statements or are both false statements, the two conditionals that the biconditional joins are true at the same time. This makes the biconditional (conjunction) a true statement.

Definition of a Biconditional

The conjunction

$$(p \to q) \wedge (q \to p)$$

is called a **biconditional** and is abbreviated by writing $p \leftrightarrow q$. The statement $p \leftrightarrow q$ is read "p if and only if q." Table 2.8 shows that the biconditional $p \leftrightarrow q$ is true when p and q are both true or are both false. When statements p and q have opposite truth values, $p \leftrightarrow q$ is false.

Since $p \leftrightarrow q$ and $q \leftrightarrow p$ refer to the conjunction of the same conditional and its converse, they are equivalent statements. Thus, the order in which p and q appear in a biconditional does not matter.

Table 2.8 TRUTH TABLE FOR BICONDITIONAL $p \leftrightarrow q$

p	q	$p \leftrightarrow q$
T	T	T
T	F	F
F	T	F
F	F	T

Examples

1. Let p represent the statement "$3 + 7 = 10$," q represent the statement "A triangle has three sides," and r represent the statement "9 is a prime number." Express each of the following biconditionals as an English sentence and determine its truth value:

(**a**) $p \leftrightarrow q$ (**b**) $p \leftrightarrow r$ (**c**) $\sim q \leftrightarrow r$

Solution:

(**a**) $3 + 7 = 10$ if and only if a triangle has three sides. **True** since both p and q are true.

(**b**) $3 + 7 = 10$ if and only if 9 is a prime number. **False** since p (true) and r (false) have different truth values.

(**c**) A triangle does not have three sides if and only if 9 is the prime number. **True** since both $\sim q$ and r are false.

2. Construct a truth table to show that the statement

$$(p \to q) \leftrightarrow (\sim p \vee q)$$

is a tautology.

Solution:

(1)	(2)	(3)	(4)	(5)	(6)
p	q	$p \to q$	$\sim p$	$\sim p \vee q$	$(p \to q) \leftrightarrow (\sim p \vee q)$
T	T	T	F	T	T
T	F	F	F	F	T
F	T	T	T	T	T
F	F	T	T	T	T

Notice that in each row the truth values of columns (3) and (5) agree, so that T is entered in column (6). Since column (6) shows that the biconditional is true for all combinations of truth values of p and q,

$$(p \to q) \leftrightarrow (\sim p \vee q)$$

is a tautology.

3. In Example 2, let p represent "The machine is a computer," and q represent "The machine cannot make a mistake." What statement is logically equivalent to the statement "If the machine is a computer, then the machine cannot make a mistake"?

Solution: The statement "If the machine is a computer, then the machine cannot make a mistake" can be represented as $p \to q$. The truth table shows that $p \to q$ in column (3) is logically equivalent to $\sim p \vee q$ in column (5). The negation of p is the statement "The machine is not a computer." Therefore the logical equivalent of "If the machine is a computer, then the machine cannot make a mistake" is the statement **"The machine is not a computer or the machine cannot make a mistake."**

Exercise Set 2.7

1. Which statement is always true?
    (1) $p \wedge \sim p$ (2) $(p \wedge \sim p) \leftrightarrow (q \wedge \sim q)$ (3) $p \to \sim p$ (4) $p \wedge q$

2. If $p \leftrightarrow q$ is false, which statement *must* also be false?
    (1) $\sim p \wedge q$ (2) $p \wedge q$ (3) $p \to q$ (4) $p \vee q$

3. If j and k each represent a compound statement and $j \leftrightarrow k$ is always true, then which of the following statements is always true?
   (1) j and k are logically equivalent. (3) $(j \wedge k)$ is a tautology.
   (2) j and k are tautologies. (4) $(j \vee k)$ is a tautology.

4. Let p represent "x is a prime number," q represent "x is an odd number," and r represent "x is a multiple of 7." Express each of the following as an English sentence, and determine its truth value if $x = 21$:
   (a) $p \leftrightarrow q$ (b) $q \leftrightarrow r$ (c) $p \leftrightarrow \sim r$ (d) $\sim p \leftrightarrow (q \wedge r)$

5. Let p represent "He is a snob," and q represent "He likes cavier." Express each of the following as an English sentence:
   (a) $p \leftrightarrow q$ (b) $\sim p \leftrightarrow q$

6. If p represents "$x > 5$" and q represents "x is divisible by 3," which statement is true if $x = 10$?
   (1) $p \rightarrow q$ (2) $(p \vee q) \rightarrow q$ (3) $(p \wedge q) \rightarrow p$ (4) $p \leftrightarrow q$

7. Let p represent the statement "x is a prime number," and q represent the statement "$x + 2$ is a prime number." Which statement is false when $x = 15$?
   (1) $p \leftrightarrow q$ (2) $p \rightarrow \sim q$ (3) $\sim p \wedge q$ (4) $p \leftrightarrow \sim q$

8–11. For each of the following statements, (a) construct a truth table, and (b) determine whether the statement is a tautology:

8. $(p \rightarrow q) \leftrightarrow \sim(p \wedge \sim q)$ 10. $[q \wedge (\sim p \vee q)] \leftrightarrow (p \rightarrow q)$

9. $[(p \leftrightarrow q) \rightarrow p] \leftrightarrow (p \vee q)$ 11. $(p \wedge q) \rightarrow [p \vee q) \leftrightarrow (p \rightarrow q)]$

12. (a) Construct a truth table for the statement

$$(p \vee \sim q) \leftrightarrow (\sim p \rightarrow \sim q).$$

   (b) Determine whether the statement in part (a) is a tautology, and give a reason for your answer.
   (c) Let p represent the statement "Today is a holiday," and q represent the statement "I do not attend school." Using your results from part (a), determine which of the following statements is logically equivalent to the statement "Today is a holiday, or I attend school."
   (1) If today is a holiday, then I do not attend school.
   (2) If today is not a holiday, then I attend school.
   (3) If today is not a holiday, then I do not attend school.
   (4) If I do not attend school, then today is not a holiday.

13. Construct a truth table to show that the biconditional is logically equivalent to the conjunction of a conditional statement and its converse.

2.8 REASONING LOGICALLY

=== KEY IDEAS ===

Knowing the truth value of a compound statement, and of each of its component statements except one, may allow a conclusion to be drawn regarding the truth value of the remaining component statement.

Drawing Conclusions

If $p \vee q$ is true and p is false, we may conclude that q must be true. If it is given that p is true and the conditional $p \rightarrow q$ is also true, what can be said about the truth value of q? Since the hypothesis and the conditional are true, the conclusion cannot be false; that is, q must be true.

Next, suppose that $\sim p$ is true and $p \rightarrow q$ is true. What conclusion can be drawn about the truth value of q? No conclusion is possible. Since $\sim p$ is true, p is false. If the hypothesis is false, then the conditional is true regardless of the truth value of the conclusion.

Examples

1. If $p \leftrightarrow q$ is false, draw a conclusion regarding the truth value of each of the following:

(a) $p \wedge q$ (b) $p \vee q$ (c) $p \rightarrow q$

Solution:

(a) $p \wedge q$ is **false**. Since the biconditional is false, either p or q must be false, so that the conjunction must be false.

(b) $p \vee q$ is **true**. Since the biconditional is false, either p or q must be true, so that the disjunction is true.

(c) No conclusion. Since the biconditional is false, either p is true and q is false ($p \rightarrow q$ is false), or p is false and q is true ($p \rightarrow q$ is true). Therefore no conclusion can be drawn regarding the truth value of $p \rightarrow q$.

2. If $p \rightarrow q$ is true and q is false, which statement is false?

(1) $\sim p$ (2) $p \leftrightarrow q$ (3) $p \vee q$ (4) $q \rightarrow p$

Solution: Suppose p is true. Since q is false, $p \rightarrow q$ is false. This contradicts the given that $p \rightarrow q$ is true. Hence, p is false. Since p and q are both false, the disjunction $p \vee q$ is false. The correct choice is **(3)**.

3. Based on the truth values that are given for the first two sentences, write the truth value for the third sentence. If the truth value cannot be determined from the information given, write "CANNOT BE DETERMINED."

(a) 1. It rains or it is cold. (TRUE)
    2. It is cold. (FALSE)
    3. It rains. (?)

(b) 1. If it is August, then it is hot. (TRUE)
    2. It is hot. (TRUE)
    3. It is August. (?)

(c) 1. $q \rightarrow p$ (TRUE)
    2. p (FALSE)
    3. $\sim q$ (?)

(d) 1. $p \vee q$ (TRUE)
    2. $p \rightarrow q$ (TRUE)
    3. q (?)

Solution:

(a) Let A represent the statement, "It rains," and B represent the statement "It is cold." The statement, "It rains or it is cold," is the disjunction of statements A and B, represented by $A \vee B$. A disjunction is true when at least one of the disjuncts is true. Since it is given that $A \vee B$ is TRUE and B is FALSE, A must be TRUE. Hence, the statement "It rains" is **TRUE**.

(b) Let A represent the statement "It is August," and B represent the statement "It is hot." The first statement, "If it is August, then it is hot," is the conditional statement A implies B, represented by $A \rightarrow B$. A conditional statement is always true except in the single instance in which A is TRUE and B is FALSE. Since B is TRUE, $A \rightarrow B$ is TRUE when A is TRUE and $A \rightarrow B$ is also TRUE when A is FALSE.

Hence, the truth value of the statement "It is August" **CANNOT BE DETERMINED**.

(c) A conditional statement of the form $q \rightarrow p$ is always true except in the single instance in which the conclusion, statement p, is false and the hypothesis, statement q, is true. Since $q \rightarrow p$ is TRUE and p is FALSE, q cannot be TRUE. Hence, q is FALSE, so $\sim q$ is **TRUE**.

(d) Since the disjunction $p \vee q$ is true, at least one of the disjuncts is true. If p and q are both true, the conditional statement $p \rightarrow q$ is true. If p is false and q is true, the $p \rightarrow q$ is also true. If p is true and q is false, $p \rightarrow q$ is false. Thus, the given truth values of the first two sentences hold, provided q is **TRUE**.

Exercise Set 2.8

1–4. If p ↔ q *is false, draw a conclusion regarding the truth value of each statement, provided it can be determined.*

1. $p \wedge q$ **2.** $p \vee q$ **3.** $p \to q$ **4.** $q \leftrightarrow p$

5. Given the true statements "If we water the lawn in the morning, then it will rain in the afternoon," and "It didn't rain in the afternoon," which statement logically follows?
(1) We didn't water the lawn in the morning.
(2) We watered the lawn in the morning.
(3) It rained in the morning.
(4) It rained in the afternoon.

6. If $B \to {\sim}C$ and C are both true statements, then which conclusion must be true?
(1) B (2) ${\sim}B$ (3) ${\sim}C$ (4) $B \to C$

7. Given the true statements "Mark goes shopping, or he goes to the movies" and "Mark doesn't go to the movies," which statement *must* also be true?
(1) Mark goes shopping.
(2) Mark doesn't go shopping.
(3) Mark doesn't go shopping, and he doesn't go to the movies.
(4) Mark stays home.

8. If ${\sim}r \to s$ and ${\sim}s$ are true, which statement *must* be true?
(1) r (2) ${\sim}r$ (3) $r \to s$ (4) $s \wedge {\sim}s$

9. Given the true statement, $[(s \vee t) \wedge {\sim}s]$, which statement *must* also be true?
(1) t (2) ${\sim}t$ (3) s (4) $s \wedge {\sim}t$

10. If $p \vee {\sim}q$ is true and $p \leftrightarrow q$ is false, then:
(1) p is true, q is false.   (3) both p and q are true.
(2) p is false, q is true.   (4) both p and q are false.

11. If $p \to q$ is false, which can never be true?
(1) $q \to p$ (2) ${\sim}q \to p$ (3) $p \vee q$ (4) $p \leftrightarrow q$

12. If $p \to q$ is true and $p \leftrightarrow q$ is false, then:
(1) p is true, q is false.
(2) p is false, q is true.
(3) both p and q are true.
(4) both p and q are false.

13. Given these true premises: (1) If John gets a driver's license, then he will buy a car, and (2) John does not buy a car, which of the following conclusions is true?
(1) John gets a driver's license.
(2) John does not get a driver's license.
(3) John buys a car.
(4) No conclusion is possible.

14–17. Based on the truth values that are given for the first two sentences, write the truth value for the third sentence. If the truth value cannot be determined from the information given, write "CANNOT BE DETERMINED."

14. (1) If it is sunny, then it will not rain.   (TRUE)
     (2) It is sunny.   (FALSE)
     (3) It will rain.   (?)

15. (1) If I study, I pass math.   (TRUE)
     (2) I pass math.   (TRUE)
     (3) I study.   (?)

16. (1) It is August and it is cold.   (FALSE)
     (2) It is not cold.   (TRUE)
     (3) It is August.   (?)

17. (1) $p \vee q$ (TRUE)
     (2) $p \wedge q$ (FALSE)
     (3) $p \rightarrow q$ (?)

REGENTS TUNE-UP: CHAPTER 2

Each of the questions in this section has appeared on a previous Course I Regents Examination. Here is an opportunity for you to review Chapter 2 and, at the same time, prepare for the Course I Regents Examination.

1. Let p represent "It is summer" and q represent "I go swimming." Using p and q, write in symbolic form: "If I go swimming, then it is summer."

2. Let p represent the statement "The base angles are congruent," and let q represent the statement "A triangle is isosceles." Using p and q, write in symbolic form: "A triangle is isosceles if and only if the base angles are congruent."

3. The inverse of a statement is $p \rightarrow \sim q$. What is the statement?

4. If *p* represents "Today is Monday" and *q* represents "I am tired," write in symbolic form, using *p* and *q*: "Today is Monday, and I am not tired."

5. Which represents the inverse of the statement "If the base angles of an isosceles triangle are congruent, then the triangle is isosceles"?
(1) If the base angles of the triangle are not congruent, then the triangle is not isosceles.
(2) If the triangle is isosceles, then the base angles are congruent.
(3) If the triangle is not isosceles, then the base angles are not congruent.
(4) If the base angles of a triangle are not congruent, then the triangle is isosceles.

6. Given the statement "If a figure is a triangle, then it is a polygon," which statement *must* be true?
(1) If a figure is a polygon, then it is a triangle.
(2) If a figure is not a triangle, then it is not a polygon.
(3) If a figure is not a polygon, then it is not a triangle.
(4) If a figure is not a triangle, then it is a polygon.

7. What is the inverse of $\sim p \to q$?
(1) $p \to \sim q$ (2) $q \to \sim p$ (3) $\sim p \to \sim q$ (4) $\sim q \to \sim p$

8. Let *p* represent "*x* is a prime number," and let *q* represent "*x* is an odd number." What is true if $x = 15$?
(1) p (2) $\sim q$ (3) $p \wedge q$ (4) $p \vee q$

9. Let *p* represent "*x* is odd." Let *q* represent "$x > 8$." When *x* is 6, which is true?
(1) p (2) q (3) $p \to q$ (4) $p \wedge q$

10. If $p \leftrightarrow q$ is true, which statement *must* also be true?
(1) $p \wedge q$ (2) $p \vee q$ (3) $p \wedge \sim q$ (4) $p \to q$

11. Let *p* represent the statement "*x* is even." Let *q* represent the statement "$x \leq 12$." Which is true if $x = 20$?
(1) $p \to q$ (2) $p \wedge \sim q$ (3) $\sim p \vee q$ (4) all of these

12. (a) Construct a truth table for the statement
$$[\, p \vee (\sim p \wedge q)\,] \leftrightarrow (\, p \vee q).$$
(b) On the basis of your results from part (a), is $[\, p \vee (\sim p \wedge q)\,] \leftrightarrow (\, p \vee q)$ a tautology?

13. (a) On your answer paper, copy and complete the truth table for the statement

$$(\sim p \to q) \leftrightarrow (p \lor q).$$

p	*q*	*~p*	*~p → q*	*p ∨ q*	*(~p → q) ↔ (~p ∨ q)*

(b) Is $(\sim p \to q) \leftrightarrow (p \lor q)$ a tautology?

(c) Let *p* represent: "I do my homework."
Let *q* represent: "I get into trouble."
Which sentence is equivalent to $(p \lor q)$?
(1) If I do not do my homework, I will get into trouble.
(2) I do my homework or I do not get into trouble.
(3) If I do my homework, I get into trouble.
(4) I do my homework and I get into trouble.

14. (a) Copy and complete the truth table for the statement

$$(p \to q) \leftrightarrow (q \lor \sim p).$$

p	*q*	*p → q*	*~p*	*q ∨ ~p*	*(p → q) ↔ (q ∨ ~p)*

(b) Why is $(p \to q) \leftrightarrow (q \lor \sim p)$ a tautology?

(c) In $(p \to q) \leftrightarrow (q \lor \sim p)$, let *p* represent "We pollute the water," and let *q* represent "The fish will die." Which statement is logically equivalent to "If we pollute the water, then the fish will die"?
(1) The fish will die or we do not pollute the water.
(2) We pollute the water and the fish will die.
(3) If we do not pollute the water, then the fish will not die.
(4) If the fish die, then we pollute the water.

15. Each part consists of three sentences. Write the truth value for the third sentence in each part based on the truth values given for the first two sentences.

(a)
John will buy a car.	TRUE
John will get a raise.	FALSE
John will buy a car if and only if John will get a raise.	?

(b)
Physics is a science.	TRUE
Jan plays piano.	FALSE
Physics is a science and Jan does *not* play piano.	?

(c)
If a polygon is a square, then the polygon is a rectangle.	TRUE
Polygon *ABCD* is a square.	TRUE
Polygon *ABCD* is a rectangle.	?

(d)
I like apples or it is Tuesday.	FALSE
I like apples.	FALSE
It is Tuesday.	?

(e)
If it is snowing, the roads are slippery.	TRUE
The roads are not slippery.	TRUE
It is not snowing.	?

Answers to Odd-Numbered Exercises: Chapter 2

Section 2.1

1. {3, 4, 5, 6}   **5.** { }   **9.** {0, 1, 2}

3. {3, 6, 9, 12}   **7.** { }   **11.** {−1, 1}

Section 2.2

1. $3 + 4 \neq 4 + 3$ (F)

3. A week does not have 7 days. (F)

5. The sum of two even numbers is not an odd number. (T)

7. Parallel lines intersect. (F)

9. False.

11. (2)

Section 2.3

1. All rectangles are squares and the product of two odd numbers is an odd number. (F)

3. The product of two negative numbers is not a negative number or the product of two odd numbers is an odd number. (T)

5. False.　　**7.** True.　　**9.** (3)　　　　**11.** (2)

13. p is F, q is T, $\sim p$ is T, $\sim p \land q$ is T, and $\sim q$ is F.

15. (3)

17.

p	q	$\sim p$	$\sim q$	$\sim p \lor \sim q$	$(\sim p \lor \sim q) \land p$
T	T	F	F	F	F
T	F	F	T	T	T
F	T	T	F	T	F
F	F	T	T	T	F

Section 2.4

1. True, False　　　　**3.** True　　　　**5.** True　　**7.** (2)　　**9.** (3)

11.

p	q	$p \land q$	$\sim p$	$(p \land q) \to \sim p$
T	T	T	F	F
T	F	F	F	T
F	T	F	T	T
F	F	F	T	T

13.

p	q	$p \to q$	$\sim q$	$[(p \to q) \land \sim q]$	$\sim p$	$[(p \to q) \land \sim q] \to p$
T	T	T	F	F	F	T
T	F	F	T	F	F	T
F	T	T	F	F	T	T
F	F	T	T	T	T	T

Section 2.5

1. contrapositive　　　　**3.** true　　　**5.** $\sim q \to p$

7. Converse: If a triangle is isosceles, then the triangle has two equal sides.
Inverse: If a triangle does not have two equal sides, then the triangle is not isosceles.
Contrapositive: If a triangle is not isosceles, then the triangle does not have two equal sides.

9. Converse: $\sim q \to p$
　　Inverse: $\sim p \to q$
　　Contrapositive: $q \to \sim p$

11. (2)

13. (1)

15.

p	q	p ∧ q	~(p ∧ q)	~p	~q	~p ∨ ~q
T	T	T	F	F	F	F
T	F	F	T	F	T	T
F	T	F	T	T	F	T
F	F	F	T	T	T	T

The statements ~(p ∨ q) and ~p ∧ ~q are logically equivalent.

17.

p	q	~q	~p	~q → ~p	p ∨ ~q
T	T	F	F	T	T
T	F	T	F	F	T
F	T	F	T	T	F
F	F	T	T	T	T

The statements ~q → ~p and p ∨ ~q are not logically equivalent.

Section 2.6

1. (1)

3.

p	q	~q	p ∧ ~q	~p	~q → ~p	(p ∧ ~q) → (~q → ~p)
T	T	F	F	F	T	T
T	F	T	T	F	F	F
F	T	F	F	T	T	T
F	F	T	F	T	T	T

The statement (p ∧ ~q) → (~q → ~p) is not a tautology.

5.

p	q	p ∨ q	~p	[(p ∨ q) ∧ ~p]	[(p ∨ q) ∧ ~p] → q
T	T	T	F	F	T
T	F	T	F	F	T
F	T	T	T	T	T
F	F	F	T	F	T

The statement [(p ∨ q) ∧ ~p] → q is a tautology.

7.

p	q	p → q	p ∧ (p → q)	[p ∧ (p → q)] → q
T	T	T	T	T
T	F	F	F	T
F	T	T	F	T
F	F	T	F	T

The statement $[\,p \wedge (\,p \rightarrow q)\,] \rightarrow q$ is a tautology.

Section 2.7

1. (3)   **3.** (1)

5. (a) He is a snob if and only if he likes cavier.
   (b) He is not a snob if and only if he likes cavier.

7. (1)

9. (a)

p	q	p ↔ q	[(p ↔ q) → p]	p ∨ q	[(p ↔ q) → p] ↔ (p ∨ q)
T	T	T	T	T	T
T	F	F	T	T	T
F	T	F	T	T	T
F	F	T	F	F	T

(b) The statement $[(\,p \leftrightarrow q) \rightarrow p] \leftrightarrow (\,p \vee q)$ is a tautology.

11. (a)

p	q	p ∧ q	p ∨ q	p → q	[(p ∨ q) ↔ (p → q)]	(p ∧ q) → [(p ∨ q) ↔ (p → q)]
T	T	T	T	T	T	T
T	F	F	T	F	F	T
F	T	F	T	T	T	T
F	F	F	F	T	F	T

(b) The statement $(\,p \wedge q) \rightarrow [(\,p \vee q) \leftrightarrow (\,p \rightarrow q)$ is a tautology.

13.

p	q	p → q	q → p	(p → q) ∧ (q → p)	p ↔ q
T	T	T	T	T	T
T	F	F	T	F	F
F	T	T	F	F	F
F	F	T	T	T	T

Section 2.8
1. False
3. Cannot be determined
5. (1)
7. (1)
9. (1)
11. (4)
13. (2)
15. Cannot be determined
17. Cannot be determined

Regents Tune-Up: Chapter 2
1. $q \to p$
3. $\sim p \to q$
5. (1)
7. (1)
9. (3)
11. (2)
13. (a)

p	q	~p	~p → q	p ∨ q	(~p → q) ↔ (p ∨ q)
T	T	F	T	T	T
T	F	F	T	T	T
F	T	T	T	T	T
F	F	T	F	F	T

    (b) Yes   **(c)** (1)

15. (a) False   **(c)** True   **(e)** True
    (b) True   **(d)** False

ALGEBRA

Unit Two

CHAPTER 3
SOLVING LINEAR EQUATIONS

3.1 TRANSLATING ENGLISH TO ALGEBRA

Algebra is used to solve problems, so it is important to be able to represent commonly encountered phrases algebraically. Keep in mind that the terms *plus, sum, more than,* and *increased* or *exceeded by* involve the operation of **addition**; the words *difference, less than,* and *diminished by* involve the operation of **subtraction**; the terms *product, two times, three times,* and so forth, are related to **multiplication**; the term *quotient* refers to **division**.

Interpreting Algebraic Expressions

If n represents some unknown number, how do expressions such as $n + 5$, $n - 3$, $\frac{n}{2}$, and $3n$ compare to the original number? By identifying the arithmetic operations contained in the expression, we may express the relationship between the variables, numbers, and arithmetic operations in words.

Expression	Interpretation
$n + 5$	5 more than n *or* the sum of n and 5 *or* "...exceeds n by 5"
$n - 3$	3 less than n *or* n diminished by 3 *or* 3 subtracted from n
$\frac{n}{2}$	One half of n *or* n divided by 2 *or* the quotient of n divided by 2
$3n$	Three times n *or* the product of 3 and n *or* 3 multiplied by n

By expressing the width and the length of a rectangle in terms of the same variable, the dimensions of the rectangle can be compared.

64

RECTANGLE		COMPARING LENGTH TO WIDTH
Width	Length	Interpretation
x	$x + 4$	The length is 4 more than the width *or* The length exceeds the width by 4.
x	$2x$	The length is twice the width.
x	$2(x + 1)$	The length is twice the sum of the width and 1.
x	$2x + 1$	The length exceeds twice the width by 1.

Writing Algebraic Expressions

If y represents John's age, how would you represent John's age 5 years ago? John's age 9 years from now? John's age when he is three times as old as he is now? These situations can be represented algebraically by writing an appropriate expression that adds to y, subtracts from y, or multiplies y.

Expression	Interpretation
$y - 5$	John's age 5 years ago
$y + 9$	John's age 9 years from now
$3y$	John's age when he is three times as old as he is now
$2(y - 8)$	Twice John's age 8 years ago

Examples

1. Express, in terms of n, each of the following:
   (**a**) a number diminished by 6
   (**b**) a number increased by 7
   (**c**) four times the number
   (**d**) the quotient of a number divided by 3
   (**e**) 8 more than the number
   (**f**) 15 less than the number
   (**g**) the value, in cents, of n nickels
   (**h**) the value, in cents, of n dimes

Solution:

(**a**) $n - 6$	(**e**) $n + 8$
(**b**) $n + 7$	(**f**) $n - 15$
(**c**) $4n$	(**g**) $5n$ ($= 5$ cents $\cdot\ n$ nickels)
(**d**) $\dfrac{n}{3}$	(**h**) $10n$ ($= 10$ cents $\cdot\ n$ dimes)

2. The length of a certain rectangle is three times its width. If x represents the width of the rectangle, express in terms of x the perimeter of the rectangle.

Solution: Let x = width of the rectangle.
Then $3x$ = length of the rectangle.

The perimeter of a rectangle is the distance around the figure, which is found by finding the sum of the lengths of the four sides.

$$\begin{aligned} \text{Perimeter} &= 3x + x + 3x + x \\ &= 4x \quad\quad + 4x \\ &= \mathbf{8x} \end{aligned}$$

Exercise Set 3.1

1–6. If x *represents a number, express each of the following in terms of* x*:*

1. The sum of the number and 9.

2. 5 less than twice the number.

3. Twice the number diminished by 5.

4. 4 more than three times the number.

5. Three times the number increased by 4.

6. The quotient of two times the number divided by 3.

7–9. If the length of a certain rectangle is represented by l, *then, for each of the following situations, express the width in terms of* l*:*

7. Width is one half of the length.

8. Width is 7 less than the length.

9. Width exceeds one-third the length by 5.

10–12. If s *represents Susan's age, then represent each of the following situations in terms of* s*:*

10. Susan's age 9 years from now.

11. Susan's age 10 years ago.

12. Susan's age when she is twice the age she was 5 years ago.

13–15. In a certain rectangle, the width of the rectangle is represented by w*:*

　　(a) *Express the length in terms of* w.
　　(b) *Express the perimeter of the rectangle in terms of* w.

13. The length exceeds the width by 10.

14. The length is twice the width.

15. The length exceeds three times the width by 1.

3.2 SOLVING EQUATIONS USING INVERSE OPERATIONS

Equations that have the same solution set are **equivalent**. The equations
$$x + 1 = 4, \quad 2x = 6, \quad \text{and} \quad x = 3$$
are equivalent since {3} is the solution set of each equation. When solving an equation, our goal is to try to obtain an equivalent equation in which the variable is "isolated" on one side of the equation, so that the solution set can be read from the opposite side, as in $x = 3$.

An equation in which the greatest exponent of the variable is 1 is called a **first-degree** or **linear equation**.

Terms Related to Equations

The open sentence $2x + 1 = 7$ is an *equation* whose left side is $2x + 1$ and whose right side is 7. An **equation** is a number sentence that states that the expressions on either side of the equality symbol represent the same number.

Equations may be true or false. The equation $2x + 1 = 7$ is false when $x = 4$ since $2(4) + 1$ is *not* equal to 7. If x is replaced by 3, the equation $2x + 1 = 7$ is true since the left side becomes $2(3) + 1$, which evaluates to $6 + 1$ or 7. Thus, $x = 3$ is a *solution* or *root* of the equation. Any replacements of a variable that make an equation a true statement are called **solutions** or **roots** of that equation.

In the equation $2x + 1 = 7$ the variable is *not isolated*. In the equation $x = 3$, the variable is *isolated*. A variable in an equation is **isolated** when the equation is in the form $x = a$ where a stands for some number.

General Strategy for Solving Equations

Solving an equation means finding the root or roots of that equation. The solution process usually involves finding an equivalent equation in which the variable is isolated. If we know that the original equation and an equation that has the form $x = a$ are equivalent, then the number a is a solution of the original equation.

Isolating Variables in Equations

To isolate a variable in an equation, we must "undo" each arithmetic operation that is linked to the variable by performing its *inverse* operation on both sides of the equation. Addition and subtraction are inverse operations, as are multiplication and division. The following property is used to help isolate the variable when solving an equation.

MATH FACTS

BALANCE PROPERTY OF EQUATIONS

If the same arithmetic operation is performed on *both* sides of an equation, then the new equation is equivalent to the original equation.

Solving Equations Using Addition or Subtraction

Let a and b stand for numbers.

- To solve an equation in the form $x - a = b$, isolate x by adding a on both sides of the equation.
- To solve an equation in the form $x + a = b$, isolate x by subtracting a from both sides of the equation.

Examples

1. Solve: $x - 4 = -1$

Solution: In the equation $x - 4 = -1$, 4 is *subtracted* from x. Isolate x by *adding* 4 to both sides of the equation.

Original equation:	$x - 4 = -1$
Add 4 on each side:	$x - 4 + 4 = -1 + 4$
Simplify:	$x + 0 = 3$
Final equivalent equation:	$x = 3$

The solution is **3**.

Note: Since $x - 4 = -1$ and $x = 3$ are equivalent equations, they have the same solution set. Hence, 3 is the solution of both equations.

2. Solve: $x + 7 = 5$.

Solution: In the equation $x + 7 = 5$, 7 is *added* to x. Isolate x by *subtracting* 7 from both sides of the equation.

$$
\begin{aligned}
x + 7 - 7 &= 5 - 7 \\
x + 0 &= -2 \\
x &= -2
\end{aligned}
$$

The solution is **–2**.

Note: Since adding the opposite of a number is equivalent to subtracting that number, the equation $x + 7 = 5$ could also be solved by *adding* –7 to both sides of the equation. The solution would look like this:

$$
\begin{aligned}
x + 7 + (-7) &= 5 + (-7) \\
x + (7 - 7) &= -2 \\
x &= -2
\end{aligned}
$$

3. Solve: $-8 = y - 3$.

Solution: Isolate y by adding 3 to both sides of the equation.

$$
\begin{aligned}
-8 &= y - 3 \\
-8 + 3 &= y - 3 + 3 \\
-5 &= y + 0 \\
-5 &= y \text{ or } y = -5
\end{aligned}
$$

The solution is **–5**.

4. Solve: $b + 2.3 = -4.8$

Solution: Isolate b by subtracting 2.3 from both sides of the equation. You may find it convenient to write 2.3 underneath each number from which it is being subtracted.

Align decimal points:
$$
\begin{aligned}
b + 2.3 &= -4.8 \\
- 2.3 &= - 2.3 \\
\hline
b + 0 &= -7.1 \\
b &= -7.1
\end{aligned}
$$

The solution is **–7.1**.

Solving Equations Using Multiplication or Division

Let a and b stand for numbers with $a \neq 0$.

- To solve an equation in the form $ax = b$, isolate x by dividing both sides of the equation by a.
- To solve an equation in the form $\dfrac{x}{a} = b$, isolate x by multiplying both sides of the equation by a.

Examples

5. Solve: $-2n = 10$

Solution: In the equation, n is *multiplied* by -2. Isolate n by *dividing* both sides of the equation by -2.

$$-2n = 10$$
$$\frac{-2n}{-2} = \frac{10}{-2}$$
$$1 \cdot n = -5$$
$$n = -5$$

The solution is **–5**.

Note: Since multiplying by $-\frac{1}{2}$ is the same as dividing by -2, the original equation can also be solved by multiplying both sides of the equation by the multiplicative inverse (reciprocal) of -2, which is $-\frac{1}{2}$.

6. Solve: $\frac{y}{4} = -5$

Solution: Since y is *divided* by 4, isolate y by *multiplying* both sides of the equation by 4.

$$\frac{y}{4} = -5$$
$$4\left(\frac{y}{4}\right) = 4(-5)$$
$$1 \cdot y = -20$$
$$y = -20$$

The solution is **–20**.

7. Solve: $\frac{3t}{2} = -21$

Solution: In examples 5 and 6, we choose a number to multiply or divide the equation that would make the coefficient of the variable 1. In the given equation, t is multiplied by $\frac{3}{2}$. Since the product of a number and its multiplicative inverse (reciprocal) is 1, multiply both sides of the equation by the reciprocal of $\frac{3}{2}$, which is $\frac{2}{3}$.

$$\frac{3t}{2} = -21$$
$$\frac{2}{3} \times \left(\frac{3t}{2}\right) = \frac{2}{3} \times (-21)$$
$$\left(\frac{2}{3} \times \frac{3}{2}\right)t = 2 \times (-7)$$

$$1 \cdot t = -14$$
$$t = -14$$

The solution is **–14**.

Checking Roots

To check that a number is a root of an equation, follow these steps:

Step 1. Return to the original equation and replace the variable with the value(s) obtained as the solution.

Step 2. Compare the values of each side of the equation. If these values are the same, then the number is a root of the equation.

Examples

8. Determine whether 3 is a root of the equation $2x - 11 = 5$.

Solution: Replace x by 3.

$$\frac{2(3) - 1 = 5}{6 - 1 \Big|}$$
$$5 = 5\checkmark$$

9. Solve *and* check: $\frac{2}{5}n = 14$.

Solution: When the coefficient of the variable is a fraction, the variable may be isolated by multiplying both sides of the equation by the *reciprocal* of the fraction.

$$\left(\frac{5}{2}\right)\left(\frac{2n}{5}\right) = (14)\left(\frac{5}{2}\right)$$

$$n = \frac{\overset{7}{\cancel{14}}}{1} \cdot \frac{5}{\underset{1}{\cancel{2}}}$$

$$n = 35$$

$$Check: \frac{2n}{5} = 14$$

$$\frac{2}{5}(35)$$

$$\frac{2}{\cancel{5}}(\overset{7}{\cancel{35}})$$

$$14 = 14\checkmark$$

Solving Word Problems

If a certain number is increased by 5, the result is –8. What is the number? To solve this problem:

Think. . .	Take This Action. . .
1. What must I find?	Represent the unknown with a variable: Let n = unknown number.
2. What are the facts?	Translate the facts into an equation: A number increased by 5 is –8 n $+5$ $=$ -8
3. What is the answer?	Solve the equation: $n + 5 - 5 = -8 - 5$ $n + 0 = -13$ $n = -13$
4. Is my answer correct?	Check the answer in the statement of the problem: $-13 + 5 = -8$ True.

Exercise Set 3.2

1–35. Solve for the variable and check.

1. $x + 5 = -9$

2. $-3x = 12$

3. $-7 = x - 1$

4. $\dfrac{n}{2} = 8$

5. $-0.5w = -35$

6. $-4 = x + 2$

7. $-10 = \dfrac{t}{-3}$

8. $40 = -8s$

9. $n + 8 = -7$

10. $-7.2 = 3 + x$

11. $x - \dfrac{1}{2} = \dfrac{5}{2}$

12. $4n = -3.2$

13. $y + 0.6 = 3$

14. $\dfrac{a}{-3} = -6$

15. $1.2 = m - 1.3$

16. $t + 2 = \dfrac{4}{3}$

17. $\dfrac{2x}{3} = 16$

18. $2 = y + 0.6$

19. $-21 = \dfrac{7x}{8}$

20. $\dfrac{x}{5} = 1.7$

21. $\dfrac{3x}{4} = -12$

22. $y - \dfrac{1}{3} = -2$

23. $\dfrac{r}{3} = -8.4$

24. $-10 = \dfrac{-5a}{3}$

25. $3x = 1\dfrac{1}{8}$

26. $\dfrac{x}{4} = 2\dfrac{1}{3}$

27. $-2x = 1\dfrac{1}{5}$

28. $x + 1.9 = -7.6$

29. $\dfrac{a}{-4} = 1.6$

30. $b + \dfrac{1}{3} = 2\dfrac{1}{6}$

31. $\dfrac{-3h}{5} = -15$

32. $8k = -\dfrac{7}{2}$

33. $1\dfrac{1}{3}y = 8$

34. $0.7t = 4.9$

35. $2\dfrac{1}{4}p = -18$

36. A number diminished by 7 is –3. What is the number?

37. The quotient of a number divided by –4 is 5. What is the number?

38. 19 exceeds a number by 25. What is the number?

39. –14 is 8 less than a number. What is the number?

40. Two thirds of a number is 15. What is the number?

3.3 SOLVING EQUATIONS USING MORE THAN ONE OPERATION

$\triangle$
KEY IDEAS

To solve an equation in which the variable is involved in two or more arithmetic operations, perform inverse operations one at a time, starting with addition (or subtraction). If the equation contains parentheses, remove these first.

Solving Equations with Two Arithmetic Operations

In the equation $2n + 5 = -11$ the variable is involved in two operations: multiplication ($2n$) and addition ($2n + 5$). In solving this equation, undo the addition before the multiplication.

73

Examples

1. Solve and check: $2n + 5 = -11$.

Solution: $2n + 5 - 5 = -11 - 5$

$$2n + \quad 0 = -16$$

$$\frac{2n}{2} = \frac{-16}{2}$$

$$n = \textbf{-8}$$

Check: $\quad 2n + 5 = -11$

$$2(-8) + 5$$

$$-16 + 5$$

$$-11 = -11\checkmark$$

2. Solve and check: $\frac{x}{3} - 2 = 13$.

Solution: $\frac{x}{3} - 2 + 2 = 13 + 2$

$$\frac{x}{3} + 0 \quad = 15$$

$$\overset{1}{\cancel{3}}\left(\frac{x}{\cancel{3}}\right) = 3(15)$$

$$x = \textbf{45}$$

Check: $\frac{x}{3} - 2 = 13$

$$\frac{45}{3} - 2$$

$$15 - 2$$

$$13 = 13\checkmark$$

3. The EZ-Car Rental Agency charges \$35 for the first day, and \$22 for each additional day. If John's car rental bill was \$167, for how many days did John rent the car?

Solution: Let d = number of additional days after the first day that John rented the car.

Charge for first day		Charge for each additional day		Total charge
35	+	22d	=	167

$$35 + 22d - 35 = 167 - 35$$

$$22d = 132$$

$$\frac{22d}{22} = \frac{132}{22}$$

$$d = 6$$

John rented the car for a total of $1 + 6$ or **7 days**.

Check: $\$35 + 6(\$22) = \$35 + \$132 = \$167$.

Solving Problems Involving Averages

The average of a set of values is found by dividing the sum of the values by the number of values in the set.

Examples

4. Bonnie receives grades of 79, 83, and 86 on her first three mathematics exams. What grade must she receive on her next exam in order to have an average grade of 85 for the four exams?

Solution: Let x = Bonnie's next exam grade.

$$\frac{79 + 83 + 86 + x}{4} = 85$$

$$\frac{248 + x}{4} = 85$$

Multiply each side by 4: $\cancel{4}\left(\frac{248 + x}{\cancel{4}}\right) = 4(85)$

$$248 + x = 340$$
$$x = 340 - 248 = 92$$

Bonnie must receive **92** on her next exam.

5. Bill drives at 55 m.p.h. for 2 hours. At what rate of speed must he drive for the next hour so that his average speed for the trip is 50 m.p.h.?

Solution: Let x = Bill's average rate of driving during the third hour.

	Average rate	×	Time	=	Distance
First 2 hours	55 m.p.h.		2		110
Next hour	x		1		x

The table shows that in 3 hours of driving Bill travels a *total* of $110 + x$ miles.

$$\text{Average rate } = \frac{\text{Total distance}}{\text{Total time}}$$

$$50 = \frac{110 + x}{3}$$
$$150 = 110 + x$$
$$150 - 110 = x$$
$$40 = x \text{ } or \text{ } x = \textbf{40 m.p.h.}$$

Solving an Equation with Decimals

To clear an equation of one or more decimal terms, multiply each member of the equation by the power of 10 that will change the decimal number having the greatest number of decimal places to an integer.

Example

6. Solve: $0.03x - 0.7 = 0.8$

Solution:

Method 1	Method 2
The coefficient of x contains two decimal places, which is the greatest number of decimal positions in the equation. Multiply each member of the equation by 10^2, or 100.	$0.03x - 0.7 = 0.8$ $0.03x - 0.7 + 0.7 = 0.8 + 0.7$ $0.03x = 1.5$ $\dfrac{0.03x}{0.03} = \dfrac{1.5}{0.03}$
$100(0.03x) - 100(0.7) = 100(0.8)$ $3x - 70 = 80$ $3x = 80 + 70$ $\dfrac{3x}{3} = \dfrac{150}{3}$ $x = \mathbf{50}$	$x = \dfrac{1.50}{0.03} \cdot \dfrac{100}{100}$ $= \dfrac{150}{3}$ $x = \mathbf{50}$

Solving an Equation with Fractions

To clear an equation of all of its fractions, multiply both sides of the equation by the least common denominator (LCD) of all of its fractions.

Examples

7. Solve and check: $\dfrac{3y}{4} - \dfrac{7}{12} = \dfrac{1}{6}$

Solution: The LCD is 12 since 12 is the smallest positive number into which each of the denominators 4, 12, and 6, divide evenly. Hence, multiply each term on both sides of the equation by 12.

$$12\left(\frac{3y}{4}\right) - 12\left(\frac{7}{12}\right) = 12\left(\frac{1}{6}\right)$$
$$9y - 7 = 2$$
$$9y = 2 + 7$$
$$\frac{9y}{9} = \frac{9}{9}$$
$$y = \mathbf{1}$$

The solution is **1**. The check is left for you.

Note: This equation can also be solved by adding $\frac{7}{12}$ to both sides of the equation and then isolating y by multiplying both sides of the equation by $\frac{4}{3}$, the reciprocal of the coefficient of y.

Solving an Equation with Parentheses

If an equation contains parentheses, remove them by applying the distributive property of multiplication over addition (or subtraction).

8. Solve and check: $3(1 - 2x) = -15$.

Solution:
$$3(1 - 2x) = -15$$
$$3 - 6x = -15$$
$$3 - 6x - 3 = -15 - 3$$
$$\frac{-6x}{-6} = \frac{-18}{-6}$$
$$x = \mathbf{3}$$

Check: $3(1 - 2x) = -15$
$$3(1 - 2 \cdot 3)$$
$$3(1 - 6)$$
$$3(-5)$$
$$-15 = -15$$

Exercise Set 3.3

1–40. Solve for the variable and check.

1. $0.3x = 12$

2. $0.4y + 5 = 1$

3. $3x - 1 = -16$

4. $-2x + 7 = -13$

5. $32 = 3w + 5$

6. $\frac{x}{5} + 8 = 10$

7. $\frac{y}{3} + 5 = 3$

8. $3(2x - 5) = 9$

9. $0.2x + 0.3 = 8.1$

10. $0.3y + 5.2 = 4$

11. $-2(1 - t) = 16$

12. $7(1 - x) - 1 = 20$

13. $1 + 4(7 - 2x) = 5$

14. $4(2x + 3) = -4$

15. $2(3x - 5) - 12 = 0$

16. $-17 = 2(5 + m)$

17. $3 = 4x + 19$

18. $19 - (1 - x) = 18$

19. $7 - 3(x - 1) = -17$

20. $\frac{3x}{4} - 2 = -17$

21. $-3(9 - 7p) = -13$

22. $1.04x + 8 = 60$

23. $\frac{3t}{2} + 5 = 17$

24. $0 = 18 - 2(h + 1)$

25. $8 - 5r = -7$

26. $-(8x - 3) = 19$

27. $0.4(x - 10) = 5.6$

28. $-12 + 3k = -6$

29. $\frac{3}{4}s - 1 = 8$

30. $\frac{2y}{3} - \frac{1}{2} = \frac{15}{2}$

31. $\frac{3n}{2} - \frac{5}{6} = \frac{2}{3}$

32. $\frac{3p}{4} + \frac{7}{8} = 2$

33. $\frac{2x - 5}{-3} = -7$

34. $\frac{5y}{7} - \frac{11}{14} = 1$

35. $0.8(0.4r + 1) = 0.16$

36. $6\left(\frac{x}{2} + 1\right) = -9$

37. $13 - \frac{3x}{4} = -8$

38. $2\left(\frac{y}{4} + 7\right) = 3$

39. $3(9 - 7p) = 27$

40. $0.25(3x - 5) = \frac{5}{2}$

41. The sum of three times a number and 5 is 32. What is the number?

42. When 2 is subtracted from one third of a number, the result is 3. What is the number?

43. 37 exceeds three times the sum of a number and 5 by 1. What is the number?

44. 45 is 9 greater than twice the difference obtained by subtracting 7 from a number. What is the number?

45. A video store rents video tapes at the following rate: $3 for the first day and $2 for each additional day the tape is out. If Bill returns a video tape and is charged $13, for how many days is he being charged?

46. If the length of a side of a square is represented by $2x - 1$, find x if the perimeter of the square is 36. (The perimeter of a square is equal to four times the length of a side.)

47. The length of a certain rectangle is represented by $3x + 5$. If its width is 11 and its perimeter is 80, find the value of x and the length of the rectangle. (The perimeter of a rectangle is twice the sum of its length and width.)

48. If the average of the set of numbers $\{45, 67, 54, 51, x\}$ is 60, find x.

49. Carol's average driving speed for a 4-hour trip was 48 m.p.h. During the first 3 hours her average rate was 50 m.p.h. What was her average rate for the last hour of the trip?

50. A car travels at an average rate of 40 m.p.h. for the first 5 hours and at a uniform rate of speed for the next 3 hours. If the average rate of travel during the 8-hour trip is 43 m.p.h., what is the average rate for the last 3 hours of the trip?

3.4 SOLVING EQUATIONS INVOLVING MORE THAN ONE VARIABLE TERM

△
KEY IDEAS
△△

An equation in which the variable appears in more than one term can be solved by first collecting and then combining like terms. If the variable appears on both sides of the equation, rearrange terms so that like terms involving the variable are on the same side of the equation and constant terms (numbers without variable factors) are on the opposite side.

Variable Terms on the Same Side

To solve an equation in which the variable appears in more than one term on the same side of the equation, first combine like terms. Then solve the resulting equation as usual.

Examples

1. Solve and check: $5x + 3x = -24$.

Solution: Begin by combining the like terms, $5x$ and $3x$, which gives $8x$.

Thus,
$$8x = -24$$
$$\frac{8x}{8} = \frac{-24}{8}$$
$$x = -3$$

Check: $5x + 3x = -24$
$$5(-3) + 3(-3) \,\vert$$
$$-15 + (-9) \,\vert$$
$$-24 = -24 \checkmark$$

2. Solve and check: $\dfrac{x}{4} - \dfrac{5x}{8} = 3$.

Solution: The LCD is 8 since it is the smallest positive number into which 4 and 8 divide evenly. Multiply each member on both sides of the equation by 8:

$$8\left(\frac{x}{4}\right) - 8\left(\frac{5x}{8}\right) = 8(3)$$
$$2x - 5x = 24$$
$$-3x = 24$$
$$\frac{-3x}{-3} = \frac{24}{-3}$$
$$x = -8$$

The solution is **-8**. The check is left for you.

3. Solve and check: $3(2y + 5) - 8y = 1$.

Solution: Remove the parentheses by multiplying each term inside of the parentheses by 3. Then collect like terms.

$$
\begin{aligned}
3(2y) + 3(5) - 8y &= 1 \\
6y + 15 - 8y &= 1 \\
(6y - 8y) + 15 &= 1 \\
-2y &= 1 - 15 \\
\frac{-2y}{-2} &= \frac{-14}{-2} \\
y &= 7
\end{aligned}
$$

The solution is **7**. The check is left for you.

4. The sum of two numbers is 25. When twice the larger number is subtracted from three times the smaller, the difference is 5. Find the numbers.

Solution: Let x = smaller of the two numbers.
Then $25 - x$ = larger of the two numbers.

$$
\begin{aligned}
3x - 2(25 - x) &= 5 \\
\text{Remove parentheses:} \quad 3x - 50 + 2x &= 5 \\
\text{Combine like terms:} \quad (3x + 2x) - 50 &= 5 \\
5x - 50 &= 5 \\
5x - 50 + 50 &= 5 + 50 \\
5x &= 55 \\
\frac{5x}{5} &= \frac{55}{5} \\
x &= 11
\end{aligned}
$$

so that $25 - x = 25 - 11 = 14$.

The smaller number is **11**, and the larger number is **14**. The check is left for you.

5. A soda machine contains 20 coins. Some of the coins are nickels, and the rest are quarters. If the value of the coins is $4.40, find the number of coins of each kind.

Solution: Let x = number of nickels.
Then $20 - x$ = number of quarters.

$$
\underbrace{\textit{Value of nickels}}_{} \quad + \quad \underbrace{\textit{Value of quarters}}_{} \quad = \quad \underbrace{\$4.40}_{}
$$

$$
0.05x \quad + \quad 0.25(20 - x) \quad = \quad \$4.40
$$

To clear the equation of decimals, multiply each member of both sides of the equation by 100.

$$100(0.05x) + 100[0.25(20 - x)] = 100(4.40)$$
$$5x + 25(20 - x) = 440$$
$$5x + 500 - 25x = 440$$
$$-20x + 500 = 440$$
$$-20x = 440 - 500$$
$$\frac{-20x}{-20} = \frac{-60}{-20}$$
$$x = 3$$
$$\text{and } \mathbf{20 - x = 17}$$

There are **3 nickels** and **17 quarters**. The check is left for you.

Variable Terms on Different Sides

If variable terms appear on opposite sides of an equation, work toward collecting variables on the same side of the equal sign and constant terms (numbers) on the other side.

Examples

6. Solve and check: $5w + 14 = 3(w - 8)$.

Solution: $\qquad\qquad\qquad\qquad 5w + 14 = 3(w - 8)$

Apply the distributive property: $\qquad\quad 5w + 14 = 3w - 24$
Subtract $3w$ from each side: $\qquad -3w + 5w + 14 = -3w + 3w - 24$
Simplify: $\qquad\qquad\qquad\qquad\quad\; 2w + 14 = -24$
Subtract 14 from each side: $\qquad 2w + 14 - 14 = -24 - 14$
Divide each side by 2: $\qquad\qquad\qquad\; \dfrac{2w}{2} = \dfrac{-38}{2}$
$$w = \mathbf{-19}$$

$\qquad$ *Check*: $\qquad\qquad 5w + 14 = 3(w - 8)$
$$5(-19 + 14 \mid 3(-19 - 8)$$
$$-95 + 14 \mid 3(-27)$$
$$-81 = -81 \checkmark$$

7. Solve and check: $\dfrac{x+1}{2} = \dfrac{5x - 3}{3} + 5$.

Solution: Clear the equation of its fractions by multiplying each member on both sides of the equation by 6, the LCD of 2 and 3.

$$^3\cancel{6}\left(\frac{x+1}{\cancel{2}}\right) = {}^2\cancel{6}\left(\frac{5x-3}{\cancel{3}}\right) + 6(5)$$

$$3(x + 1) = 2(5x - 3) + 30$$
Remove the parentheses: $\qquad 3x + 3 = 10x - 6 + 30$
Combine like terms: $\qquad\quad\; 3x + 3 = 10x + 24$

81

Subtract 3 from each side:

$$3x + 3 - 3 = 10x + 2 - 3$$
$$3x = 10x + 21$$

Subtract $10x$ from each side:

$$3x - 10x = 10x - 10x + 21$$
$$-7x = 21$$

Divide each side by -7:

$$\frac{-7x}{-7} = \frac{21}{-7}$$
$$x = -3$$

The solution is **–3**. The check is left for you.

8. Solve and check: $3(4b - 7) = 2(3b + 11) + 5$.

Solution: On each side of the equation remove the parentheses by multiplying each term inside the parentheses by the number that is outside the parentheses. Then collect the variable terms on the left side of the equation, and the constant terms on the right side of the equation.

$$3(4b - 7) = 2(3b + 11) + 5$$
$$3(4b) + 3(-7) = 2(3b) + 2(11) + 5$$
$$12b - 21 = 6b + 22 + 5$$
$$12b - 6b = 27 + 21$$
$$6b = 48$$
$$\frac{6b}{6} = \frac{48}{6}$$
$$b = 8$$

The solution is **8**. The check is left for you.

9. In 7 years Maria will be twice as old as she was 3 years ago. What is Maria's present age?

Solution: Let $x =$ Maria's present age.
Then $x + 7 =$ Maria's age 7 years from now,
and $x - 3 =$ Maria's age 3 years ago.

$$x + 7 = 2(x - 3)$$
$$x + 7 = 2x - 6$$
$$x + 7 - 7 = 2x - 6 - 7$$
$$x = 2x - 13$$
$$x - 2x = 2x - 13 - 2x$$
$$-x = -13$$
$$\frac{x}{-1} = \frac{-13}{-1}$$
$$x = 13$$

Maria's present age is **13 years**. The check is left for you.

Exercise Set 3.4

1–40. Solve for the variable and check.

1. $3x + 4x = -28$

2. $5x - 2x = -27$

3. $0.8x + 0.1x = -36$

4. $7t = t - 42$

5. $w - 0.4w = -4.2$

6. $0.54 - 0.07y = 0.2y$

7. $6h - 14 = 3h + 13$

8. $3(x - 4) - 6 = 0$

9. $2(a - 5) = 4$

10. $3(x + 4) = x$

11. $3s + 2(8 - s) = 3$

12. $7(5 - w) + w = -1$

13. $7 - (3n - 5) = n$

14. $9b = 2b - 3(8 - b)$

15. $3(x - 2) + 2(2 - x) = 0$

16. $1.2x - 0.35 = 0.5x + 5.25$

17. $4(b - 3) = 9 - 3b$

18. $4c - (c + 7) = 5$

19. $5(2x + 1) - (6x - 7) = 0$

20. $1.6(3x - 1) + 0.2x = 8.4$

21. $3(5 - 2n) = 2n - 9$

22. $0.03t = 0.07 + 0.6t$

23. $-(8 - 3x) = 7x + 12$

24. $1 + 7w = 5(7 - 2w)$

25. $7x + 3(x - 2) = -16$

26. $3(4 - x) = 2(x + 11)$

27. $2(d - 7) = 3(7 - d)$

28. $5(8 + x) = 3(x + 6)$

29. $-(11 + 5m) = 4(7 - 2m)$

30. $\dfrac{a - 2}{3} = a - 6$

31. $\dfrac{n - 5}{2} = 3n - 5$

32. $5(6 - q) = -3(q + 2)$

33. $13 - 2(x - 7) = x$

34. $7(2p - 1) = 4(1 - 2p)$

35. $3(7 - 2n) = 6(n + 2)$

36. $5(n + 2) - 2(n + 2) = -9$

37. $y - 6 = 3(2y + 9) + y$

38. $2(7 - 3x) = 3(x + 1) - 7$

39. $2(x - 1) + 3(x + 7) = 4$

40. $0.7(x - 0.2) + 0.3(x - 0.2) = 0.4$

41–46. Clear the equation of its fractions and then solve for the variable.

41. $\dfrac{x}{3} + \dfrac{x}{4} = 2$

42. $\dfrac{2y}{3} - \dfrac{y}{5} = 4$

43. $n + \dfrac{3}{2}n = 15$

44. $\dfrac{5k}{3} - \dfrac{3k}{2} = k + 5$

45. $\dfrac{x + 1}{4} + \dfrac{2x + 5}{3} = \dfrac{1}{12}$

46. $\dfrac{c + 2}{2} - 4 = \dfrac{c - 1}{3}$

47. The length of a rectangle exceeds twice its width by 5. If the perimeter of the rectangle is 52, find the dimensions of the rectangle.

48. A 72-inch board is cut into two pieces so that the larger piece is seven times as long as the shorter piece. Find the length of each piece.

49. How old is David if his age 6 years from now will be twice his age 7 years ago?

50. Three years ago Jane was one-half as old as she will be 2 years from now. What is Jane's present age?

51. The product of 3 and 1 less than a number is the same as twice the number increased by 14. What is the number?

52. Allan has nickels, dimes, and quarters in his pocket. The number of nickels is 1 more than twice the number of quarters. The number of dimes is 1 less than the number of quarters. If the value of the change in his pocket is 85 cents, how many of each coin does Allan have?

53. Twice the sum of a number and 9 is the same as four times the difference of the number and 6. What is the number?

54. Howard has 14 coins in his pocket, consisting of nickels and dimes. If the value of the coins is 90 cents, how many of each coin does he have?

55. Six times the sum of a number and 7 is the same as the number diminished by 13. What is the number?

56. The average of four numbers is 32. The largest of the four numbers is 5 more than one of the other numbers, and exceeds twice the smallest number by 10. The remaining number is one-half as great as the largest of the four numbers. Find the four numbers.

3.5 SOLVING CONSECUTIVE INTEGER PROBLEMS

KEY IDEAS

Types of Integers	Examples
Consecutive	$8, 9, 10,\ldots$ $n, n + 1, n + 2,\ldots$
Consecutive *even*	$14, 16, 18,\ldots$ $n, n + 2, n + 4,\ldots$
Consecutive *odd*	$11, 13, 15,\ldots$ $n, n + 2, n + 4,\ldots$

Consecutive Integer Problems

A list of integers is **consecutive** if each integer after the first is 1 more than the integer that comes before it.

Consecutive integers: $\{\ldots, -4, -3, -2, -1, 0, 1, 2, 3, 4, \ldots, n, n+1, \ldots\}$

A list of odd integers is consecutive if each integer after the first is 2 more than the odd integer that comes before it. Similarly, consecutive even integers also differ by 2.

Consecutive odd integers: $\{\ldots, -5, -3, -1, 1, 3, 5, \ldots, n, n+2, \ldots\}$
Consecutive even integers: $\{\ldots, -6, -4, -2, 0, 2, 4, 6, \ldots, n, n+2, \ldots\}$

Examples

1. If x represents an *even* integer, express each of the following in terms of x:
  (**a**) the next three consecutive *even* integers
  (**b**) the next larger *odd* integer
  (**c**) the next *smaller* even integer

Solution:
  (**a**) Given that x is an even integer, then $x + 2$, $x + 4$, and $x + 6$ are the next three consecutive *even* integers.
  (**b**) Suppose that $x = 6$; then 7 is the next larger odd integer. In general, if x is an even integer, then $x + 1$ must represent the next larger odd integer.
  (**c**) Suppose that $x = 6$; then 4 is the next smaller even integer. In general, if x is an even integer, then $x - 2$ must represent the next smaller even integer.

2. Find three consecutive odd integers such that twice the sum of the second and the third is 43 more than three times the first.

Solution: Let x = first odd integer.
  Then $x + 2$ = second consecutive odd integer.
  and $x + 4$ = third consecutive odd integer.

Twice the sum of the second and third is 43 more than 3 times the first integer.

$$
\begin{aligned}
2[(x + 2) + (x + 4)] &= 43 & &+ 3x \\
2(2x + 6) &= 43 & &+ 3x \\
4x + 12 &= 43 & &+ 3x \\
4x + 12 - 3x &= 43 & &+ 3x - 3x \\
x + 12 &= 43 & & \\
x + 12 - 12 &= 43 & &- 12 \\
x &= 31 & &
\end{aligned}
$$

85

$$\text{Then } x + 2 = 33$$
$$x + 4 = 35$$

The three consecutive odd integers are **31**, **33**, and **35**.

Check: Is $\underline{2(33 + 35) = 43 + 3(31)}$?
$$\underline{2(68) \mid 43 + 93}$$
$$136 = 136\checkmark$$

Exercise Set 3.5

1. If $n + 2$ represents an odd integer, express each of the following in terms of n:
   (a) the next three consecutive odd integers
   (b) the next larger *even* integer
   (c) the next *smaller* odd integer

2. Find three consecutive integers whose sum is 60.

3. Find four consecutive odd integers whose sum is –72.

4. Find four consecutive even integers whose sum is 124.

5. Find four consecutive integers whose sum is 15 less than 5 times the first.

6. Find four consecutive integers such that the sum of the first and the fourth is 29.

7. Find four consecutive integers such that the sum of the second and the fourth is 26.

8. The lengths of the sides of a triangle are consecutive even integers. Find the length of each side if the perimeter of the triangle is 24.

9. Find four consecutive odd integers such that the sum of three times the second integer and the last integer is 104.

10. Find four consecutive odd integers such that their sum is 1 less than five times the smallest.

11. The length and the width of a rectangle are consecutive odd integers. If the perimeter of the rectangle is 48, find the dimensions of the rectangle.

12. The lengths of the sides of a triangle are consecutive even integers. The perimeter of the triangle is the same as the perimeter of a square whose side is 5 less than the shortest side of the triangle. Find the lengths of the sides of the triangle.

13. Find the largest of five consecutive even integers if their average is –12.

14. Find the smallest of four consecutive odd integers if their average is 35.

3.6 SOLVING PERCENT PROBLEMS

Before you solve percent problems, here are two facts about percents that you should know:

1. A percent compares a number to 100. For example, 35% means "35 out of 100."

2. A percent may be expressed as an equivalent fraction by writing the amount of percent over 100:

$$35\% = \frac{35}{100} = 0.35.$$

Types of Percent Problems

There are three basic types of percent problems that you should know how to solve.

1. Finding a percent of a given number.

Example: What is 15% of 80?

$$n = 0.15 \times 80$$
$$n = \mathbf{12}$$

2. Finding a number when a percent of it is given.

Example: 30% of what number is 12?

$$0.30 \times n = 12$$
$$0.30n = 12$$
$$10(0.3n) = 10(12)$$
$$3n = 120$$
$$n = \frac{120}{3}$$
$$n = \mathbf{40}$$

3. Finding what percent one number is of another.

Example: What percent of 30 is 9?

$$\frac{p}{100} \times 30 = 9$$

$$\frac{p}{100} \times 30 = 9$$

87

$$\frac{\overset{3}{\cancel{30}p}}{\underset{10}{\cancel{100}}} = 9$$

$$p = \frac{10 \cdot 9}{3} = 30$$

9 is 30% of 30

Examples

1. If the rate of sales tax is 7.5%, what is the amount of sales tax on a purchase of $40.00?

Solution: Amount of tax = 7.5% of $40
$$= 0.075 \times 40$$
$$= 3$$

The amount of sales tax is **$3.00**.

2. A pair of sneakers that regularly sells for $35 is on sale for $28. What is the percent of the discount?

Solution: The dollar amount of the discount is $35 − $28 = $7.

Think: "What percent of 35 is 7?"

$$\frac{p}{100} \times 35 = 7$$

$$\frac{35p}{100} = 7$$

$$p = \frac{(100)\overset{1}{\cancel{7}}}{\underset{5}{\cancel{35}}} = 20$$

The percent of the discount is **20%**.

3. A sum of money was invested for 1 year at an annual rate of simple interest of 8%. If the income from the investment was $46, how much money was invested?

Solution: Let A = amount of money invested at 8%.

$$\underbrace{8\%}_{\substack{\downarrow\downarrow \\ 0.08 \times}} \text{ of } \underbrace{\text{the amount of money invested}}_{\substack{\downarrow \\ A}} \text{ is } \underbrace{\$46}_{\substack{\downarrow \\ = 46}}$$

$$0.08A = 46$$

Multiply each side by 100: $\quad 8A = 4600$

Divide each side by 8: $\qquad A = \dfrac{4600}{8}$

$$A = \mathbf{\$575}$$

Exercise Set 3.6

1. What is 25% of 44?

2. 6 is what percent of 30?

3. 12 is 15% of what number?

4. 18 is what percent of 24?

5. 21 is 37.5% of what number?

6. 18 is what percent of 45?

7. 20 is 62.5% of what number?

8. A sweater that regularly costs $32 is on sale for $24. What is the percent of discount?

9. If the sales tax rate is 7%, what is the amount of sales tax on a purchase of $40?

10. The larger of two numbers exceeds three times the smaller by 2. If the smaller number is 30% of the larger, find the two numbers.

11. If the two numbers are in the ratio of 2 : 5, what percent of the larger is the smaller?

12. If 50% of a number is 20, then what is 75% of the same number?

13. 30% of 50 is what percent of 60?

14. 40% of 60 is equal to 60% of what number?

15. 128 is what percent of 96?

16. A shirt is on sale at 20% off the list price. If the sale price of the shirt is $24, what is the list price?

3.7 WRITING RATIOS AND SOLVING PROPORTIONS

KEY IDEAS

The **ratio** of two numbers a and b $(b \neq 0)$ is the quotient of the numbers, and is written as

$$\frac{a}{b} \text{ or } a : b \text{ (read as "} a \text{ is to } b \text{")}.$$

A **proportion** is an equation that states that two ratios are equal. To *solve* a proportion for an unknown member, set the cross-products equal and solve the resulting equation.

Solving Problems Involving Ratios

Comparisons of quantities are often expressed in terms of a ratio. If John is 20 years old and Glen is 10 years old, then John is twice as old as Glen. The ratio of John's age to Glen's age is 2 : 1 since

$$\frac{\text{John's age}}{\text{Glen's age}} = \frac{20 \text{ years}}{10 \text{ years}} = \frac{2}{1} \text{ or } 2 : 1.$$

Examples

1. The length and the width of a rectangle are in the ratio of 3 : 1.
  (**a**) What is the ratio of the width to the perimeter?
  (**b**) If the perimeter of the rectangle is 56 cm, find the dimensions of the rectangle.

Solution: Since the length and width are in the ratio of 3:1, the length is three times as great as the width.

Let x = width of the rectangle.
Then $3x$ = length of the rectangle.

$$\begin{aligned} \text{Perimeter} &= \text{Sum of the lengths of the four sides} \\ &= x + 3x + x + 3x \\ &= 8x \end{aligned}$$

(**a**) $\dfrac{\text{Width}}{\text{Perimeter}} = \dfrac{x}{8x} = \dfrac{1}{8} \text{ or } \mathbf{1 : 8}.$

(**b**) Perimeter = $8x = 56$

$$\frac{8x}{8} = \frac{56}{8}$$

$$x = 7 \text{ and } 3x = 3(7) = 21$$

The width of the rectangle is **7**, and the length is **21**.

Check: Is the perimeter 56?
  Yes, since $P = 2(7) + 2(21) = 14 + 42 = 56.$

2. Two numbers are in the ratio of 6 : 1, and their difference is 40. What is the *smaller* number?

Solution: Since the numbers are in a 6 : 1 ratio, the larger number is six times as great as the smaller number.

$$\begin{aligned} \text{Let } x &= \text{ smaller number.} \\ \text{Then } 6x &= \text{ larger number.} \\ 6x - x &= 40 \end{aligned}$$

90

$$\frac{5x}{5} = \frac{40}{5}$$
$$x = 8.$$

The smaller number is **8**.

Check: The larger number is $6 \times 8 = 48$. Does $48 - 8 = 40$? Yes.

3. The lengths of the three sides of a triangle are in the ratio 2 : 3 : 4. If the perimeter of the triangle is 45 inches, what is the length of the longest side of the triangle?

Solution: Let $2x$ = the shortest side of the triangle
$3x$ = the next longest side of the triangle
$4x$ = the longest side of the triangle.

The perimeter of a triangle is the sum of the lengths of its three sides. Thus,

$$2x + 3x + 4x = 36$$
$$9x = 45$$
$$\frac{9x}{9} = \frac{45}{9}$$
$$x = 5$$

The length of the longest side of the triangle is $4x = 4(5) = $ **20 inches**.

Solving Proportions

In a proportion, the cross-products are equal. For example,

$$\frac{2}{6} = \frac{4}{12} \text{ and } 6 \cdot 4 = 2 \cdot 12$$
$$\checkmark$$
$$24 = 24$$

MATH FACTS

EQUAL CROSS-PRODUCTS RULE

If $\dfrac{a}{b} = \dfrac{c}{d}$, then $b \times c = a \times d$

where b and d cannot be equal to 0. The terms b and c are called **means**, and the terms a and d are the **extremes**.

Examples

4. Solve for x and check: $\dfrac{2}{3} = \dfrac{x+9}{21}$.

Solution: Cross-multiply and then simplify.

$$3(x + 9) = 2 \cdot 21$$
$$3x + 27 = 42$$
$$3x + 27 - 27 = 42 - 27$$
$$3x = 15$$
$$\frac{3x}{3} = \frac{15}{3}$$
$$x = 5$$

Check: $\dfrac{2}{3} = \dfrac{x+9}{21}$

$$\frac{5+9}{21}$$
$$\frac{14}{21}$$
$$\frac{14 \div 7}{21 \div 7}$$
$$\frac{2}{3} = \frac{2}{3} \checkmark$$

5. The ratio of the number of girls to the number of boys in a certain mathematics class is 3 : 5. If there is a total of 32 students in the class, how many are girls and how many are boys?

Solution:

Method 1	Method 2
Let x = number of girls.	Let $3x$ = number of girls.
Then $32 - x$ = number of boys.	Then $5x$ = number of boys.

Method 1

$$\frac{\text{Girls}}{\text{Boys}} = \frac{3}{5} = \frac{x}{32 - x}$$

Cross-multiply:

$$5x = 3(32 - x)$$
$$5x = 96 - 3x$$
$$5x + 3x = 96 - 3x + 3x$$
$$\frac{8x}{8} = \frac{96}{8}$$
$$x = 12$$
$$32 - x = 32 - 12 = 20$$

Method 2

$$3x + 5x = 32$$
$$8x = 32$$
$$\frac{8x}{8} = \frac{32}{8}$$
$$x = 4$$

Therefore $3x = 3(4) = 12$
and $5x = 5(4) = 20$

There are **12 girls** and **20 boys** in the class.

6. The denominator of a fraction is 5 more than the numerator. If the numerator is decreased by 7 and the denominator is not changed, the new fraction is equal to $\dfrac{1}{3}$. Find the original fraction.

Solution: Let x = the numerator of the original fraction.
Then $x + 5$ = the denominator of the original fraction.

If the numerator of the original fraction is decreased by 7, then the new fraction is $\frac{x-7}{x+5}$. Since the new fraction is equal to $\frac{1}{3}$, form the proportion.

$$\frac{x-7}{x+5} = \frac{1}{3}$$

In a proportion, the cross-products are equal. Hence,
$$
\begin{aligned}
3(x - 7) &= 1(x + 5) \\
3x - 21 &= x + 5 \\
3x - 21 + 21 &= x + 5 + 21 \\
3x &= x + 26 \\
3x - x &= x - x + 26 \\
2x &= 26 \\
x &= \frac{26}{2} = 13
\end{aligned}
$$

The original fraction is $\dfrac{x}{x+5} = \dfrac{13}{13+5} = \dfrac{13}{18}$.

Exercise Set 3.7

1–10. Solve each proportion for the variable.

1. $\dfrac{2}{6} = \dfrac{8}{x}$

2. $\dfrac{9}{x} = \dfrac{3}{4}$

3. $\dfrac{x}{2} = \dfrac{50}{4}$

4. $\dfrac{1}{x+1} = \dfrac{10}{5}$

5. $\dfrac{2}{3} = \dfrac{2-x}{12}$

6. $\dfrac{2x-5}{3} = \dfrac{9}{4}$

7. $\dfrac{x+5}{4} = \dfrac{x+2}{3}$

8. $\dfrac{4}{x+3} = \dfrac{1}{x-3}$

9. $\dfrac{10-x}{5} = \dfrac{7-x}{2}$

10. $\dfrac{4}{11} = \dfrac{x+6}{2x}$

11. Jane's age exceeds Sue's age by 5. If the ratio of Jane's age to Sue's age is 3 : 2, how old is each girl?

12. Find the number that must be added to both the numerator and the denominator of the fraction $\dfrac{7}{12}$ in order for the resulting fraction to have a value of $\dfrac{3}{4}$.

13. In Jill's purse the ratio of the number of dimes to the number of nickels is 3 : 4. If the value of these coins is $3.00, how many of each coin does Jill have?

14. The lengths of the sides of a triangle are in the ratio of 1 : 3 : 4. If the perimeter of the triangle is 72, what is the length of the shortest side?

15. John is twice as old as Steve, and Jeffrey is three times as old as John. If the sum of their ages is 63 years, how old is Jeffrey?

16. A class consisted of 14 boys and 19 girls. On a certain day all the boys were present and some girls were absent, so that the girls present made up only 30% of the class attendance. How many girls were absent?

17. Three numbers are in the ratio of 1 : 3 : 5. If their average is 18, find the smallest number.

REGENTS TUNE-UP: CHAPTER 3

Each of the questions in this section has appeared on a previous Course I Regents Examination. Here is an opportunity for you to review Chapter 3 and, at the same time, prepare for the Course I Regents Examination.

---

1. Solve for x: $4x - 3 = 41$.

2. Solve for a: $3a + 0.2 = 5$.

3. Solve for x: $4(2x - 1) = 20$.

4. Solve for x: $0.03x = 36$.

5. What percent of 200 is 14?

6. Solve for x: $2(x + 3) = x + 7$.

7. Solve for x: $5 + 3(x + 2) = 14$.

8. Solve for a: $\dfrac{a + 2}{12} = \dfrac{5}{3}$.

9. Solve for y: $0.02y - 1.5 = 8$.

10. The ratio of Tom's shadow to Emily's shadow is 3 to 2. If Tom is 180 centimeters tall, how many centimeters tall is Emily?

11. The length of a rectangle is 3 more than three times its width. The perimeter of the rectangle is 62. Find the length and the width of the rectangle.

12. Solve for t: $3t + 5(6 - t) = 4$.

13. Solve for x: $\dfrac{2}{3}x - 2 = 16$.

14. Using the letter n to represent a number, express "four less than twice this number" in terms of n.

15. Solve for x: $0.06x + 0.3x = 7.2$.

16. Solve for x: $\dfrac{x}{3} + \dfrac{x}{2} = 5$.

17. The length of a side of a square is represented by $(3x - 1)$. If the perimeter of the square is 68, find the value of x.

18. What is the value in cents of n nickels and d dimes?
(1) $n + d$ (2) $5n + 10d$ (3) $0.05n + 0.10d$ (4) $15nd$

19. A booklet contains 30 pages. If nine pages in the booklet have drawings, what percent of the pages in the booklet have drawings?
(1) 30%   (2) 9%   (3) 3%   (4) $\frac{3}{10}$%

20. If x represents a number, which expression represents a number that is 5 less than three times x?
(1) $5x - 3$ (2) $5 - 3x$ (3) $3x - 5$ (4) $3 - 5x$

21. The equation $5x + 10 = 55$ has the same solution set as the equation:
(1) $x = 45$ (2) $x + 10 = 11$ (3) $5x = 65$ (4) $5x + 15 = 60$

22. If x represents the smallest of three consecutive odd integers, then the largest would be represented by
(1) $x + 2$ (2) $x + 3$ (3) $x + 4$ (4) $x + 5$

23. Three numbers are in the ratio $2 : 3 : 5$. If the smallest number is multiplied by 8, the result is 32 more than the sum of the second and third numbers. Find the numbers.

24. Find three consecutive odd integers such that the sum of the first and twice the second is 6 more than the third.

25. The denominator of a fraction is 4 less than twice the numerator. If 3 is added to both the numerator and the denominator, the new fraction is equal to $\frac{2}{3}$. Find the original fraction.

Answers to Odd-Numbered Exercises: Chapter 3

Exercise Set 3.1

1. $x + 9$	**7.** $\frac{1}{2}$	**13.** (a) $w + 10$ (b) $4w + 20$
3. $2x - 5$	**9.** $\frac{1}{3} + 5$	**15.** (a) $3w + 1$ (b) $8w + 2$
5. $3x + 4$	**11.** $s - 10$	

Exercise Set 3.2

1. -14	**9.** -15	**17.** 24	**25.** $\frac{3}{8}$	**33.** 6
3. -6	**11.** 3	**19.** -24	**27.** $-\frac{3}{5}$	**35.** -8
5. 70	**13.** 2.4	**21.** -16	**29.** -6.4	**37.** -20
7. 30	**15.** 2.5	**23.** -25.2	**31.** 25	**39.** -6

Exercise Set 3.3

1. 40	**11.** 9	**21.** $\frac{2}{3}$	**31.** 1	**41.** 9
3. −5	**13.** $\frac{7}{2}$	**23.** 8	**33.** 13	**43.** 7
5. 9	**15.** $\frac{11}{3}$	**25.** 3	**35.** −2	**45.** 6
7. −6	**17.** −4	**27.** 24	**37.** 28	**47.** $x = 8$, length = 29
9. 4	**19.** 9	**29.** 12	**39.** 0	**49.** 42

Exercise Set 3.4

1. −4	**15.** 2	**29.** 13	**43.** 6
3. −40	**17.** 3	**31.** 1	**45.** −2
5. −7	**19.** −3	**33.** 9	**47.** width = 7, length = 19
7. 9	**21.** 3	**35.** $\frac{3}{4}$	**49.** 20
9. 7	**23.** −5	**37.** $-\frac{11}{2}$	**51.** 17
11. −13	**25.** −1	**39.** −3	**53.** 21
13. 6	**27.** 7	**41.** $\frac{24}{7}$	**55.** −11

Exercise Set 3.5

1. (a) $n + 4, n + 6, n + 8$ **3.** −21, −19, −17, and −15   **9.** 23, 25, 27, 29
   (b) $n + 3$ **5.** 21, 22, 23, 24   **11.** 11 and 13
   (c) n **7.** 11, 12, 13, 14   **13.** −8

Exercise Set 3.6

1. 11	**5.** 56	**9.** $2.80	**13.** 25
3. 80	**7.** 32	**11.** 40	**15.** $133\frac{1}{3}$

Exercise Set 3.7

1. 24	**9.** 5	**15.** 42
3. 25	**11.** 15 and 10	**17.** 6
5. −6	**13.** 18 dimes, 24 nickels	
7. 7		

Regents Tune-Up: Chapter 3

1. 11	**7.** 1	**13.** 27	**19.** (1)	**25.** $\frac{11}{18}$
3. 3	**9.** 475	**15.** 20	**21.** (4)	
5. 7	**11.** 7 and 24	**17.** 6	**23.** 8, 12, 20	

CHAPTER 4

SOLVING LITERAL EQUATIONS AND INEQUALITIES

4.1 SOLVING LITERAL EQUATIONS

KEY IDEAS

A **formula** is an equation that uses mathematical operations to explain how the value of one variable is related to the value(s) of one or more other variables. A formula is an example of a *literal equation*. A **literal equation** is any equation that contains more than one variable.

Evaluating Formulas

The value of a specified variable in a formula can be found if the values of the other variables that appear in the formula are given. For example, the formula to convert from degrees Fahrenheit to degrees Celsius is the equation $C = \frac{5}{9}(F - 32)$. To find the Celsius temperature that is equivalent to 68 degrees Fahrenheit, write the equation, replace F with 68, and simplify:

Write the equation: $\qquad C = \frac{5}{9}(F - 32)$

Substitute 68 for F: $\qquad = \frac{5}{9}(68 - 32)$

Evaluate, following order of operations: $\qquad = \frac{5}{9}(36)$

Divide 36 by 9: $\qquad = \frac{5}{\cancel{9}}(\cancel{36})^{4}$

$\qquad = 5 \cdot 4$

$\qquad = 20$

A temperature of 68 degrees Fahrenheit is equivalent to **20 degrees Celsius**.

Solving for a Given Variable

Sometimes it is helpful to solve a literal equation for a specified variable, so that it becomes easy to see how this variable depends on the other members of the equation. We solve for a given variable in an equation that contains two or more different variables by *isolating* the specified variable. Isolating a

97

variable so that it stands alone on one side of the equation is accomplished in much the same way that an equation having a single variable is solved.

Examples

1. The perimeter P of a rectangle is given by the formula $P = 2l + 2w$. Solve for the length, l.

Solution: $P = 2l + 2w$

Subtract $2w$ from each side: $P - 2w = 2l + 2w - 2w$
Simplify: $P - 2w = 2l$

Divide each side by the coefficient of l: $\dfrac{P - 2w}{2} = \dfrac{2l}{2}$

Simplify: $\dfrac{P - 2w}{2} = l \text{ or } l = \dfrac{P - 2w}{2}$

2. Solve for x: $ax = bx + c$.

Solution: $ax = bx + c$
Subtract bx from each side of the equation: $ax - bx = bx + c - bx$
Apply the distributive property and simplify: $x(a - b) = c$

Divide by the coefficient of x: $\dfrac{x(a - b)}{(a - b)} = \dfrac{c}{(a - b)}$

Since division by 0 is undefined, $a \neq b$. $x = \dfrac{c}{a - b}$

Exercise Set 4.1

1–5. Given the formula $C = \dfrac{5}{9}(F - 32)$, *find C for each value of* F.

1. $F = 32$ **2.** $F = -4$ **3.** $F = 50$ **4.** $F = 0$ **5.** $F = 212$

6. In the formula $C = \dfrac{5}{9}(F - 32)$, at what temperature does $F = C$?

  (1) $+40$ (2) -40 (3) $+8$ (4) -8

7. In the formula $C = \pi D$, find C if $\pi = \dfrac{22}{7}$ and $D = 14$.

8. In the formula $A = \pi r^2$, find A if $\pi = 3.14$ and $r = 10$.

9. In the formula $z = xy^2$, find z if $x = 4$ and $y = -3$.

10–15. Solve for x.

10. $p = 4x$

11. $c = ax + b$

12. $rx + sx = t$

13. $a(x + b) = c$

14. $ax - c = d - bx$

15. $\dfrac{x}{2} - \dfrac{x}{3} = a$

16. Solve for t: $rt = d$.

17. Solve for h: $A = \dfrac{1}{2} bh$.

18. Solve for w: $P = 2(l + w)$.

19. Solve for F: $C = \dfrac{5}{9}(F - 32)$.

20. Solve for b: $x = \dfrac{a + b + c}{3}$.

21. Solve for h: $A = 2l + 2w + 2h$.

4.2 SOLVING INEQUALITIES USING INVERSE OPERATIONS

KEY IDEAS

To solve an inequality, isolate the variable, using addition, subtraction, multiplication, and division as is done when solving an equation. Be careful! If both sides of an inequality are multiplied or divided by the same *negative* number, then an equivalent inequality results *only if the direction of the inequality is* **reversed**.

Graphing Inequalities

The set of values that satisfies an inequality of the form $x \geq a$ can be graphed by darkening a and all points to the right of a on the number line. All points to the left of a satisfy the inequality $x < a$. For example, the graph of $x \geq 2$ is

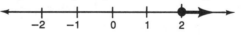

The darkened circle around 2 indicates that 2 is included in the set of values that satisfies $x \geq 2$.

The graph of $x < -1$ is

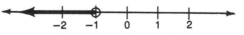

The open circle around -1 indicates that the set of values that satisfies $x < -1$ does *not* include -1.

Solving Inequalities

To solve an inequality algebraically, express the inequality as an equivalent inequality that has the variable on one side of the inequality sign and a constant (number) on the other side. The three properties that follow are useful when working toward isolating a variable in an inequality. Note that each property is true for both the "is greater than (>)" and the "is less than (<)" inequality relation.

99

1. *Addition/Subtraction Property.* An equivalent inequality results whenever the same number is added *or* subtracted on each side of an inequality. For example, to solve $x - 1 > 2$, proceed as follows:

$$\text{Add 1 to each side: } \quad x - 1 + 1 > 2 + 1$$
$$\text{Simplify:} \qquad\qquad\qquad\quad \boldsymbol{x > 3}$$

2. *Multiplication/Division Property I.* An equivalent inequality results whenever each side of an inequality is multiplied by or divided by the same *positive* number. For example, to solve $\frac{x}{2} \geq -3$, proceed as follows:

$$\text{Multiply each side by 2: } \quad 2\left(\frac{x}{2}\right) \geq 2(-3)$$
$$\text{Simplify:} \qquad\qquad\qquad\qquad \boldsymbol{x \geq -6}$$

3. *Multiplication/Division Property II.* Whenever each side of an inequality is multiplied by *or* divided by a negative number, an equivalent inequality results, provided that the direction of the original inequality sign is reversed (from "greater than" to "less than," or vice versa). For example, to solve $-3x > 12$, proceed as follows:

$$\text{Divide each side by } -3: \quad \frac{-3x}{-3} \ \square \ \frac{12}{-3}$$
$$\text{Simplify:} \qquad\qquad\qquad\quad x \ \square \ -4$$
$$\text{Reverse the inequality sign:} \quad \boldsymbol{x < -4}$$

Examples

1. If the replacement set is $\{1, 3, 5, 7\}$, find the solution set of $2x + 4 > 10$.

Solution:

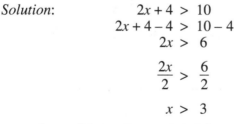

$$2x + 4 > 10$$
$$2x + 4 - 4 > 10 - 4$$
$$2x > 6$$
$$\frac{2x}{2} > \frac{6}{2}$$
$$x > 3$$

The members of the replacement set that satisfy this inequality are 5 and 7. The *solution set is* **{5, 7}**.

2. Solve: $4 > 1 - x$.

Solution:	$4 > 1 - x$
Subtract 1 from each side:	$4 - 1 > -1 + 1 - x$
Simplify:	$3 > -x$
*See note below.	$-x < 3$
Divide each side by -1:	$\frac{-x}{-1} \ \square \ \frac{3}{-1}$
Reverse the inequality sign:	$\boldsymbol{x > -3}$

Note: Switching the left and right sides of an inequality results in an equivalent inequality, provided that the direction of the inequality symbol is reversed. In general, $a > b$ and $b < a$ are equivalent inequalities.

3. Find and graph the solution set: $\qquad 1 - 2x \le x + 13$.

Solution: $\qquad\qquad\qquad\qquad 1 - 2x \le x + 13$

Subtract x from each side: $\qquad\quad 1 - 2x - x \le x + 13 - x$

Combine like terms: $\qquad\qquad\quad 1 - 3x \le 13$

Subtract 1 from each side: $\qquad\quad 1 - 3x - 1 \le 13 - 1$

Divide each side by -3: $\qquad\qquad \dfrac{-3x}{-3} \,\square\, \dfrac{12}{-3}$

Reverse the inequality sign: $\qquad\qquad\qquad x \ge -4$

The graph of $x \ge -4$ is

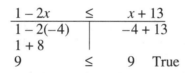

The solution set is $x \ge -4$.

Check: Although it is not possible to check each number of the solution set, you should choose at least one representative value of the solution set and verify that this number makes the original inequality a true statement. Since -4 is a member of the solution set, replace x by -4 in the original inequality.

$$
\begin{array}{c|c}
1 - 2x & \le & x + 13 \\
\hline
1 - 2(-4) & & -4 + 13 \\
1 + 8 & & \\
9 & \le & 9 \quad \text{True}
\end{array}
$$

4. If the replacement set is the set of integers, find the *smallest* value of y that makes the inequality $1 > 3 - 2(y - 4)$ a true statement.

Solution: $\qquad\qquad\qquad\qquad\qquad 1 > 3 - 2(y - 4)$

Apply the distributive property: $\qquad 1 > 3 - 2y + 8$

Simplify: $\qquad\qquad\qquad\qquad\quad 1 > -2y + 11$

Subtract 11 from each side: $\qquad 1 - 11 > -2y + 11 - 11$

Simplify: $\qquad\qquad\qquad\qquad\quad -10 > -2y$

Divide each side by -2: $\qquad\qquad \dfrac{-10}{-2} \,\square\, \dfrac{-2y}{-2}$

Reverse the inequality sign: $\qquad\quad 5 < y \text{ or } y > 5$

Since the replacement set is the set of integers, the solution set is $\{5, 7, 8, 9, ...\}$. The *smallest* member of the solution set is **6**.

Check: Replace x by 6 in the *original* inequality.

$$1 > 3 - 2(y - 4)$$
$$3 - 2(6 - 4)$$
$$3 - 2(2)$$
$$3 - 4$$
$$1 > -1 \qquad \text{True}$$

Exercise Set 4.2

1–21. Find and graph the solution set for each of the following inequalities:

1. $x + 8 > 5$

2. $y - 1 \le -3$

3. $9 - 2x > 1$

4. $\dfrac{n}{-3} < 2$

5. $3(x + 1) > -6$

6. $\dfrac{x}{3} - \dfrac{x}{2} < 2$

7. $\dfrac{y}{3} - 1 \le -2$

8. $2x > 4x - 1$

9. $7 \le 3x - 2$

10. $1 \ge 3 - \dfrac{x}{5}$

11. $3x + 12 \ge 5x + 4$

12. $1 - 5x < -9$

13. $4 - \dfrac{x}{2} \ge 1$

14. $\dfrac{x}{2} \le 3 + \dfrac{x}{6}$

15. $7(9 - 2x) > -(x + 2)$

16. $9 > 2(5 - y) - 3$

17. $3(1 - x) + x > 0$

18. $11 \le 3(1 - 7x) - 6$

19. $\dfrac{x}{6} - 3 > \dfrac{x}{2}$

20. $2 + \dfrac{n}{4} \ge n + \dfrac{3}{4}$

21. $0.1x - 0.02x \ge 8$

22. If the replacement set is the set of integers, what is the largest value of x that makes the inequality $7x - 6 < 8$ a true statement?

23–28. If the replacement set is $\{-2, -1, 0, 1, 2\}$, find the solution set for each of the following inequalities.

23. $1 - 3x \ge -5$

24. $2(x + 1) < x$

25. $4x + 5 - x \ge 2$

26. $0 > x + 1$

27. $1 - 3(3 - x) \le 2$

28. $3(x + 1) < 2(x - 1)$

4.3 SOLVING COMPOUND INEQUALITIES

KEY IDEAS

The inequality $x \geq 3$ is a *compound inequality* since $x \geq 3$ means $x > 3$ OR $x = 3$. A **compound inequality** is an inequality that represents the disjunction or conjunction of two inequalities. The inequality $5 < x < 9$ means 5 is less than x and, at the same time, x is less than 9. Thus, $5 < x < 9$ is a *compound inequality* since it means $5 < x$ AND $x < 9$.

The graph of a compound inequality of the form $a \leq x \leq b$ is a segment on the number line whose left endpoint is a and whose right endpoint is b.

Graphing Compound Inequalities

The graph of the conjunction $x \geq 1$ AND $x < 4$ shown in Figure 4.1 consists of the portion on the number line over which the graphs of the two simple inequalities $x \geq 1$ and $x < 4$ overlap.

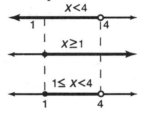

Figure 4.1   Graph of $1 \leq x < 4$

The solution interval includes all numbers 1 and 4, including 1 but not 4. The compound inequality that represents this set of numbers is $1 \leq x < 4$, read as "1 is less than or equal to x *and* x is less than 4."

Hence,

$$(x \geq 1) \wedge (x < 4) \text{ means } 1 \leq x < 4$$

The graph of the solution set of $(x \leq -1) \vee (x \geq 3)$ (see Figue 4.2) consists of all values less than or equal to -1 *or* greater than or equal to 3. Hence the graph consists of two nonoverlapping intervals since all values *between* -1 and 3 do *not* satisfy this inequality condition.

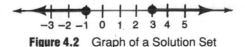

Figure 4.2   Graph of a Solution Set

103

Examples

1. Express the inequality $(x \geq -2) \wedge (x < 3)$ as a single inequality.

Solution: The "boundary" values of the interval that x can range between are -2 and 3, including -2, so that the inequality $-2 \leq x < 3$ is equivalent to $(x \geq -2) \wedge (x < 3)$.

2. If x is an *integer,* list the numbers that satisfy $-4 < x < 1$.

Solution: The solution set consists of the set of all integers between -4 and 1, but not including -4 and 1.

$$-4 < x < 1 = \{-3, -2, -1, 0\}.$$

3. Which of the following is a member of the solution of $-3 < x \leq 1$?
(1) -3 \qquad (2) -4 \qquad (3) -2 \qquad (4) 2

Solution: The solution set consists of the set of all integers between -3 and 1, including 1. Since -2 is contained in this interval, the correct choice is **(3)**.

4. Which inequality is represented by the following graph?

(1) $2 < x \leq -3$ \qquad (2) $-3 \leq x < 2$ \qquad (3) $-3 < x \leq 2$ \qquad (4) $-3 \leq x \leq 2$

Solution: The graph includes all values between -3 and 2, including -3. The correct choice is **(2)**.

Solving Compound Inequalities

To solve a compound inequality, solve each inequality that appears on either side of the logical connective. Then combine the two solution sets into a compound inequality.

Examples

5. Find and graph the solution set: $(-2x + 8 < 0) \wedge (x < 7)$.

Solution:
$$\begin{array}{ccc}
-2x + 8 < 0 & \wedge & x < 7 \\
-2x + 8 - 8 < 0 - 8 & & \\
-2x < -8 & & \\
\dfrac{-2x}{-2} \;\square\; \dfrac{-8}{-2} & & \\
x > 4 & & \\
(x > 4) & \wedge & (x < 7)
\end{array}$$

104

The graph of $(x > 4) \wedge (x < 7)$ is

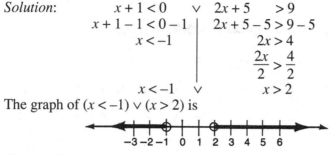

The solution is **$4 < x < 7$**.

6. Find and graph the solution set: $(x + 1 < 0) \vee (2x + 5 > 9)$.

Solution:

$$
\begin{array}{c|c}
x + 1 < 0 \quad \vee & 2x + 5 \quad > 9 \\
x + 1 - 1 < 0 - 1 & 2x + 5 - 5 > 9 - 5 \\
x < -1 & 2x > 4 \\
 & \dfrac{2x}{2} > \dfrac{4}{2} \\
x < -1 \quad \vee & x > 2
\end{array}
$$

The graph of $(x < -1) \vee (x > 2)$ is

The solution is **$(x < -1) \vee (x > 2)$**.

Exercise Set 4.3

1–4. Express each of the following inequalities as an equivalent inequality having the form $a \leq x \leq b$, *and then draw its graph.*

1. $(x > -5) \wedge (x < 3)$ **3.** $(x > -3) \wedge (x \leq 0)$

2. $(x \leq 7) \wedge (x > 2)$ **4.** $(-x \leq 4) \wedge (x + 1 \leq -1)$

5. Which inequality is represented by the accompanying graph?

(1) $-2 \leq x < 3$ (3) $x > 3$ or $x \leq 2$
(2) $-2 < x \leq 3$ (4) $x \geq 3$ or $x < -2$

6. Which inequality is represented by the graph below?

(1) $-2 < x \leq 6$ (3) $-2 < x < 6$
(2) $-2 \leq x < 6$ (4) $-2 \leq x \leq 6$

7. Which statement is represented by the graph below?

(1) $x \leq 6$ (3) $x > 6$
(2) $x < 6$ (4) $x \geq 6$

8. Which graph represents the solution set of $-2 \le x < 1$?

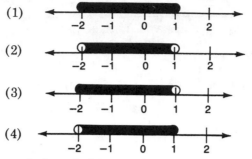

(1)

(2)

(3)

(4)

9–14. Find and graph the solution set.

9. $(x > -2) \wedge (3x + 1 > 16)$

10. $(x + 1 \ge 0) \vee (3 - x > 7)$

11. $(4x + 1 \le 9) \wedge (2x \ge x - 5)$

12. $(2 - n > 0) \vee (19 - 3n < 1)$

13. $(8 < 3x - 4) \wedge (2x - 1 \le 13)$

14. $(x - 5 \ge 2x) \vee \left(\dfrac{x}{3} - 1 \ge 0\right)$

4.4 SOLVING WORD PROBLEMS USING INEQUALITIES

KEY IDEAS

Word problems that use phrases such as *is greater than* (>), *is less than* (<), *is at most* (≤), and *is at least* (≥) require that an inequality be used to determine the set of values that satisfy the conditions of the problem.

Solving Inequalities

Word problems that involve inequalities are solved using an approach similar to the one used to solve word problems that lead to equations.

Examples

1. Find all integers such that the integer increased by 6 is greater than one-third of itself.

Solution: Let x = an integer.

Integer increased by 6 is greater than one third of the number

$$x \qquad + 6 \qquad > \qquad \frac{x}{3}$$

$$x + 6 > \frac{x}{3}$$

Multiply each side by 3: $\qquad 3(x + 6) > 3\left(\frac{x}{3}\right)$

Remove parentheses: $\qquad\qquad 3x + 18 > x$

Subtract x from each side: $\quad 3x - x + 18 > x - x$

Simplify: $\qquad\qquad\qquad\quad 2x + 18 > 0$

Subtract 18 from each side: $\qquad 2x > -18$

Divide each side by 2: $\qquad\qquad \dfrac{2x}{2} > \dfrac{-18}{2}$

$$x < -9$$

Since x can be any integer greater than –9, the set {**–8, –7, –6, –5, . . .**} represents the set of *all* integers having the property that, when they are increased by 6, they are greater than one-third of the original number.

Check: When –8 from the solution set is used, $-8 + 6 = -2$.
Is $-2 > -\dfrac{8}{3}\left(=-2\dfrac{2}{3}\right)$? Yes.

2. Steve is 5 years older than Peter. If the sum of their ages is at most 37 years, what is the *oldest* that Peter can be?

Solution: $\qquad$ Let x = Peter's age.
$\qquad\qquad$ Then $x + 5$ = Steve's age.
$\qquad\qquad x + (x + 5) \le 37 \qquad$ (The sum of their ages cannot exceed 37.)
$\qquad\qquad\qquad 2x + 5 \le 37$
$\qquad\qquad 2x + 5 - 5 \le 37 - 5$
$\qquad\qquad\qquad \dfrac{2x}{2} \le \dfrac{32}{2}$
$\qquad\qquad\qquad\quad x \le 16$

The *oldest* that Peter can be is **16 years**.

Check: If Peter is 16, then Steve is $16 + 5 = 21$. Is $16 + 21 \le 37$? Yes.

3. What is the smallest positive integer that can be added to both the numerator and the denominator of the fraction $\dfrac{2}{7}$ so that the value of the resulting fraction *is at least* $\dfrac{1}{2}$?

Solution: Let x = a positive integer.

$$\frac{2 + x}{7 + x} \ge \frac{1}{2}$$

Since the numerator and the denominator of both fractions are positive numbers, an equivalent inequality results from cross-multiplying.

$$2(2 + x) \ge 1(7 + x)$$

Remove the parentheses: $\qquad 4 + 2x \ge 7 + x$

Subtract x from each side: $\quad 4 + 2x - x \ge 7 + x - x$

Simplify: $4 + x \geq 7$

Subtract 4 from each side: $4 + x - 4 \geq 7 - 4$

Simplify: $x \geq 3$

If x is an integer, then the solution set to the inequality is $\{3, 4, 5, 6, \ldots\}$. Therefore **3** is the *smallest* integer value that can be added to both the numerator and the denominator of $\frac{2}{7}$ so that the value of the resulting fraction is at least $\frac{1}{2}$.

Check: $\dfrac{2+3}{7+3} = \dfrac{5}{10} = \dfrac{1}{2}.$ Is $\dfrac{1}{2} \geq \dfrac{1}{2}$? Yes.

Exercise Set 4.4

1. The sum of three consecutive odd integers is at most 75. What are the largest possible integers that may comprise this set of three?

2. One number is five times another number. The sum of the two numbers is greater than 45. Find the smallest possible values for the two numbers if both are integers.

3. A certain classroom has six rows, and each row must have the same number of student desks. What is the minimum number of desks that must be placed in each row so that the classroom can accommodate at least 32 students and each student has his or her own desk?

4. George is twice as old as Edward, and Edward's age exceeds Robert's age by 4 years. If the sum of their ages is at least 56 years, what is the minimum age of each person?

5. The length and the width of a rectangle are in the ratio 4 : 1. If the perimeter of the rectangle is at most 90, what are the largest possible dimensions of the rectangle?

6. Susan has seven more nickels than dimes in her pocketbook. If the total value of these coins is at least $2.00, what is the least number of nickels that she has?

7. Jerry is saving to buy a VCR that costs $345. Jerry has already saved $37. What is the least amount of money Jerry must save each week so that at the end of 11 weeks he has enough money (excluding tax) to buy the VCR?

8. What is the smallest positive integer that, when added to both the numerator and the denominator of the fraction $\frac{3}{11}$, makes the value of the resulting fraction at least $\frac{1}{2}$?

9. The length of each side of a certain triangle is 7 less than twice the length of a side of a certain square, which is a whole number. What are the smallest possible dimensions of the triangle if the perimeter of the triangle is at least as great as the perimeter of the square?

10. What is the greatest positive integer that can be added to both the numerator and the denominator of the fraction $\dfrac{11}{45}$ so that the value of the resulting fraction does not exceed $\dfrac{1}{3}$?

11. A portion of a wire 60 inches in length is bent to form a rectangle whose length exceeds twice its width by 1. What are the smallest possible dimensions of the rectangle if no more than 4 inches of the wire is unused?

REGENTS TUNE-UP: CHAPTER 4

Each of the questions in this section has appeared on a previous Course I Regents Examination. Here is an opportunity for you to review Chapter 4 and, at the same time, prepare for the Course I Regents Examination.

---

1. Solve for y in terms of a, b, and x: $ay - bx = 2$.

2. Solve for E in terms of I and R: $\dfrac{E}{I} = R$.

3. Solve for b in terms of V and h: $V = \dfrac{bh}{3}$.

4. Given the inequality $8x \geq 3(x - 5)$.
   (a) Solve for x.
   (b) Choose one value of x from your solution in part (a), and show that it makes the inequality $8x \geq 3(x - 5)$ true.

5. One member of the solution set of $-5 < x \leq -1$ is:
   (1) -1 (2) -5 (3) 3 (4) -6

6. An expression equivalent to $3x - 2 < 7$ is:
   (1) $x > \dfrac{5}{3}$ (2) $x < \dfrac{5}{3}$ (3) $x > 3$ (4) $x < 3$

7. The expression $6 \leq x + 4$ is equivalent to:
   (1) $x \geq 2$ (2) $x \leq 2$ (3) $x \leq -2$ (4) $x \geq 10$

8. The largest possible value of x in the solution set of $2x + 1 \leq 7$ is:
   (1) 6 (2) 2 (3) 3 (4) 4

9. Which inequality is represented by the graph?

   (1) $x < 4$ (3) $-4 \leq x < 4$
   (2) $-4 < x \leq 4$ (4) $-4 \leq x \leq 4$

10. Patty needs $80 to buy a bicycle. She has already saved $35. If she saves $10 a week from her earnings, what is the least number of weeks she must work to have enough money to buy the bicycle?
(1) 5   (2) 8   (3) 3   (4) 4

11. Which inequality is the solution set of the graph shown below?

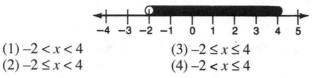

(1) $-2 < x < 4$ (3) $-2 \leq x \leq 4$
(2) $-2 \leq x < 4$ (4) $-2 < x \leq 4$

12. An architect wants to design a rectangular room so that its length is 8 meters more than its width, and its perimeter is greater than 56 meters. If each of the dimensions of the room must be a whole number of meters, what are the *smallest* possible measures, in meters, of the length and width?

13. One number is four times another. The sum of the two numbers is less than 12. Find the largest possible values for the two numbers if both are integers.

14. Find the three largest consecutive integers whose sum is less than 86.

15. One integer is 3 more than twice another integer. The sum of these integers is greater than 24. Find the *smallest* values for these integers.

16. Brian has $78 and wants to purchase tapes through a music club. Each tape costs $7.50. The music club will add a total postage and handling charge of $3.50 to his order. What is the greatest number of tapes he can purchase?

17. The cost of a telephone call from Wilson, N.Y., to East Meadow, N.Y., is $0.60 for the first three minutes plus $0.17 for each *additional* minute. What is the greatest number of whole minutes of a telephone call if the cost cannot exceed $2.50?

Answers to Odd-Numbered Exercises: Chapter 4

Section 4.1

1. 0   **7.** 44   **13.** $\dfrac{c - ab}{a}$ **19.** $\dfrac{9}{5}C + 32$

3. 10   **9.** 36   **15.** $6a$ **21.** $\dfrac{A - 2l - 2w}{h}$

5. 100   **11.** $\dfrac{p}{4}$ **17.** $\dfrac{2A}{b}$

Section 4.2

1. $x > -3$ **7.** $y \leq -3$ **13.** $x \leq 6$ **19.** $x < -9$ **25.** $\{-1, 0, 1, 2\}$

3. $x < 4$ **9.** $x \geq 3$ **15.** $x < 5$ **21.** $x \geq 100$ **27.** $\{-2, -1, 0, 1, 2\}$

5. $x > -3$ **11.** $x \leq 4$ **17.** $x < \dfrac{3}{2}$ **23.** $\{-2, -1, 0, 1, 2\}$

Section 4.3

1. $-5 < x < 3$ **5.** (1)   **9.** $x > 5$ **13.** $4 < x \leq 7$

3. $-3 < x \leq 0$ **7.** (2)   **11.** $-5 \leq x \leq 2$

Section 4.4

1. 23, 25, 27   **7.** 28

3. 6   **9.** 15

5. length = 36, width = 9   **11.** length = 19, width = 9

Regents Tune–Up: Chapter 4

1. $\dfrac{ay - 2}{b}$ **5.** (1)   **9.** (3)   **13.** 2 and 8   **17.** 14

3. $\dfrac{3V}{h}$ **7.** (1)   **11.** (4)   **15.** 7 and 17

CHAPTER 5

OPERATIONS WITH POLYNOMIALS

5.1 CLASSIFYING POLYNOMIALS

KEY IDEAS

The algebraic expression $5x^2 - 3x + 13$ is an example of a *trinomial* since it contains *three* terms: $5x^2$, $3x$, and 13. The expression $5x^2 - 3x$ is a *binomial* since it contains *two* terms. The expression $5x^2$ is a *monomial* since it contains *one* term. The general name given to an algebraic expression that contains one monomial or the sum (or difference) of *more than one* monomial term is *polynomial.*

Monomials

A **monomial** is a single number, a variable, or the product of numbers and variables with positive exponents. The *coefficient* of the monomial $3x^2y$ is 3 and the *literal* or variable factor is x^2y. Here are some additional examples.

Monomial	Numerical Coefficient	Literal Factor
$-7x^2y^3$	-7	x^2y^3
y	1 (since $y = 1y$)	y
$-a^2$	-1 (since $-a^2 = -1a^2$)	a^2
$\dfrac{x^3}{2}$	$\dfrac{1}{2}\left(\text{Since } \dfrac{x^3}{2} = \dfrac{1}{2} \cdot x^3\right)$	x^3

    Like monomials have the same literal (variable) factors. For example, $2xy$ and $5xy$ are *like* monomials since they differ in only their numerical coefficients. The monomials x^2y and xy^2 are *unlike* terms since the variable bases do *not* have the same exponents.

    Like monomials can be added or subtracted by combining their numerical coefficients and using the same literal factor:

$$2xy + 5xy = (2 + 5)xy = 7xy.$$

Examples

    1. Write as a monomial: *3mn − 7mn + mn*

Solution: $3mn - 7mn + mn = (3mn - 7mn) + mn$

$$= -4mn + 1mn$$
$$= -3mn$$

2. Express as a binomial: $x - (3x - 4)$

Solution: Treat the negative sign in front of the parentheses as -1 and remove the parentheses by multiplying it by each term inside the parentheses. Thus,

$$x - (3x - 4) = x - 1(3x - 4)$$
$$= x + (-1)(3x) + (-1)(-4)$$
$$= x - 3x + 4$$
$$= -2x + 4$$

Note: The parentheses in the expression $-(3x - 4)$ can also be removed by rewriting each term with its opposite sign.

Polynomials

A **polynomial** is a monomial or the sum (or difference) of unlike monomial terms. A **binomial** is a polynomial having *two* monomial terms, while a **trinomial** is a polynomial having *three* monomial terms. The polynomial $3x^2 + x - y$ is a polynomial in *two* variables since it has two different variables, x and y. The polynomial $3x^2 + x - 7$ is a polynomial in *one* variable since x is the only variable that it contains.

A polynomial in one variable is in **standard form** when its terms are arranged so that the exponents decrease in value as the polynomial is read from left to right. To write a polynomial in standard form, it may be necessary to rearrange its terms. For example,

> *Given polynomial:* $\qquad 8x^3 + 6x^4 - 2x + 5$
>
> *Polynomial in standard form:* $6x^4 + 8x^3 - 2x + 5$

The **degree** of a polynomial in one variable is the greatest exponent of the variable. If the polynomial consists of a single number, its degree is 0. Here are three more examples.

Polynomial	Standard Form	Degree
1. $4x + 2x^3 - x^2$	$2x^3 - x^2 + 4x$	3
2. $5 - 3x$	$-3x + 5$	1 $(3x = 3x^1)$
3. 2	2	0 $(2 = 2 \cdot 1 = 2x^0)$

Simplifying Polynomials

A polynomial is in *simplest* form when no two of its terms can be combined. For example, the first and last terms of the polynomial $5c^4d - 3c + c^4d$ are like terms, so that the polynomial can be simplified as follows:

$$5c^4d - 3c + c^4d = (5c^4d + 1c^4d) - 3c$$
$$= \mathbf{6c^4d - 3c}.$$

Exercise Set 5.1

1. All of the following are like monomials *except:*
(1) $-4m^3n^2p^5$ (2) $12m^3n^2p^5$ (3) $7m^2n^3p^5$ (4) $n^2m^3p^5$

2. Write each of the following polynomials in standard form, and determine its degree:
(a) $x^2 - x^3 + x - 12$ **(b)** $-(6x - x^4 + 3x - 8)$

3. If $3(2x - 1) = 5x$, what is the value of the polynomial $x^3 - 2x^2 + 10$?

4. If $x - 2 = 10 - 3x$ and $2y - 1 = 1$, what is the value of the binomial $3x - 2y$?

5–10. Simplify each of the following expressions:

5. $3(2a^3 - 5a + 1) + 15a$ **8.** $a^2b - 2ab + 3a^2b$

6. $-(3x^2 - 7x - 9)$ **9.** $2y - 5(3y - 8y^2 + 1)$

7. $5ax - x^2 + 6ax$ **10.** $3x^2y^3 - 7x^2y^3 + 5x^2y^3$

11. If $2x$ represents the width of a rectangle and $5x$ represents the length, express the perimeter of the rectangle as a monomial in terms of x.

5.2   MULTIPLYING AND DIVIDING MONOMIALS

$$\widehat{\text{KEY IDEAS}}$$

Powers of the *same* nonzero base are *multiplied* by *adding* their exponents and keeping the same base. Powers of the *same* nonzero base are *divided* by *subtracting* their exponents and keeping the same base. For example,

$$y^{10}y^3 = y^{10+3} = \mathbf{y^{13}} \quad \text{and} \quad \frac{y^{10}}{y^3} = y^{10-3} = \mathbf{y^7}.$$

Multiplying Monomials

To **multiply** monomials, group and then multiply like factors. For example, to multiply $3x^5$ by $2x^3$, proceed as follows:

Group like factors: $\qquad\qquad\qquad\qquad (3x^5)(2x^3) = (3 \cdot 2)(x^5 x^3)$
Multiply numerical factors: $\qquad\qquad\qquad\qquad\qquad = 6(x^5 x^3)$
Multiply variable factors with the same $\qquad\qquad\qquad = 6x^{5+3}$
   base by adding their exponents: $\qquad\qquad\qquad\quad = 6x^8$

Example

1. Multiply: $\dfrac{1}{2}a^2b$ and $-\dfrac{3}{4}ab^3$

    Solution: $\quad \left(\dfrac{1}{2}a^2b\right)\left(-\dfrac{3}{4}ab^3\right) = \left(\dfrac{1}{2}\right)\left(-\dfrac{3}{4}\right)(a^2b)\,(ab^3)$

$$= \left(-\dfrac{3}{8}\right)(a^2 \cdot a)\,(b \cdot b^3)$$

$$= -\dfrac{3}{8}\,a^3b^4$$

Power Law of Exponents

The properties of exponents can be used to develop a shortcut method for raising a power to another power. For example,

$$(x^5)^4 = x^5 \cdot x^5 \cdot x^5 \cdot x^5 = x^{5+5+5+5} = x^{20}.$$

with $5 \cdot 4 = 20$.

In general, to raise a power to another power, *multiply* their exponents.

$$(a^m)^p = a^{mp}$$

Similarly, to raise a product of variables to a power, raise each factor of the product to the power. For example,

$$(ab)^3 = a^3b^3 \quad \text{and} \quad (-4y)^2 = (-4)^2y^2 = 16y^2.$$

In general,

$$(a^m b^n)^p = a^{mp}b^{np}.$$

Examples

2. Simplify: $(-2a^4)^3$

115

Solution: $(-2a^4)^3 = (-2)^3(a^4)^3 = -8a^{12}$

3. Simplify: $(w^3y^2)^5$

Solution: $(w^3y^2)^5 = (w^3)^5(y^2)^5 = w^{3 \cdot 5}\,y^{2 \cdot 5} = w^{15}y^{10}$

Dividing Monomials

As the following example illustrates, monomials are **divided** using a procedure similar to that for multiplying monomials. To divide $\frac{24a^5b^4c}{8a^2b}$, proceed as follows:

Group the quotients of like variables:
$$\frac{24a^5b^4c}{8a^3b} = \left(\frac{24}{8}\right)\left(\frac{a^5}{a^3}\right)\left(\frac{b^4}{b^1}\right)c$$

Divide numerical coefficients:
$$= 3\left(\frac{a^5}{a^3}\right)\left(\frac{b^4}{b^1}\right)c$$

Divide factors having the same base by
subtracting their exponents:
$$= 3a^{5-3}b^{4-1}c$$
$$= 3a^2b^3c$$

Examples

4. Divide: $\dfrac{20y^3}{4y^7}$.

Solution: $\dfrac{20y^3}{4y^7} = \dfrac{5y^3}{y^7} = 5y^{-4} = \dfrac{5}{y^4}$

5. Divide $-15m^2np^4$ by $3m^5p^4$.

Solution: Write the quotient in fractional form and simplify:
$$\frac{-15m^2np^4}{+3m^5p^4} = \left(\frac{-15}{3}\right)\left(\frac{m^2}{m^5}\right)(n)\left(\frac{p^4}{p^4}\right)$$
$$= (-5)(m^{2-5})(n)(p^{4-4})$$
$$= (-5)(m^{-3})\,(n)(p^0)$$

Replace p^0 by 1:
$$= (-5)(m^{-3})n$$

Replace m^{-3} by $\dfrac{1}{m^3}$:
$$= \frac{-5n}{m^3}$$

Exercise Set 5.2

1–14. Multiply.

1. $(5y^2)(-3y)$

2. $(-p^2q)(-7pq^3)$

3. $(-3ax)(-5ay)$

4. $4y^3(-3y^2)$

5. $y^2 \cdot y^3 \cdot y^4$

6. $(-x^4)(-x^2)(-x)$

7. $(7q^2w^3)(-3qw)$

8. $(x^2y^3)(-2xy^2)$

9. $(-3m^5n^2)(-9mn^3)$

10. $(-0.2ab)(0.35a^3b^2)$

11. $(0.4y^3)(-0.15y^2)$

12. $(rs^2)(r^2s)(r^2s^2)$

13. $(-3x)(7x^2)(-2x)$

14. $\left(\frac{1}{2}x^2y\right)\left(\frac{2}{5}x^3y^2\right)$

15–23. Simplify.

15. $b^2(-b^3)$

16. $x^4(-x)^3$

17. $(-a)^5(-a^5)$

18. $(x^2y^4)^6$

19. $(-2ab^2)^5$

20. $-a^2(-a)^2$

21. $(-b)(-b)^4$

22. $3y^2 + (3y)^2$

23. $(-2a^3) + (-2a)^3$

24–33. Divide.

24. $\dfrac{21p^8}{3p^5}$

25. $\dfrac{-9b^7}{18b^3}$

26. $\dfrac{y^2}{y^6}$

27. $\dfrac{3x^7}{-27x^7}$

28. $\dfrac{72c^6}{8c^{11}}$

29. $\dfrac{8a^2b^3}{12ab^2}$

30. $\dfrac{12r^3}{3rs^4}$

31. $(-12x^3y^2)^2 \div (48xy)$

32. $(1.05x^5y^3) \div (0.35x^2y^2)$

33. $\dfrac{(3x-y)^8}{(3x-y)^7}$

5.3 ADDING AND SUBTRACTING POLYNOMIALS

KEY IDEAS

To find the *sum* of two or more polynomials, add their like terms. To find the *difference* of two polynomials, change the operation from subtraction to addition by replacing each term of the polynomial that is being subtracted with its opposite. Then combine like terms.

Adding Polynomials

To add two polynomials, collect like terms and then simplify. For example,

$$(4x^2 + 3y) + (5x^2 - 2y + 7) = (4x^2 + 5x^2) + (3y - 2y) + 7$$
$$= 9x^2 + y + 7$$

Sometimes it is easier to add polynomials by writing them on separate lines, one underneath the other, so that like terms are aligned in the same vertical columns. The numerical coefficients of like terms can then be added mentally. For example, the sum of $(4x^2 + 3y)$ and $(5x^2 - 2y + 7)$ may be found as follows:

Write the first polynomial, using 0 as a placeholder: $\qquad 4x^2 + 3y + 0$
Write the second polynomial, placing like terms in the $\qquad +$
same columns: $\qquad \underline{5x^2 - 2y + 7}$
Combine like terms in each column: $\qquad 9x^2 + y + 7$

Example

1. The lengths of the sides of a triangle are represented by the binomials $3x + 7y$, $2x - 15y$, and $4x + 21y$. Express the perimeter of this triangle as a binomial.

Solution: The perimeter of a triangle is found by adding the lengths of its sides.

$$\begin{array}{r} 3x + 7y \\ 2x - 15y \\ \underline{4x + 21y} \\ \text{Perimeter} = 9x + 13y \end{array}$$

Subtracting Polynomials

To subtract a polynomial from another polynomial, add the *opposite* of each term of the polynomial that is being subtracted.

Examples

2. Subtract $3a - 2b - 9c$ from $5a + 7b - 4c$.

Solution:
$(5a + 7b - 4c) - (3a - 2b - 9c) = (5a + 7b - 4c) + (-3a + 2b + 9c)$
Group like terms: $= (5a - 3a) + (7b + 2b) + (-4c + 9c)$
Combine like terms: $= \mathbf{2a + 9b + 5c}$

3. How much greater is $29a^2 + 13b + 3c$ than $17a^2 - 6b + 8c$?

Solution: Subtract $(17a^2 - 6b + 8c)$ from $(29a^2 + 13b + 3c)$. It is sometimes convenient to perform the subtraction by writing the polynomials vertically, aligning like terms in the same columns. Write the polynomial being subtracted underneath the other polynomial. Change to an addition example by replacing the sign of each term of the bottom polynomial with its opposite sign. Then add like terms.

<u>Original Subtraction Example</u>

$$29a^2 + 13b + 3c$$
$$-\ \underline{+17a^2 - 6b + 8c}$$

<u>Equivalent Addition Example</u>

$$29a^2 + 13b + 3c$$
$$\underline{-17a^2 + 6b - 8c}$$
$$\mathbf{12a^2 + 19b - 5c}$$

Exercise Set 5.3

1–9. Add.

1. $(5y - 8) + (3y - 2)$

2. $(2x - 5y + 4c) + (-3x + 2y - 3c)$

3. $(7p^3 - 4p^2 + 9) + (-5p^3 + 6p - 7)$

4. $(-x^3 + 7x^2 - 9) + (3x^3 + x^2 - 6x)$

5. $(3x^2 - 5x) + (2x^2 + 7) + (9x - 10)$

6. $\left(\frac{3}{7}a^2 + 1\right) + \left(\frac{5}{14}a^2 - 3\right)$

7. $\left(\frac{1}{3}x^2 + 2x\right) + \left(\frac{1}{2}x^2 - 5x\right)$

8. $\left(\frac{7}{10}b^2 - 2.9b\right) + \left(\frac{4}{25}b^2 + 0.8b\right)$

9. $(y - 3.75y^2) + \left(-\frac{1}{4}y + 0.88y^2\right)$

10. Find the average of $3x^2 - 9$, $2x^2 + 2$, and $4x^2 + 1$.

11–13. Subtract.

11. $(6x + 5) - (3x - 2)$

12. $(4n^2 + 11) - (10n^2 + 7)$

13. $(3y^4 - 2y^3 - 9) - (4y^3 + 3y^2 - 7y)$

14. Subtract $3x - 2$ from $4x + 3$.

15. From $5x^2 - 2x + 3$, subtract $3x^2 + 4x + 3$.

16. From $3x^2 - 4$, subtract $x^2 + 2x - 7$.

17. From $3x^2 - 7x + 12$, subtract $x^2 - 7x - 3$.

18. From the sum of $5x^2 + 8$ and $-2x^2 - 4$, subtract $3x^2 + 7$.

19. From the sum of $2x^3 + 6x^2 - 3$ and $x^3 - 9x + 7$, subtract $8x^3 + x^2 - 5x + 2$.

20. How much greater is $x^4 + 6x^2 + 12$ than $8x^2 - 9$?

21. What polynomial must be added to $4x^3 - 5x^2 + 13$ in order for their sum to be 0?

5.4 MULTIPLYING AND DIVIDING POLYNOMIALS

KEY IDEAS

The methods used for *multiplying* and *dividing* polynomials are based on the distributive property, as illustrated in the following set of arithmetic examples:

1. $\dfrac{15 + 9}{3} = \dfrac{15}{3} + \dfrac{9}{3} = 5 + 3 = \mathbf{8}$

2. $(30 + 1)(10 + 2) = \begin{array}{r} 30 + 1 \\ \times\, 10 + 2 \end{array}$

Multiply 30 and 1 by 10:	$300 + 10$
Multiply 30 and 1 by 2:	$60 + 2$
Add the columns:	$300 + 70 + 2 = \mathbf{372}$

Multiplying a Polynomial by a Monomial

To **multiply** a polynomial by a monomial, multiply *each* term of the polynomial by the monomial. For example,

$$4x^2(2x^3 + xy - 3y^2) = 4x^2(2x^3) + 4x^2(xy) + 4x^2(-3y^2)$$
$$= 8x^{2+3} + 4x^{2+1}y - 12x^2y^2$$
$$= \mathbf{8x^5 + 4x^3y - 12x^2y^2}$$

Examples

1. The length of a rectangle is represented by $4x^2 - 5x + 1$, and its width is represented by $3x$. Represent the area of the rectangle as a trinomial.

Solution: Area =   Length ×   Width
$$= (4x^2 - 5x + 1) \times \quad (3x)$$
$$= (4x^2)3x - 5x(3x) + 1(3x)$$
$$= \mathbf{12x^3 - 15x^2 + 3x}$$

2. Multiply: $3ab^2(5a^2 - 2ab + 4a^3)$.

Solution:
$$3ab^2(5a^2 - 2ab + 4a^3) = 3ab^2(5a^2) + 3ab^2(-2ab) + 3ab^2(4a^3)$$
$$= 15(aa^2)b^2 - 6(aa)(b^2b^1) + 12(aa^3)b^2$$
$$= \mathbf{15a^3b^2 - 6a^2b^3 + 12a^4b^2}$$

3. Multiply: $0.8m(0.7m + 5n^3)$.

Solution: $0.8m(0.7m + 5n^3) = 0.8m^1(0.7m^1) + 0.8m(5n^3)$
$$= (0.8)(0.7)m^2 + (0.8)(5)mn^3$$
$$= \mathbf{0.56m^2 + 4mn^3}$$

Dividing a Polynomial by a Monomial

To **divide** a polynomial by a monomial, divide each term of the polynomial by the monomial. For example,

$$\frac{72x^3 - 32x^2}{8x} = \frac{72x^3}{8x} - \frac{32x^2}{8x}$$
$$= \left(\frac{72}{8}\right)\left(\frac{x^3}{x}\right) - \left(\frac{32}{8}\right)\left(\frac{x^2}{x}\right)$$
$$= 9x^{3-1} - 4x^{2-1}$$
$$= \mathbf{9x^2 - 4x}$$

Multiplying a Polynomial by a Polynomial

To **multiply** a polynomial by another polynomial, write the second polynomial underneath the first polynomial. Then multiply each term of the second polynomial by the polynomial above it using a procedure much like the one used to multiply two multidigit whole numbers. For example, the product of $(2x + 7)$ and $(x + 3)$ may be found as follows:

$$
\begin{array}{r}
2x \;+\; 7 \\
x \;+\; 3 \\
\hline
2x^2 \;+\; 7x \\
+\; 6x \;+\; 21 \\
\hline
2x^2 \;+\; 13x + 21
\end{array}
$$

$$x(2x + 7) =$$
$$3(2x + 7) =$$
Add like terms by column:

The product of $(2x + 7)$ and $(x + 3)$ is $\mathbf{2x^2 + 13x + 21}$.

The product of $(2x + 7)$ and $(x + 3)$ can also be obtained by repeated application of the distributive property.

$$
\begin{aligned}
(2x + 7)(x + 3) &= (2x + 7)x + (2x + 7)3 \\
&= (2x^2 + 7x) + (6x + 21) \\
&= 2x^2 + 13x + 21
\end{aligned}
$$

Exercise Set 5.4

1–15. Multiply.

1. $3x(x^2 - 5x + 7)$

2. $5y^2(y^3 + 8y - 1)$

3. $-4a(a^3 + 6a - 9)$

4. $0.06c^3(0.5c^2 - 0.8)$

5. $(3x + 7)(2x - 9)$

6. $(11a - 3)(8a + 7)$

7. $(r^2 + 5r + 8)(r + 3)$

8. $(3s^2 - 4s + 7)(5s + 2)$

9. $(x + y)(x^2 - xy + y^2)$

10. $(9x^3 - 2x^2 + 7)(4x - 1)$

11. $(5w + 8)(5w - 8)$

12. $(2c - 7)^2$

13. $(3y^2 - 9)(3y^2 + 9)$

14. $(a^2 + ab + b^2)(a - b)$

15. $(n + 4)^3$

16. Express as a trinomial in terms of x the area of a square whose side has a length represented by $5x - 2$.

17. To the product of $(4x - 1)$ and $(3x - 5)$, add $2x^2 - 8x - 9$.

18. From the product of $(2x + 3)$ and $(x - 6)$, subtract $-3x^2 + 5x + 4$.

19. From the product of $(x^2 - 4)$ and $(2x + 1)$, subtract $4x^3 - 2x^2 + 7x - 2$.

20–27. Divide.

20. $\dfrac{32a^5 - 8a^2}{4a}$

21. $\dfrac{15p^3 - 45p^2 + 9p}{3p}$

22. $\dfrac{t^4 + t^3 + 5t^2}{t^2}$

23. $\dfrac{18r^4 - 27r^3s^2}{9r^2}$

24. $\dfrac{30y^6 + 5y^3 - 10y^2}{-5y^2}$

25. $\dfrac{h^3k^2 + h^2k^3 - 7hk}{hk}$

26. $\dfrac{(a^{2b})^2 - (ab)^3}{ab}$

27. $\dfrac{0.14a^3 - 1.05a^2b}{0.7a}$

28–29. Solve for x.

28. $(x + 6)(x - 6) = x(x + 4)$ **29.** $(x - 3)^2 = x(x + 12)$

30. A square and a rectangle have the same area. The length of the rectangle is 8 more than a side of the square, and the width of the rectangle is 4 less than a side of the square. Find the length of a side of the square.

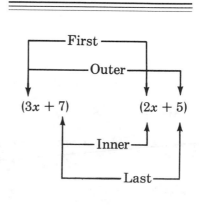

31. In the accompanying figure, a border 4 cm in width surrounds a rectangle whose length exceeds its width by 5. If the area of the rectangular border is 216 cm², what are the dimensions of the inner rectangle?

5.5 MULTIPLYING BINOMIALS USING FOIL

KEY IDEAS

When a pair of binomials such as $(3x + 7)$ and $(2x + 5)$ are written next to each other on the same line, special pairs of terms may be identified by their position: $3x$ and $2x$ are the _First_ terms of each binomial; $3x$ and 5 are the _Outermost_ terms; 7 and $2x$ are the _Innermost_ terms; and 7 and 5 are the _Last_ terms of each binomial. The first letters of these four words form the word **FOIL**.

Using FOIL to Multiply Binomials

A shortcut method for multiplying a pair of binomials is based on FOIL. The letters of FOIL tell us the products that must be formed in order to obtain the product of two binomials: form the product of the First terms, the product of the Outer terms, the product of the Inner terms, and the product of the Last terms. Then write the sum of these products. The following horizontal format can be used:

$$
\begin{array}{cccc}
\text{F} & \text{O} & \text{I} & \text{L}
\end{array}
$$

$$
(3x + 7)(2x + 5) = (3x \cdot 2x) + [(3x \cdot 5) + (7 \cdot 2x)] + (7 \cdot 5)
$$
$$
= \quad 6x^2 \quad + \quad [15x + \quad 14x] \quad + \quad 35
$$
$$
= 6x^2 + 29x + 35
$$

Example

Use the FOIL method to find each of the following products:
(a) $(x - 2)(x - 6)$ (c) $(5x - 9)(2x + 3)$
(b) $(y + 3)^2$ (d) $(w - 6)(w + 6)$

Solution:

$$
\begin{array}{cccc}
\text{F} & \text{O} & \text{I} & \text{L}
\end{array}
$$
(a) $(x - 2)(x - 6) = x^2 + [(-6x) + (-2x)] + 12 = x^2 - 8x + 12$

(b) Rewrite the square of the binomial as the product of two identical binomials.

$$
\begin{array}{cccc}
\text{F} & \text{O} & \text{I} & \text{L}
\end{array}
$$
$$
(y + 3)^2 = (y + 3)(y + 3) = y^2 + [3y + 3y] + 9 = y^2 + 6y + 9
$$

$$
\begin{array}{cccc}
\text{F} & \text{O} & \text{I} & \text{L}
\end{array}
$$
(c) $(5x - 9)(2x + 3) = 10x^2 + [15x + (-18x)] - 27 = 10x^2 - 3x - 27$

$$
\begin{array}{cccc}
\text{F} & \text{O} & \text{I} & \text{L}
\end{array}
$$
(d) $(w - 6)(w + 6) = w^2 + [6w + (-6w)] - 36$
$$
= w^2 + 0w - 36
$$
$$
= w^2 - 36
$$

Notice that the product of two binomials is *not* always a trinomial.

Exercise Set 5.5

1–20. Use FOIL to find each of the following products:

1. $(x - 4)(x - 7)$ 4. $(t - 8)(t + 8)$

2. $(y + 3)(y + 10)$ 5. $(2x + 1)(x + 4)$

3. $(a + 2)(a + 2)$ 6. $(3x + 1)(2x + 5)$

7. $(2y - 1)(4y - 3)$

8. $(a + 7)(a - 7)$

9. $(n + 9)(n + 9)$

10. $\left(x - \dfrac{1}{4}\right)\left(x + \dfrac{1}{4}\right)$

11. $(5 - 6p)(8 - 3p)$

12. $(10 - x)(10 + x)$

13. $(0.6y - 5)(0.4y + 8)$

14. $(x + y)(x - y)$

15. $(1 - 4x)(9 - 5x)$

16. $(3 - 7z)(7 + 3z)$

17. $(p + 3q)(2p - 7q)$

18. $(a + 0.8)(a - 0.8)$

19. $(4w - 3y)(2w - 5y)$

20. $(2r + 9s)(3r - 11s)$

21–23. Express each of the following products as a trinomial:

21. $(x - 7)^2$

22. $(4t + 7)^2$

23. $2(3x - 1)^2$

24–27. Find the value of p.

24. $(x + p)(x - 9) = x^2 - 81$

25. $(x + p)^2 = x^2 + 4x + 4$

26. $(x + 3)(x + p) = x^2 + 2x - 3$

27. $(px - 3)(px + 3) = 25x^2 - 9$

28. In the accompanying diagram, the width of the inner rectangle is represented by x and the length by $2x - 1$. The width of the outer rectangle is represented by $x + 3$ and the length by $x + 5$.

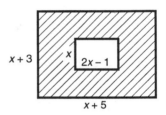

   (a) Express the area of the shaded region as a trinomial in terms of x.

   (b) If the perimeter of the outer rectangle is 24, what is the value of x?

29. The length of a rectangle exceeds the width by 3. If the length is increased by 2 and the width is decreased by 1, the area remains the same. What are the original dimensions of the rectangle?

30. The length of a rectangle is twice its width. If the length is decreased by 4 and the width is increased by 3, the area remains the same. What are the original dimensions of the rectangle?

31. The square of the larger of two consecutive integers exceeds the product of the two consecutive integers by 10. Find the smaller of the two integers.

32. In a set of three consecutive odd integers, the product of the first and the third integer exceeds the product of the first and the second integer by 14. Find the smallest integer in the set.

REGENTS TUNE-UP: CHAPTER 5

Each of the questions in this section has appeared on a previous Course I Regents Examination. Here is an opportunity for you to review Chapter 5 and, at the same time, prepare for the Course I Regents Examination.

1. Find the sum of $4a + 2b - c$ and $3a - 5b - 2c$.

2. From $5x^2 - 3x + 2$ subtract $3x^2 - 4x - 1$.

3. Express $3x^2(2x - 5)$ as a binomial.

4. Express as a trinomial: $(3x - 1)(x + 2)$.

5. Express as a trinomial: $(5x^2 + 2x - 3) - (2x^2 - 3x + 7)$.

6. Subtract $4m - h$ from $4m + h$.

7. The length of a rectangle is represented by $2x + 3$ and the width by $x + 2$. Express the area of the rectangle as a trinomial in terms of x.

8. Express $\dfrac{15x^2}{-3x}$ in simplest form.

9. Find the product of $8y^3$ and $3y^6$.

10. Perform the indicated operations and express the result as a trinomial: $3x(x + 1) + 4(x - 1)$.

11. If the product $(2x + 3)(x + k)$ is $2x^2 + 13x + 15$, find the value of k.

12. The product of $(3x^5)$ and $(4x^2)$ is:
    (1) $7x^7$ (2) $12x^7$ (3) $7x^{10}$ (4) $12x^{10}$

13. The expression $18x^6 \div 3x^3$ is equivalent to:
    (1) $15x^2$ (2) $15x^3$ (3) $6x^2$ (4) $6x^3$

14. The product of $9x^3$ and $2x^4$ is:
    (1) $11x^7$ (2) $11x^{12}$ (3) $18x^7$ (4) $18x^1$

15. The expression $(3x^2y^3)^2$ is equivalent to:
    (1) $9x^4y^6$ (2) $9x^4y^5$ (3) $3x^4y^6$ (4) $6x^4y^6$

16. The quotient of $\dfrac{-4a^6b^2}{2a^2b}$ is:

    (1) $2a^3b^2$ (2) $-2a^4b^2$ (3) $-2a^4b$ (4) $-6a^3b$

17. The length of a rectangle is represented by $x - 5$ and the width by $x + 2$. The area of the rectangle is represented by:
    (1) $x^2 + 3x - 10$ (2) $2x - 3$ (3) $x^2 - 3x - 10$ (4) $4x - 6$

18. The width of a rectangle is represented by $2x$ and the length is represented by $x^2 - x + 3$. Which expression represents the area of the rectangle?
(1) $2x^3 - x + 3$　(2) $x^2 + x + 3$　(3) $2x^2 + 2x + 6$　(4) $2x^3 - 2x^2 + 6x$

19. If $14x^3 - 35x^2 + 7x$ is divided by $7x$, the quotient is
(1) $2x^2 - 5x$　(2) $2x^2 - 5x + 1$　(3) $2x^3 - 5x^2 + x$　(4) $2x^2 - 5x + x$

20. The width of a rectangle is 3 less than its length. If l represents the length, which expression represents the area of the rectangle?
(1) $l^2 - 3l$　(2) $l^2 - 3$　(3) $3l^2$　(4) $4l - 6$

21. The expression $\dfrac{16y^3 + 4y^2 + 2y}{-2y}$, $y \neq 0$, is equivalent to
(1) $-8y^2 - 2y - 1$　(2) $-8y^2 - 2y$　(3) $-2y$　(4) $8y^2 - 2y - 1$

ANSWERS TO ODD-NUMBERED EXERCISES: CHAPTER 5

Section 5.1
1. (3) **5.** $6a^3 + 3$ **9.** $40y^2 - 13y - 5$
3. 19 **7.** $11ax - x^2$ **11.** $14x$

Section 5.2
1. $-15y^3$ **11.** $0.06y^5$ **21.** $-b^5$ **31.** $3x^5y^3$
3. $15a^2xy$ **13.** $42x^4$ **23.** $-10a^3$ **33.** $3x-y$
5. y^9 **15.** $-b^5$ **25.** $-\dfrac{b^4}{2}$
7. $-21q^3w^4$ **17.** a^{10} **27.** $-\dfrac{1}{9}$
9. $27m^6n^5$ **19.** $-32a^5b^{10}$ **29.** $\dfrac{2ab}{3}$

Section 5.3
1. $8y - 10$ **13.** $3y^4 - 6y^3 - 3y^2 + 7y - 9$
3. $2p^3 - 4p^2 + 6p + 2$ **15.** $2x^2 - 6x$
5. $5x^2 + 4x - 3$ **17.** $2x^2 + 15$
7. $\dfrac{5}{6}x^2 - 3x$ **19.** $-5x^3 + 5x^2 - 4x + 2$
9. $\dfrac{3}{4}y - 2.87y^2$ **21.** $-4x^3 + 5x^2 - 13$
11. $3x + 7$

Section 5.4

1. $3x^3 - 15x^2 + 21x$
3. $5y^5 + 40y^3 - 5y^2$
5. $5x^2 - 13x - 63$
7. $r^3 + 8r^2 + 23r + 24$
9. $x^3 + y^3$
11. $25w^2 - 64$
13. $9y^4 - 81$
15. $n^3 + 12n^2 + 48n + 64$

17. $14x^2 - 31x - 4$
19. $-2x^3 + 3x^2 - 15x - 2$
21. $5p^2 - 15p + 3$
23. $2r^2 - 3rs^2$
25. $h^2k + hk^2 - 7$
27. $0.2a^2 - 1.5ab$
29. $\dfrac{1}{2}$
31. width = 7, length = 12

Section 5.5

1. $x^2 - 11x + 28$
3. $a^2 + 4a + 4$
5. $2x^2 + 9x + 4$
7. $8y^2 - 10y + 3$
9. $n^2 + 18n + 81$

11. $18p^2 - 63p + 40$
13. $0.24y^2 + 2.8y - 40$
15. $20x^2 - 41x + 9$
17. $2p^2 - pq - 21q^2$
19. $8w^2 - 26wy + 15y^2$

21. $x^2 - 14x + 49$
23. $18x^2 - 12x + 2$
25. 2
27. 5
29. width = 5, length = 8
31. 9

Regents Tune-Up: Chapter 5

1. $7a - 3b - 3c$
3. $6x^3 - 15x^2$
5. $3x^2 + 5x - 10$

7. $2x^2 + 7x + 6$
9. $24y^9$
11. 5

13. (4)
15. (1)
17. (3)

19. (2)
21. (1)

CHAPTER 6
FACTORING

6.1 FINDING THE GREATEST COMMON FACTOR (GCF)

$$\bigwedge \text{KEY IDEAS}$$

Factoring and multiplying are opposite processes. Factoring an expression means writing the expression as the product of two or more other expressions, each of which is called a **factor** of the product. The numbers 21 and 28 have 7 as a *common* factor since $21 = 3 \cdot 7$ and $28 = 4 \cdot 7$. The monomials $8y^2$ and $12y^3$ have $4y^2$ as a common factor since $8y^2 = 2 \cdot 4y^2$ and $12y^3 = 3y \cdot 4y^2$.

Finding the Greatest Common Factor

The product of the greatest numerical and variable factors that two or more monomials have in common, if any, is called their *Greatest Common Factor* (GCF). To find the GCF of $21a^5$ and $-14a^3$, follow these steps:

Steps	Example
1. Find the largest number that divides evenly into the numerical coefficient of each term.	**1.** The largest number that divides evenly into 21 and −14 is 7.
2. Determine the greatest power of each variable base that is common to all of the monomials.	**2.** The greatest power of a that is contained in a^5 and a^3 is a^3.
3. Form the GCF by multiplying the numerical and variable factors obtained in steps 1 and 2.	**3.** The GCF of $21a^5$ and $-14a^3$ is $7a^3$.

Factoring a Polynomial by Removing the GCF

Since 5 is a factor of 30, we can find the corresponding factor by dividing 30 by 5 to obtain 6. Thus, $30 = 5 \cdot 6$. If we know the GCF of the terms of a polynomial, then we can find the corresponding polynomial factor by dividing the polynomial by the GCF. To factor $10x^2 + 15x$, follow these steps:

Steps	Example
1. Find the GCF of the terms of the polynomial.	**1.** The GCF of $10x^2$ and $15x$ is $5x$.
2. Divide each term of the polynomial by the GCF. The quotient is the other factor.	**2.** Since $$\frac{10x^2 + 15x}{5x} = \frac{10x^2}{5x} + \frac{15x}{5x} = 2x + 3,$$ the other factor is $2x + 3$.
3. Write the polynomial as the product of the two factors obtained in steps 1 and 2.	**3.** Thus, $10x^2 + 15x = 5x(2x + 3)$.
4. Check that the factorization is correct by multiplying the two factors together and then comparing the result with the original polynomial.	**4.** Since $$5x(2x + 3) = 5x(2x) + 5x(3)$$ $$= 10x^2 + 15x,$$ the factorization is correct.

In each of the examples that follow you should check that you have factored the polynomial correctly by multiplying the two factors and then comparing the resulting product with the original polynomial.

Examples

1. Factor: $6kp^2 - 21p^2$

Solution: Compare the numerical and the variable factors of $6kp^2$ and $21p^2$. Since 3 is the largest number that divides 6 and 21 evenly and p^2 is the greatest power of p in both $6kp^2$ and $21p^2$, the GCF of $6kp^2$ and $21p^2$ is $3p^2$. To find the other factor, divide the given polynomial by $3p^2$:

$$\frac{6kp^2 - 21p^2}{3p^2} = \frac{6kp^2}{3p^2} - \frac{21p^2}{3p^2}$$
$$= 2k - 7$$

Thus, $6kp^2 - 21p^2 = 3p^2(2k - 7)$.

2. Factor: $-2rs - 2$

Solution: The GCF of $-2rs$ and -2 is -2. Find the other factor by dividing the polynomial by -2:

$$\frac{-2rs - 2}{-2} = \frac{-2rs}{-2} + \left(\frac{-2}{-2}\right) = rs + 1$$

Thus, $-2rs - 2 = -2(rs + 1)$.

3. Factor: $24x^3y + 30x^2y^3$

Solution: The GCF of the terms of the polynomial is $6x^2y$ since 6 is the largest number that divides both 24 and 30 evenly, and x^2y is the factor with the greatest degree of x and y that is contained in $24x^3y$ and $30x^2y^3$. Find the other factor by dividing the polynomial by $6x^2y$:

$$\frac{24x^3y + 30x^2y^3}{6x^2y} = \frac{24x^3y}{6x^2y} + \frac{30x^2y^3}{6x^2y}$$
$$= 4x + 5y^2$$

Thus, $24x^3y + 30x^2y^3 = \mathbf{6x^2y\ (4x + 5y^2)}$.

4. Factor: $12x + 7y$

Solution: The terms of the given polynomial have no common factors other than 1, so the polynomial cannot be factored.

5. Factor: $x(x - 3) + 7(x - 3)$

Solution: We know that $xA + 7A$ can be factored as $A(x + 7)$. The given expression has the same form as $xA + 7A$ where A represents $x - 3$. Thus, $x(x - 3) + 7(x - 3) = \mathbf{(x - 3)(x + 7)}$.

Exercise Set 6.1

1–16. Factor each of the following polynomials so that one of the factors is the GCF:

1. $5x^2 + 11x$

2. $6a^3 - 9a^2$

3. $4p^2q + 12pq^2$

4. $7n^2 + 7t^2$

5. $x^3 + x^2 + x$

6. $14x - 7x^2$

7. $n^4 - 2n^3 + 5n^2$

8. $4s^3 - 12s^2 + 8s$

9. $3y^7 - 6y^5 + 12y^3$

10. $-3a - 3b$

11. $8u^5w^2 - 40u^2w^5$

12. $-14t^3 - 21t^5$

13. $(p - q)^3 + (p - q)^2$

14. $(x + a)^2 - a(x + a)$

15. $3b(b + 2) - b(b + 2)$

16. $18x^3y - 12x^2y^2 + 24x^4y$

17. If the area of a rectangle is $14x^3 - 21x^2$ and the width is $7x^2$, what is the length?

18. If the area of a rectangle is $45h^4 - 18h^2$ and the length is $5h^2 - 2$, what is the width?

19. Factor out the common binomial factor.
    (a) $2y(y + 2) - 3(y + 2)$ (b) $x(3x - 7) + 2(7 - 3x)$

6.2 MULTIPLYING CONJUGATE BINOMIALS AND FACTORING THEIR PRODUCTS

KEY IDEAS

Binomial pairs such as $(x + 3)$ and $(x - 3)$, $(m + 7)$ and $(m - 7)$, and $(2y + 5)$ and $(2y - 5)$ are examples of *conjugate binomials*. **Conjugate binomials** are binomials that take the sum and difference of the *same* two terms.

Multiplying Conjugate Binomials

Observe the pattern in the following examples, in which pairs of conjugate binomials are multiplied:

$$
\begin{array}{ccccccccc}
 & \overbrace{\text{F}} & & \overbrace{\text{O}} & & \overbrace{\text{I}} & & \overbrace{\text{L}} & \\
(x + 3)(x - 3) = & x^2 & - & 3x & + & 3x & - & 9 & = x^2 - 9 \\
(m + 7)(m - 7) = & m^2 & - & 7m & + & 7m & - & 49 & = m^2 - 49 \\
(2y + 5)(2y - 5) = & 4y^2 & - & 10y & + & 10y & - & 25 & = 4y^2 - 25
\end{array}
$$

Notice that the sum of the outer and inner products will always be equal to 0, so that the product always lacks a "middle" term. *The product of a pair of conjugate binomials is a binomial formed by taking the difference of the squares of the first and last terms of each of the original binomials.*

MATH FACTS

Multiplying Conjugate Binomials

$$(a + b)(a - b) = a^2 - b^2$$

132

Example

1. Find the product of $(x - 3y)$ and $(x + 3y)$.

Solution: $(x - 3y)(x + 3y) = x^2 - (3y)^2 = \boldsymbol{x^2 - 9y^2}$

Factoring the Difference of Two Squares

By reversing the pattern observed in multiplying two conjugate binomials, we obtain a method for factoring a binomial that represents the difference of two squares. To illustrate, since $(x + 5)(x - 5) = x^2 - 25$, then $x^2 - 25$ can be factored by observing that

$$x^2 - 25 = (x)^2 - (5)^2 = (x + 5)(x - 5).$$

To factor a binomial that is the *difference* of two squares, write the product of the corresponding pair of conjugate binomials.

MATH FACTS

Factoring the Difference of Two Squares

$$p^2 - q^2 = (p + q)(p - q)$$

Examples

2. Factor: $n^2 - 100$

Solution: $n^2 - 100 = (n)^2 - (10)^2 = \boldsymbol{(n + 10)(n - 10)}$

3. Factor: $4a^2 - 25b^2$

Solution: $4a^2 - 25b^2 = (2a)^2 - (5b)^2 = \boldsymbol{(2a + 5b)(2a - 5b)}$

4. Factor: $0.16y^2 - 0.09$

Solution: $0.16y^2 - 0.09 = (0.4y)^2 - (0.3)^2 = \boldsymbol{(0.4y + 0.3)(0.4y - 0.3)}$

5. Factor: $p^2 - \dfrac{36}{49}$

Solution: $p^2 - \dfrac{36}{49} = (p)^2 - \left(\dfrac{6}{7}\right)^2 = \boldsymbol{\left(p + \dfrac{6}{7}\right)\left(p - \dfrac{6}{7}\right)}$

Exercise Set 6.2

1–16. Multiply.

1. $(x + 2)(x - 2)$

2. $(y - 10)(y + 10)$

3. $(3a - 1)(3a + 1)$

4. $(-n + 6)(-n - 6)$

5. $(4 - x)(4 + x)$

6. $(1 - 2y)(1 + 2y)$

7. $(0.8n - 7)(0.8n + 7)$

8. $\left(x - \dfrac{2}{3}\right)\left(x + \dfrac{2}{3}\right)$

9. $\left(2y - \dfrac{1}{5}\right)\left(2y + \dfrac{2}{10}\right)$

10. $\left(0.2p - \dfrac{3}{8}\right)\left(0.2p + \dfrac{3}{8}\right)$

11. $(0.5n + 0.3)(0.5n - 0.3)$

12. $(a - 7b)(a + 7b)$

13. $(x^2 - 8)(x^2 + 8)$

14. $(y^3 - z)(y^3 + z)$

15. $(m^2 - n^2)(m^2 + n^2)$

16. $(2w - 3)^2$

17–32. Factor.

17. $x^2 - 144$

18. $y^2 - 0.49$

19. $25 - a^2$

20. $p^2 - \dfrac{1}{9}$

21. $16a^2 - 36$

22. $64x^2 - 1$

23. $h^2 - k^2$

24. $121w^2 - 25z^2$

25. $\dfrac{4}{9}x^2 - 49$

26. $0.36y^2 - 0.64x^2$

27. $0.09h^2 - 0.04$

28. $\dfrac{1}{4}y^2 - \dfrac{1}{9}$

29. $100a^2 - 81b^2$

30. $9x^2 - 25y^2$

31. $n^4 - 49$

32. $x^6 - y^4$

6.3 FACTORING QUADRATIC TRINOMIALS

KEY IDEAS

Using FOIL, $(x + 2)(x + 5) = x^2 + 7x + 10$. Thus, $(x + 2)$ and $(x + 5)$ are the binomial factors of the *quadratic trinomial* $x^2 + 7x + 10$. Not all quadratic trinomials have two binomial factors. We can attempt to factor a quadratic trinomial as the product of two binomials by using a systematic trial and error method that is based on applying the FOIL method in reverse.

Terms of a Quadratic Polynomial

A quadratic polynomial in a single variable takes the general form

$$ax^2 + bx + c,$$

where a cannot be equal to 0.

Quadratic Polynomial	SPECIAL TERMS		
	Quadratic	Linear	Constant
$2x^2 + 5x + 3$	$2x^2$	$5x$	3
$x^2 - 2x$	x^2	$-2x$	0
$x^2 - 16$	x^2	$0x$	-16

Factoring $ax^2 + bx + c$ ($a = 1$)

FOIL can be used to verify that the product of $(x + 2)$ and $(x + 5)$ is $x^2 + 7x + 10$. The binomial factors of the quadratic trinomial $x^2 + 7x + 10$ may, therefore, be written as follows:

$$x^2 + 7x + 10 = (x + 2)(x + 5).$$

product of 2 and 5

sum of 2 and 5

Observe that there is a relationship between the terms of the binomial factors and the values of the coefficients of the terms of the quadratic trinomial. For example, the product of 2 and 5 is 10 (c term), and the sum of 2 and 5 is 7 (coefficient of b term). This suggests a convenient method by which similar types of quadratic trinomials may be expressed in factored form. When attempting to factor a quadratic trinomial of the form $x^2 + bx + c$, follow these steps:

Step 1. Write the two binomial factors: $x^2 + bx + c = (x + ?)(x + ?)$

Step 2. Using trial and error, find the missing terms by thinking, "What *two* numbers when multiplied together give c and when added together give b?"

Step 3. If two such numbers exist, call them p and q, then write the trinomial in factored form as $(x + p)(x + q)$.

Step 4. Check that you have factored the polynomial correctly by multiplying the two factors and then comparing the resulting product with the original polynomial.

Examples

1. Factor: $x^2 + 11x + 18$

Solution:

Step 1. Write $x^2 + 11x + 18 = (x + ?)(x + ?)$

Step 2. Find the missing numbers. The factors of 18 are 1 and 18, 2 and 9, 3 and 6. Use 2 and 9 as the factors of 18 since $2 + 9 = 11$.

Step 3. Write the trinomial in factored form:

$$x^2 + 11x + 18 = (\textbf{\textit{x}} + \textbf{2})(\textbf{\textit{x}} + \textbf{9}).$$

Step 4. Check by multiplying the factors using FOIL.

$$(x + 2)(x + 9) = x^2 + 9x + 2x + 18$$
$$= x^2 + 11x + 18$$

2. Factor: $y^2 - 7y + 12$

Solution:

Step 1. $y^2 - 7y + 12 = (y + ?)(y + ?)$

Step 2. Since the factors of +12 must have the same sign and add up to −7, both factors must be negative. The possible factors of +12 are limited to −1 and −12, −2 and −6, −3 and −4. Use −3 and −4 as the factors of +12 since $(-3) + (-4) = -7$.

Step 3. Thus, $y^2 - 7y + 12 = (\textbf{\textit{y}} - \textbf{3})(\textbf{\textit{y}} - \textbf{4})$.

Step 4. The check is left for you.

3. Factor: $n^2 - 5n - 14$

Solution:

Step 1. $n^2 - 5n - 14 = (n + ?)(n + ?)$

Step 2. Since the factors of −14 must have opposite signs, these are the possible pairs of factors: −1 and 14, 1 and −14, −2 and 7, 2 and −7. Use 2 and −7 as the factors of −14 since $2 + (-7) = -5$.

Step 3. Thus, $n^2 - 5n - 14 = (\textbf{\textit{n}} + \textbf{2})(\textbf{\textit{n}} - \textbf{7})$.

Step 4. The check is left for you.

Factoring $ax^2 + bx + c$ $(a \neq 1)$

When factoring a quadratic trinomial in which the coefficient of the quadratic term is *not* 1, you must take into account the effect of the coefficient a in forming the products of the "first" and "outer" terms. For example, to factor $2x^2 - 7x + 6$, proceed as follows:

Step 1. Determine the possible first terms of the binomial factors. The only positive factors of the numerical coefficient of $2x^2$ are 2 and 1. Therefore the first term of the binomial factors must be $2x$ and x:

$$2x^2 - 7x + 6 = (2x \ \square)(x \ \square).$$

Step 2. Determine the possible last terms of the binomial factors. Since $+6$ is the last term of the quadratic polynomial, the last terms of the binomial factors must be factors of 6, *and* they must have the *same* sign. Since the middle term is negative, the last terms of each binomial factor must be negative. This means that the last terms of the binomial factors are restricted to the following pairs:

$$-1 \text{ and } -6, \quad -2 \text{ and } -3.$$

Step 3. The factors of $+6$ found in step 2 can be used to form the possible binomial factors of $2x^2 - 7x + 6$. For each possible pair, test whether the sum of the outer product and the inner product gives a middle term of $-7x$.

Possible Binomial Factors	Outer + Inner Products = $-7x$?
$(2x - 1)(x - 6)$	$-12x + (-x) = -13x$
$(2x - 6)(x - 1)$	$-2x + (-6x) = -8x$
$(2x - 3)(x - 2)$	$-4x + (-3x) = -7x$ *Success!*

Therefore $2x^2 - 7x + 6 = (\mathbf{2x - 3})(\mathbf{x - 2})$.

You should verify that this is the correct factorization by using FOIL to multiply the two factors.

Factoring $ax^2 + bx + c$ by Decomposing bx

If a quadratic trinomial of the form $ax^2 + bx + c$ can be factored, then its binomial factors can be arrived at by rewriting its middle term, bx, as the sum of two other terms and then factoring. The numerical coefficients of these two terms must be chosen so that their sum is b and their product is $a \cdot c$. To factor $3x^2 + 10x + 8$ using this method, first find two integers whose sum is $+10$ (the coefficient of x), and whose product is 24 (the product of 3 and 8). Since these integers are 4 and 6, write

$$3x^2 + 10x + 8 = 3x^2 + \underbrace{4x + 6x}_{+10x} + 8$$

Group the first and last
pairs of terms: $= (3x^2 + 4x) + (6x + 8)$

Factor out the GCF of each
pair of terms: $= x(3x + 4) + 2(3x + 4)$

Factor out the common
binomial factors: $= (3x + 4)(x + 2)$

Thus, $3x^2 + 10x + 8 = \mathbf{(3x + 4)(x + 2)}$.

Example

4. Factor $4x^2 + 3x - 7$

Solution: Method 1: Use trial and error.

Step 1. List possible pairs of binomial factors of the quadratic polynomial whose first terms are factors of the quadratic coefficient a. Since the product of the first terms of each factor must be $4x^2$, possible binomial factors must take the form

$$(2x \ ?)(2x \ ?) \ or \ (4x \ ?)(x \ \ ?).$$

Step 2. Find all possible pairs of factors. The possible pairs of factors of -7 are 1 and -7, and -1 and 7.

Step 3. Form all possible pairs of binomial factors. Find the pair of factors whose outer and inner products have a sum of $3x$. You should verify that the binomials $(4x + 7)$ and $(x - 1)$ satisfy these conditions.

Therefore $4x^2 + 3x - 7 = \mathbf{(4x + 7)(x - 1)}$.

Method 2: Decompose the middle term and then factor.

Step 1. Find two integers whose sum is $+3$ and whose product is $(4)(-7)$ or -28. The integers that have this property are -4 and $+7$.

Step 2. Decompose the middle term of the original quadratic trinomial.
$$4x^2 + 3x - 7 = 4x^2 \underbrace{- 4x + 7x}_{+3x} - 7$$

Step 3. Group the first pair of terms and the last pair of terms together. Then factor out the GCF of each pair of terms.
$$4x^2 + 3x - 7 = (4x^2 - 4x) + (7x - 7)$$
$$= 4x(x - 1) + 7(x - 1)$$

Step 4. Factor out the common binomial factor.

$$4x^2 + 3x - 7 = 4x(x - 1) + 7(x - 1)$$
$$= \boldsymbol{(x - 1)\,(4x + 7)}$$

Since the order in which the factors in a product are written does not matter, we could also write $4x^2 + 3x - 7 = (4x + 7)\,(x - 1)$. You should verify that this is the correct factorization by using FOIL to multiply the two factors.

Exercise Set 6.3

1–24. Factor.

1. $x^2 + 8x + 15$

2. $x^2 - 10x + 21$

3. $x^2 + 4x - 21$

4. $y^2 + 6y + 9$

5. $n^2 + 3n - 88$

6. $a^2 - 4a - 45$

7. $w^2 - 13w + 42$

8. $b^2 + 3b - 40$

9. $t^2 - 7t - 60$

10. $y^2 - 9y + 8$

11. $s^2 - s - 56$

12. $x^2 - 19x + 90$

13. $y^2 - 2y + 1$

14. $a^2 + a - 20$

15. $2a^2 + 5a - 3$

16. $2q^2 - q - 15$

17. $3x^2 + 2x - 21$

18. $5s^2 + 14s - 3$

19. $5t^2 + 18t - 8$

20. $3n^2 + 29n - 44$

21. $7x^2 + 52x - 32$

22. $-x^2 + x + 12$

23. $3h^2 + h - 4$

24. $x^4 - 3x^2 - 10$

25. Factor $4x^2 + 4x - 15$ by decomposing the middle term.

26. Factor $3x^2 + 10x - 48$ by decomposing the middle term.

27. If $6x + 5$ is a factor of $12x^2 - 14x - 20$, what is the other binomial factor?

28. If $5x - 8$ is a factor of $30x^2 - 38x - 16$, what is the other binomial factor?

29. If the binomial factors of $4x^2 - 36x + 81$ are identical, what is each factor?

30. Factor $3y^2 + 11y - 42$ as the product of two binomials.

6.4   FACTORING COMPLETELY

KEY IDEAS

A polynomial is *factored completely* when *each* of its factors cannot be factored further. Sometimes it is necessary to apply more than one factoring technique in order to factor a polynomial completely.

A Strategy for Factoring Completely

To factor a polynomial completely, proceed as follows:
    1. Factor out the GCF, if any.
    2. If there is a binomial, determine whether it can be factored as the difference of two squares.
    3. If there is a quadratic trinomial, determine whether it can be factored as the product of two binomials by using the reverse of FOIL.

Examples

    1. Factor completely: $3x^3 - 75x$

Solution: First factor out the GCF of $3x$.

$$3x^3 - 75x = 3x(x^2 - 25)$$
$$= 3x(x - 5)(x + 5)$$

    2. Factor completely: $t^3 + 6t^2 - 16t$

Solution: First factor out the GCF of t.

$$t^3 + 6t^2 - 16t = t(t^2 + 6t - 16)$$
$$= t(t + 8)(t - 2)$$

    3. Factor completely: $x^4 - y^4$

Solution: Factor as the difference of two squares.

$$x^4 - y^4 = (x^2)^2 - (y^2)^2$$

Factor $(x^2 - y^2)$:
$$= (x^2 - y^2)(x^2 + y^2)$$
$$= (x - y)(x + y)(x^2 + y^2)$$

Exercise Set 6.4

1–12. Factor completely.

    1. $2y^2 - 50$

    2. $b^3 - 49b$

    3. $x^3 + x^2 - 56x$

    4. $8w^3 - 32w$

    5. $-x^2 - 7x - 10$

    6. $3y^2 - 9y + 6$

    7. $12s^3 - 2s^2 - 4s$

    8. $10y^3 + 50y^2 - 500y$

    9. $3t^4 + 12t^3 - 15t^2$

    10. $p^4 - 1$

    11. $9a^2w^2 - 12a^2w + 4a^2$

    12. $-5t^2 + 5$

    13. Factor completely: $x^2(x + 3) - 4(x + 3)$

    14. Factor completely: $6x^2 + 9x - 60$

6.5 SOLVING QUADRATIC EQUATIONS BY FACTORING

$$\triangle \atop \text{KEY IDEAS}$$

Equations like $x^2 = 25$, $y^2 - 4y - 21 = 0$, and $x^2 + 4x = 5$ are called **quadratic equations** since in each equation the greatest exponent of any variable is 2. Some quadratic equations can be solved by arranging their terms so the resulting equations have the standard form $ax^2 + bx + c = 0$ and then factoring the quadratic trinomial, if possible.

Zero Product Rule

If a and b are real numbers and $ab = 0$, then at least one of these numbers is 0. This property of real numbers is called the **zero product rule**. Thus, if $x(x-4) = 0$ then either $x = 0$, $x - 4 = 0$, or both factors are equal to 0. If $x - 4 = 0$, then $x = 4$. Thus, the *two* roots of the equation $x(x - 4) = 0$ are $x = 0$ and $x = 4$.

Solving Quadratic Equations by Factoring

A **quadratic equation** is in *standard form* when the equation is written as

$$ax^2 + bx + c = 0,$$

where $a \neq 0$. If $a = 0$, then the equation reduces to

$$bx + c = 0,$$

which is a *linear* (first-degree) *equation*. If the left side of the quadratic equation $ax^2 + bx + c = 0$ can be factored, then the zero product rule allows us to solve the quadratic equation by setting each factor equal to 0. To illustrate, the quadratic equation $x^2 + 4x = 5$ can be solved by factoring, using the following procedure:

Steps	Example
1. Write the quadratic equation in standard form.	**1.** $x^2 + 4x - 5 = 0$
2. Factor the quadratic trinomial.	**2.** $(x + 5)(x - 1) = 0$
3. Set each factor equal to 0.	**3.** $(x + 5) = 0$ or $(x - 1) = 0$
4. Solve each linear equation.	**4.** $x = -5$ or $x = 1$
5. Write the solution set.	**5.** $\{-5, 1\}$

Each member of the solution set should be checked by substituting for x in the *original* equation.

$$\text{Let } x = -5.$$
$$\frac{x^2 + 4x}{(-5)^2 + 4(-5)} = 5$$
$$25 - 20$$
$$5 = 5$$

$$\text{Let } x = 1.$$
$$\frac{x^2 + 4x}{(1)^2 + 4(1)} = 5$$
$$1 + 4$$
$$5 = 5$$

When solving quadratic equations, keep in mind that:

- The quadratic equation must be expressed in standard form *before* you attempt to factor the quadratic polynomial.
- Every quadratic equation has *two* solutions, although they may not be numerically different. Each solution is called a **root** of the equation.
- The solution set of a quadratic equation can be checked by verifying that each different root makes the *original* equation a true statement.
- Not every quadratic equation can be solved by factoring. However, in this course, only quadratic equations that can be solved by factoring will be considered.

Examples

1. Solve for x: $2x^2 - 15 = 7x$

Solution:

Write the quadratic equation, which
must be put into standard form: $2x^2 - 15 = 7x$
Subtract $7x$ from each side: $2x^2 - 15 - 7x = 7x - 7x$
Simplify: $2x^2 - 7x - 15 = 0$
Factor the quadratic polynomial: $(2x + 3)(x - 5) = 0$
Set each factor equal to 0: $2x + 3 = 0$ or $x - 5 = 0$
Solve each equation: $2x = -3$ or $x = 5$
$$x = \frac{-3}{2}$$

The solution set is $\left\{\dfrac{-3}{2}, 5\right\}$. The check is left for you.

2. Solve for p: $p^2 = 3p$

Solution: $p^2 - 3p = 0$

$p(p - 3) = 0$

$p = 0$ or $p - 3 = 0$

$p = 3$

The solution set is $\{0, 3\}$. The check is left for you.

Note: A common error in solving this type of equation is to begin by dividing each side of the original equation by p. This is incorrect since we would then be dividing by 0, one of the solutions of the equation.

3. Solve for a: $a^2 + 6a + 9 = 0$

Solution:

$(a + 3)(a + 3) = 0$

$a + 3 = 0$ or $a + 3 = 0$

$a = -3$ or $a = -3$

The two roots are equal, so the root does not have to be written twice in the solution set.

The solution set is $\{-3\}$. The check is left for you.

4. Solve for y: $y^2 - 49 = 0$

Solution:

$(y - 7)(y + 7) = 0$

$y - 7 = 0$ or $y + 7 = 0$

$y = 7$ or $y = -7$

The solution set is $\{-7, 7\}$. The check is left for you.

5. Solve for a: $6a^2 + 18a + 12 = 0$

Solution: Observe that 6 is a common factor of each term of the equation. Therefore the equation may be simplified *before* attempting to factor the quadratic polynomial by dividing each term of the equation by 6.

$$\frac{6a^2}{6} + \frac{18a}{6} + \frac{12}{6} = \frac{0}{6}$$

$$a^2 + 3a + 2 = 0$$

$$(a + 2)(a + 1) = 0$$

$$a + 2 = 0 \text{ or } a + 1 = 0$$

$$a = -2 \text{ or } a = -1$$

The solution set is $\{-2, -1\}$. The check is left for you.

6. One positive number is 4 more than another. The sum of the squares of the two numbers is 40. Find the numbers.

Solution:

$$\text{Let } x = \text{smaller number.}$$
$$\text{Then } x + 4 = \text{larger number.}$$
$$x^2 + (x + 4)^2 = 40$$
$$x^2 + (x + 4)(x + 4) = 40$$
$$x^2 + x^2 + 8x + 16 = 40$$
$$2x^2 + 8x + 16 = 40$$
$$2x^2 + 8x + 16 - 40 = 0$$
$$\frac{2x^2}{2} + \frac{8x}{2} - \frac{24}{2} = \frac{0}{2}$$
$$x^2 + 4x - 12 = 0$$
$$(x + 6)(x - 2) = 0$$
$$x + 6 = 0 \ \text{ or } \ x - 2 = 0$$
$$x = -6 \ \text{ or } \ x = 2$$

Reject –6 since numbers must be positive.

$$x + 4 = 2 + 4 = 6$$

The two numbers are **2** and **6**. Check the answer in the statement of the problem. Is the sum of the squares of 2 and 6 equal to 40? Yes, since $2^2 + 6^2 = 4 + 36 = 40$.

7. Solve for n: $\dfrac{n - 4}{2} = \dfrac{3n}{n + 4}$

Solution: Cross-multiply. Then write the quadratic equation in standard form.

$$(n - 4)(n + 4) = 2(3n)$$
$$n^2 - 16 = 6n$$
$$n^2 - 6n - 16 = 0$$
$$(n - 8)(n + 2) = 0$$
$$n - 8 = 0 \ \text{ or } \ n + 2 = 0$$
$$n = 8 \ \text{ or } \ n = -2$$

The solution set is $\{-2, 8\}$. The check is left for you.

Exercise Set 6.5

1–22. Find the solutions for each of the following equations:

1. $(x + 2)(x - 8) = 0$

2. $y^2 + 3y + 2 = 0$

3. $x^2 + 14x + 49 = 0$

4. $x^2 - 7x = 0$

5. $x^2 - 5x + 4 = 0$

6. $x^2 - x = 12$

7. $a^2 + a = 90$

8. $x(3x - 1) = 0$

9. $q^2 - 6q = 27$

10. $11n - n^2 = 0$

11. $2r^2 - 5r - 3 = 0$

12. $6 = t^2 - t$

13. $y^3 - 9y = 0$

14. $x^2 + 4x = 60$

15. $y^2 + 9y = 36$

16. $3x^2 + 7x = 6$

17. $2n^2 + 9n = 5$

18. $2x^2 + 48 = 20x$

19. $9x^2 - 12x + 4 = 0$

20. $2b^2 - 18 = 5b$

21. $6x = x^2$

22. $35 + w^2 = 12w$

23–25. Cross-multiply and then solve for x.

23. $\dfrac{x+5}{x+1} = \dfrac{x-1}{4}$　　**24.** $\dfrac{x-3}{x-2} = \dfrac{x+3}{2x}$　　**25.** $\dfrac{x-2}{x} = \dfrac{x+4}{3x}$

26–32. Solve each of the following algebraically, and check:

26. One positive number is 5 more than another. The sum of their squares is 53. Find both numbers.

27. The square of a certain positive number is 10 more than three times the number. Find the number.

28. One positive number is 8 more than another. The sum of their squares is 130. Find both numbers.

29. The square of a positive number decreased by four times the number is 12. Find the positive number.

30. The sum of the squares of two consecutive positive integers is 52. Find the integers.

31. Find the smallest of three consecutive odd integers such that two times the product of the first and third integers exceeds 19 times the second integer by 25.

32. Find three consecutive odd integers such that three times the square of the first integer is twelve more than the product of the second and third integers.

6.6 SOLVING WORD PROBLEMS INVOLVING QUADRATIC EQUATIONS

KEY IDEAS

The solutions to many types of word problems lead to quadratic equations. A few representative types are discussed in this section.

Geometry-Related Problems

It is possible that a root of a quadratic equation may *not* satisfy the conditions of a word problem. For example, it may be necessary to reject any negative roots that arise in solving a quadratic equation in which the variable represents a measurement such as length or width.

Examples

1. The length of a rectangle exceeds the width by 6. If the area is 55, find the dimensions of the rectangle.

Solution: Let x = width of the rectangle.

Then $x + 6$ = length of the rectangle.

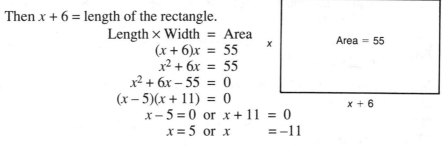

$$\text{Length} \times \text{Width} = \text{Area}$$
$$(x + 6)x = 55$$
$$x^2 + 6x = 55$$
$$x^2 + 6x - 55 = 0$$
$$(x - 5)(x + 11) = 0$$
$$x - 5 = 0 \text{ or } x + 11 = 0$$
$$x = 5 \text{ or } x = -11$$

Reject $x = -11$ since the width cannot be a negative number.

The width of the rectangle is **5**, and the length is $5 + 6$ or **11**.

Check: The length (11) exceeds the width (5) by 6, and the area is 55 since $11 \times 5 = 55$.

2. The length of a rectangular garden is twice its width. The garden is surrounded by a rectangular concrete walk having a uniform width of 4 feet. If the area of the garden and the walk is 330 square feet, what are the dimensions of the garden?

Solution: Let x = width of the garden.

Then $2x$ = length of the garden.

In the accompanying diagram the innermost rectangle represents the garden. Since the walk has a uniform width, the width of the larger (outer) rectangle is $4 + x + 4 = x + 8$. The length of the larger rectangle is $4 + 2x + 4 = 2x + 8$. The area of the larger rectangle is given as 330. Hence:

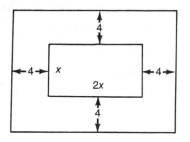

$$\text{Length} \times \text{Width} = \text{Area}$$
$$(2x + 8)(x + 8) = 330$$
$$2x^2 + 24x + 64 = 330$$
$$2x^2 + 24x + 64 - 330 = 0$$
$$2x^2 + 24x - 266 = 0$$
$$\frac{2x^2}{2} + \frac{24x}{2} - \frac{266}{2} = \frac{0}{2}$$
$$x^2 + 12x - 133 = 0$$
$$(x - 7)(x + 19) = 0$$
$$x - 7 = 0 \ \text{ or } \ x + 19 = 0$$
$$x = 7 \ \text{ or } \ x \quad = -19.$$

Reject -19 since the width must be a positive number.
The width of the garden is **7 feet** and the length is **14 feet**.

Number-Related Problems

In some problems, two or more numbers are related in a way that allows them to be represented using the same variable. Additional information about these unknown numbers is included in the problem so that an equation involving the variable can be written and then solved.

Examples

3. The difference between two positive numbers is 5, and the product of the numbers is 36. Find the two numbers.

Solution: Let x = smaller positive number.
 Then $x + 5$ = larger positive number.

$$\underbrace{\textit{Product of numbers}}\ \underbrace{\textit{is}}\ \underbrace{36}$$
$$x(x + 5) = 36$$
$$x^2 + 5x = 36$$
$$x^2 + 5x - 36 = 0$$

$$(x-4)(x+9) = 0$$
$$x-4=0 \text{ or } x+9 = 0$$
$$x=4 \text{ or } x \quad = -9$$

Reject -9 since the problem states that the numbers are positive.
The smaller number $= x = 4$ and the larger number $= x + 5 = 9$. The check is left for you.

4. If the second of three positive consecutive integers is added to the product of the first and the third, the result is 71. Find the integers.

Solution: Let $x =$ first positive integer.
Then $x + 1 =$ next larger consecutive positive integer.
and $x + 2 =$ third of three positive consecutive integers.

Second integer	+	product of first and third	is 71
$x + 1$	+	$x(x + 2)$	$= 71$
$x + 1$	+	$x^2 + 2x$	$= 71$
		$x^2 + 3x + 1$	$= 71$
		$x^2 + 3x + 1 - 71$	$= 0$
		$x^2 + 3x - 70$	$= 0$
		$(x - 7)(x + 10)$	$= 0$
		$x - 7 = 0$ or $x + 10 = 0$	
		$x = 7$ or x	$= -10$.

Reject -10 since the integers must be positive.
The three consecutive integers are 7, 8, and 9. The check is left for you.

Exercise Set 6.6

1. Find a positive number that is 42 less than its square.

2. Find two consecutive positive integers such that the square of the first decreased by 25 equals three times the second.

3. Find three consecutive positive odd integers such that the square of the smallest exceeds two times the largest by 7.

4. The sum of two positive integers is 31. If the sum of the squares of these numbers is 625, find the smaller of the numbers.

5. Find three consecutive positive integers such that the square of the first is equal to the third.

6. The sum of a number and the square of its additive inverse is 30. What is the number?

7. When the first of three consecutive positive integers is multiplied by the third, the result is 1 less than six times the second. Find the three integers.

8. Find three positive consecutive odd integers such that the square of the smallest is 9 more than the sum of the other two.

9. Find three consecutive positive integers such that the product of the first and the third integer is 15.

10. The side of a certain square is 3 feet longer than that of another square. The sum of the areas is 117 square feet. Find the length of the *smaller* square.

11. If a side of a square is doubled and an adjacent side is diminished by 3, a rectangle is formed whose area is numerically greater than the area of the square by twice the original side of the square. Find the dimensions of the original square.

12. One positive number exceeds another by 5. The sum of their squares is 193. Find both numbers.

13. The sum of two positive numbers is 12. The sum of their squares is 90. Find both numbers.

14. One positive number is 1 more than twice the other number. The difference of their squares is 40. Find both numbers.

15. A rectangular picture 30 cm wide and 50 cm long is surrounded by a frame having a uniform width. If the combined area of the picture and the frame is 2016 sq. cm, what is the width of the frame?

16. The art staff of a high school is determining the dimensions of paper to be used in a school publication. The area of each sheet is to be 432 sq cm. The staff has agreed on margins of 3 cm on each side and 4 cm on top and bottom. If the printed matter is to occupy 192 cm on each page, what must be the overall length and width of the paper?

17. A rectangular picture 24 inches by 32 inches is surrounded by a border of uniform width. If the area of the border is 528 square inches less than the area of the picture, find the width of the border.

18. The perimeter of a certain rectangle is 24 inches. If the length is doubled and the width is tripled, then the area is increased by 160 square inches. Find the dimensions of the original square.

Factoring

REGENTS TUNE-UP: CHAPTER 6

Each of the questions in this section has appeared on a previous Course I Regents Examination. Here is an opportunity for you to review Chapter 6 and, at the same time, prepare for the Course I Regents Examination.

1. Factor: $x^2 + x - 30$.

2. The area of a rectangle is represented by $x^2 + 2x$ and the length by $x + 2$. Express the width of the rectangle in terms of x.

3. Factor: $x^2 - 7x + 10$.

4. Factor: $16y^2 - 9$.

5. Factor: $x^2 - y^2$.

6. Express $2x^2 - x - 3$ as the product of two binomials.

7. The larger of two numbers is 24 less than the square of the smaller. If the larger number is divided by the smaller one, the quotient is −5. Find the larger of the two numbers.

8. One positive integer is 3 less than a second positive integer. The sum of the squares of the two integers is 65. Find both positive integers.

9. Find three consecutive even integers such that the square of the first is 80 less than the square of the third.

10. A garden is in the shape of a square. The length of one side of the garden is increased by 3 feet, and the length of an adjacent side is increased by 2 feet. The garden now has an area of 72 square feet. What is the length of a side of the original square garden?

11. Find two consecutive positive integers such that the square of the smaller is 1 more than four times the larger.

12. One factor of $25x^2 - 9$ is $5x - 3$. The other factor is:
    (1) $5x - 3$ (2) $5x + 3$ (3) $-5x - 3$ (4) $-5x + 3$

13. The solution set of $x^2 - 16 = 0$ is:
    (1) $\{2, -8\}$ (2) $\{4, -4\}$ (3) $\{-4\}$ (4) $\{4\}$

14. Which are the factors of $15y^2 - 5y$?
    (1) $5y - 1$ and $3y + 5$ (3) $5y$ and $3y$
    (2) $5y$ and $3y - 1$ (4) $5y - y$ and $3y + 5$

15. Which is the solution set of $x^2 - 3x - 10 = 0$?
    (1) $\{-5, 2\}$ (2) $\{5, -2\}$ (3) $\{2, 5\}$ (4) $\{-2, -5\}$

16. Which is the solution set of the equation $2x^2 + 3x - 2 = 0$?

(1) $\left\{-\dfrac{1}{2}, 2\right\}$ (2) $\left\{\dfrac{1}{2}, -2\right\}$ (3) $\left\{\dfrac{1}{2}, 2\right\}$ (4) $\left\{-\dfrac{1}{2}, -2\right\}$

ANSWERS TO ODD-NUMBERED EXERCISES: CHAPTER 6

Section 6.1
1. $x(5x + 11)$
3. $4pq(p + 3q)$
5. $x(x^2 + x + 1)$
7. $n^2(n^2 - 2n + 5)$
9. $3y^3(y^4 - 2y^2 + 4)$
11. $8u^2w^2(u^3 - 5w^3)$
13. $(p - q)^2(p - q + 1)$
15. $2b(b + 2)$
17. $7x^2(2x - 3)$
19. (a) $(y + 2)(2y - 3)$
  (b) $(3x - 7)(x - 2)$

Section 6.2
1. $x^2 - 4$
3. $9a^2 - 1$
5. $16 - x^2$
7. $0.64n^2 - 49$
9. $4y^2 - \dfrac{1}{25}$
11. $0.25n^2 - 0.09$
13. $x^4 - 64$
15. $m^4 - n^4$
17. $(x + 12)(x - 12)$
19. $(5 + a)(5 - a)$
21. $(4a + 6)(4a - 6)$
23. $(h + k)(h - k)$
25. $\left(\dfrac{2}{3}x + 7\right)\left(\dfrac{2}{3}x - 7\right)$
27. $(0.3h + 0.2)(0.3h - 0.2)$
29. $(10a + 9b)(10a - 9b)$
31. $(n^2 + 7)(n^2 - 7)$

Section 6.3
1. $(x + 5)(x + 3)$
3. $(x - 7)(x - 3)$
5. $(n + 11)(n - 8)$
7. $(w - 7)(w - 6)$
9. $(t - 12)(t + 5)$
11. $(s - 8)(s + 7)$
13. $(y - 1)(y - 1)$
15. $(2a - 1)(a + 3)$
17. $(3x - 7)(x + 3)$
19. $(5t - 2)(t + 4)$
21. $(7x - 4)(x + 8)$
23. $(3h + 4)(h - 1)$
25. $(2x - 3)(2x + 5)$
27. $(6x + 5)(2x - 4)$
29. $2x - 9$

Section 6.4
1. $2(y + 5)(y - 5)$
3. $x(x + 8)(x - 7)$
5. $-(x + 5)(x + 2)$
7. $4s(3s + 1)(s - 1)$
9. $3t^2(t + 5)(t - 1)$
11. $a^2(3w - 2)(3w - 2)$
13. $(x + 3)(x + 2)(x - 2)$

Section 6.5
1. $-2, 8$
3. -7
5. $1, 4$
7. $-10, 9$
9. $-3, 9$
11. $-\dfrac{1}{2}, 3$
13. $0, -3, 3$
15. $-12, 3$
17. $\dfrac{1}{2}, -5$
19. $\dfrac{2}{3}$
21. $0, 6$
23. $-3, 7$
25. 5
27. 5
29. 6
31. 9

Section 6.6

1. 7	**5.** 2	**9.** 3	**13.** 3, 9	**17.** 2 inches
3. 5	**7.** 5, 6, 7	**11.** 8	**15.** 3 cm	

Regents Tune-Up: Chapter 6

1. $(x + 6)(x - 5)$	**5.** $(x + y)(x - y)$	**9.** 8	**13.** (2)
3. $(x - 2)(x - 5)$	**7.** 40	**11.** 5, 6	**15.** (2)

CHAPTER 7

OPERATIONS WITH ALGEBRAIC FRACTIONS

7.1 SIMPLIFYING ALGEBRAIC FRACTIONS

$$\wedge$$
KEY IDEAS

A fraction may be simplified by dividing the numerator and the denominator by common factors (other than 1). For example,

$$\frac{10}{15} = \frac{\overset{1}{\cancel{5}} \cdot 2}{\cancel{5} \cdot 3} = \frac{2}{3}.$$

The factor of 5 appears in the numerator and the denominator, so that the quotient is 1. The fraction $\frac{2}{3}$ is in lowest terms since the numerator and the denominator do not have any common factors greater than 1.

Algebraic Fractions

An **algebraic fraction** is the quotient of two polynomials, provided that the denominator is not equal to 0. Here are some examples of algebraic fractions:

$$\frac{-3}{7}, \quad \frac{2x}{5y}, \quad \frac{a^2b}{bc}, \quad \frac{x+y}{x^2-xy}.$$

Recall that division by 0 is not permitted. Whenever a fraction with a variable denominator is written, the replacement set for the variable(s) in the denominator will be understood to *exclude* any value(s) that would make the denominator equal to 0. For example, for the fraction $\frac{3}{m-1}$, m cannot be equal to 1. If m were equal to 1, the denominator of the fraction would be equal to 0 since $1 - 1 = 0$.

Cancellation of Common Factors

Whenever a fraction contains the same *factor* in both the numerator and the denominator, the quotient of these identical *factors* is 1, so that they can be crossed out. This process is sometimes referred to as **cancellation**. For example, since $3a$ is a *factor* of both $3ab^2$ and $3ay$, then

$$\frac{\overset{1}{\cancel{3ab^2}}}{\underset{1}{\cancel{3ay}}} = \frac{b^2}{y}.$$

On the other hand, it would be incorrect to cancel $3a$ in a fraction in which $3a$ appeared as one of the *terms* of a polynomial. Here is an example of an illegal cancellation:

$$\frac{\overset{1}{\cancel{3a}} + b^2}{\cancel{3a}}.$$

No cancellation of identical *terms* in the numerator and denominator is permitted since $3a$ is *not* a *factor* of the numerator.

Writing Fractions in Lowest Terms

Algebraic fractions, like arithmetic fractions, are written in *lowest terms* by factoring out the GCF of the numerator and the denominator and then *canceling* common factors.

Examples

1. Write in lowest terms: $\dfrac{10x^6}{15x^2}$

Solution: The GCF of $10x^6$ and $15x^6$ is $5x^2$.

$$\frac{10x^6}{15x^2} = \frac{\overset{1}{\cancel{5x^2}} \cdot 2x^4}{\underset{1}{\cancel{5x^2}} \cdot 3} = \frac{2x^4}{3}$$

2. Write in lowest terms: $\dfrac{14a^5b}{35a^4b^3}$

Solution: The GCF of $14a^5b$ and $35a^4b^3$ is $7a^4b$.

$$\frac{14a^5b}{35a^4b^3} = \frac{\overset{1}{\cancel{7a^4b}} \cdot 2a}{\underset{1}{\cancel{7a^4b}} \cdot 5b^2} = \frac{2a}{5b^2}$$

Notice that the variable part of the answer may be obtained by finding the quotient of the variable factors of the numerator and denominator:

$$\frac{\overset{2}{\cancel{14}}a^5b}{\underset{5}{\cancel{35}}a^4b^3} = \frac{2a^{5-4}b^{1-3}}{5} = \frac{2ab^{-2}}{5} = \frac{2a}{5b^2}$$

3. Write in lowest terms: $\dfrac{18a^4 - 30a^3}{3a^2}$

Solution: Step 1: Factor out the GCF from the numerator.

$$\frac{18a^4 - 30a^3}{3a^2} = \frac{6a^3(3a-5)}{3a^2}$$

Step 2. Cancel any pair of factors that is common to both the numerator and the denominator.

$$\frac{18a^4 - 30a^3}{3a^2} = \frac{\overset{2a}{\cancel{6a^3}}(3a-5)}{\cancel{3a^2}}$$

Step 3. Write the remaining factors.

$$\frac{18a^4 - 30a^3}{3a^2} = 2a\,(3a-5)$$

Alternate Method: You can simplify the fraction by dividing each monomial term in the numerator by the monomial denominator. Thus,

$$\frac{18a^4 - 30a^3}{3a^2} = \frac{18a^4}{3a^2} - \frac{30a^3}{3a^2} = 6a^2 - 10a = 2a\,(3a-5)$$

4. Write in lowest terms: $\dfrac{-2x - 10}{x^2 - 25}$

Solution: First factor the numerator and the denominator.

$$\frac{-2x - 10}{x^2 - 25} = \frac{-2(x+5)}{(x+5)(x-5)}$$

Then cancel common factors: $\dfrac{-2\overset{1}{\cancel{(x+5)}}}{\underset{1}{\cancel{(x+5)}}(x-5)} = \dfrac{-2}{(x-5)}$

Exercise Set 7.1

1–24. Write each fraction in lowest terms.

1. $\dfrac{28a^5}{4a^2}$

2. $\dfrac{-52x^3y}{-13xy}$

3. $\dfrac{100c^2}{25c^5}$

4. $\dfrac{32w^7z^6}{18w^2z^9}$

5. $\dfrac{2ab^2 - 2a^2b}{4ab}$

6. $\dfrac{-x - y}{x + y}$

7. $\dfrac{10y^3 - 5y^2}{15y}$

8. $\dfrac{12x^3 - 21x^4}{9x^2}$

9. $\dfrac{0.48xy - 0.16y}{0.8y}$

10. $\dfrac{21r^2s - 7r^3s^2}{14rs}$

11. $\dfrac{-3x - 6}{x + 2}$

12. $\dfrac{y^2 + 4y}{2y + 8}$

13. $\dfrac{3x + 15}{x^2 + 5x}$ $\qquad$ **14.** $\dfrac{x^2 - 4}{x + 2}$ $\qquad$ **15.** $\dfrac{10y + 30}{y^2 - 9}$

7.2 MULTIPLYING AND DIVIDING FRACTIONS

=== KEY IDEAS ===

To **multiply** fractions, write the product of the numerators over the product of the denominators. Then write the resulting fraction in lowest terms.

$$\frac{4}{9} \cdot \frac{3}{10} = \frac{4 \cdot 3}{9 \cdot 10} = \frac{12}{90} = \frac{2}{15}.$$

Sometimes the multiplication process can be made easier by canceling pairs of common factors in the numerator and the denominator *before* multiplying the fractions. For example,

$$\frac{4}{9} \cdot \frac{3}{10} = \frac{\overset{2}{\cancel{4}}}{\underset{3}{\cancel{9}}} \cdot \frac{\overset{1}{\cancel{3}}}{\underset{5}{\cancel{10}}} = \frac{2}{15}.$$

To **divide** one fraction by another fraction, multiply the first fraction by the reciprocal of the second fraction.

Multiplying Fractions

Algebraic fractions are *multiplied* in much the same way that fractions are multiplied in arithmetic.

Examples

1. Write the product: $\left(\dfrac{2x^3}{3y^5}\right)\left(\dfrac{4x^2}{7y}\right)$

Solution: $\left(\dfrac{2x^3}{3y^5}\right)\left(\dfrac{4x^2}{7y}\right) = \dfrac{(2x^3)(4x^2)}{(3y^5)(7y)} = \dfrac{\mathbf{8}x^5}{\mathbf{21}y^6}$

2. Write the product in lowest terms: $\left(\dfrac{2a^7}{3b^2}\right)\left(\dfrac{12b^5}{5a^3}\right)$

Solution: Cancel common factors in the numerator and denominator *before* multiplying.

$$\frac{\overset{a^4}{\cancel{2a^7}}}{\underset{1}{\cancel{3b^2}}} \cdot \frac{\overset{4b^3}{\cancel{12b^5}}}{\cancel{5a^3}} = \frac{(2a^4)(4b^3)}{5} = \frac{8a^4b^3}{5}$$

Fractions that contain polynomials should be factored, is possible, and then simplified *before* they are multiplied.

Examples

3. Write the product in lowest terms: $\dfrac{a^3 - a^2b}{20b^3} \cdot \dfrac{5a + 5b}{a^2}$

Solution: Before multiplying, factor the numerators of each fraction. Cancel pairs of common factors in the numerators and denominators. Then multiply.

$$\frac{a^3 - a^2b}{20b^3} \cdot \frac{5a + 5b}{a^2} = \frac{\overset{1}{\cancel{a^2}}(a - b)}{\underset{4}{\cancel{20b^3}}} \cdot \frac{\cancel{5}(a + b)}{\cancel{a^2}}$$

$$= \frac{(a - b)(a + b)}{4b^3}$$

$$= \frac{a^2 - b^2}{4b^3}$$

4. Write the product in lowest terms: $\dfrac{12y^2}{x^2 + 7x} \cdot \dfrac{x^2 - 49}{2y^5}$

Solution: Factor the binomials, and then cancel pairs of common factors in the numerator and denominator before multiplying.

$$\frac{12y^2}{x^2 + 7x} \cdot \frac{x^2 - 49}{2y^5} = \frac{12y^2}{x(x + 7)} \cdot \frac{(x + 7)(x - 7)}{2y^5}$$

$$= \frac{\overset{6}{\cancel{12y^2}}}{x\cancel{(x + 7)}} \cdot \frac{\overset{1}{\cancel{(x + 7)}}(x - 7)}{\underset{1y^3}{\cancel{2y^5}}}$$

Write the products of the remaining factors in the numerator and the denominator:

$$= \frac{6(x - 7)}{xy^3}$$

Dividing Fractions

Algebraic fractions are *divided* in much the same way that fractions are divided in arithmetic.

Example

5. Write the quotient in lowest terms: $\dfrac{8m^2}{3} \div \dfrac{6m^3}{3m-12}$

Solution: To begin, change from division to multiplication by taking the reciprocal of the second fraction: Where possible, factor.

$$\frac{8m^2}{3} \div \frac{6m^3}{3m-12} = \frac{8m^2}{3} \cdot \frac{3m-12}{6m^3}$$

Cancel pairs of common factors in the numerator and denominator:

$$= \frac{\overset{4}{\cancel{8m^2}}}{\underset{1}{\cancel{3}}} \cdot \frac{\overset{1}{\cancel{3}}(m-4)}{\underset{3m}{\cancel{6m^3}}}$$

Multiply the remaining factors: $= \dfrac{4(m-4)}{3m}$

Exercise Set 7.2

1–12. Write the product or quotient in simplest form.

1. $\dfrac{12a^2}{5c} \cdot \dfrac{15c^3}{8a}$

2. $\dfrac{3y}{4x} \cdot \dfrac{8x^2-4x}{9y}$

3. $\dfrac{5y+10}{x^2} \cdot \dfrac{3x^2-x^3}{15}$

4. $\dfrac{8}{rs} \cdot \dfrac{r^2s-rs^2}{12}$

5. $\dfrac{3x}{5y} \div \dfrac{9x^3}{20y^2}$

6. $\dfrac{2b^2-2b}{3a} \cdot \left(\dfrac{2a}{3b}\right)^2$

7. $\dfrac{18}{x^2-y^2} \div \dfrac{9}{x+y}$

8. $\dfrac{(a+b)^2}{4} \div \dfrac{a+b}{2}$

9. $\dfrac{2x+6}{8} \div \dfrac{x+3}{2}$

10. $\dfrac{x}{7} \div \dfrac{5x}{21x-28}$

11. $\left(\dfrac{ax^2}{b^3}\right)^3 \div \left(\dfrac{a^2x}{b}\right)^2$

12. $\dfrac{a^2-b^2}{2ab} \div \dfrac{a-b}{a^2}$

7.3 CONVERTING UNITS OF MEASURE

$$\overbrace{\text{K\scriptsize EY \normalsize I\scriptsize DEAS}}$$

The process of converting from one unit of measure to another by multiplying the original units by a fractional conversion factor is called **dimensional analysis**. In dimensional analysis, units of measure are considered factors that can be canceled in the same way that pairs of like numerical or variable factors are canceled when multiplying fractions.

Multiplication Property of Conversion Factors

To change 3 hours into an equivalent number of minutes, multiply 3 hours by the conversion factor $\frac{60 \text{ min}}{1 \text{ hr}}$. Thus,

$$3 \text{ hrs} \times \frac{60 \text{ min}}{1 \text{ hr}} = 3 \times 60 \text{ min} = 180 \text{ min}$$

Since the conversion factor has the same amount of time in both its numerator and denominator, the conversion factor is numerically equal to 1. A conversion factor is always formed so that its numerical value is 1. This guarantees that multiplying by an appropriate conversion factor changes the *unit* of measure of a physical quantity, but not its amount.

Finding Conversion Factors

A conversion factor between two related units of measure is obtained from an equation that defines their relationship. For instance, 5280 feet = 1 mile. Dividing both sides of this equation by 5280 feet gives $1 = \frac{1 \text{ mi}}{5280 \text{ ft}}$. Another conversion factor is obtained by dividing both sides of the original equation by 1 mile. This gives $1 = \frac{5280 \text{ ft}}{1 \text{ mi}}$. Multiplying either conversion factor by a quantity that is expressed in the same units as its denominator produces the equivalent amount expressed in the same units as its numerator.

Examples

1. Convert 10,560 feet into an equivalent number of miles.

Solution:

$$10,560 \text{ ft} = 10,560 \text{ ft} \times \frac{1 \text{ mi}}{5280 \text{ ft}} = \frac{10,560}{5280} \text{ mi} = \textbf{2 mi}$$

159

2. Convert $\frac{3}{4}$ miles into an equivalent number of feet.

Solution:

$$\frac{3}{4} \text{ mi} = \frac{3}{4} \text{ mi} \times \frac{5280 \text{ ft}}{1 \text{ mi}} = 3 \times 1310 \text{ ft} = \textbf{3930 ft}$$

3. Convert a speed of $18 \frac{\text{km}}{\text{hr}}$ to $\frac{\text{m}}{\text{min}}$ [where km represents kilometer and m represents meter].

Solution: Since 1 km = 1000 m, to convert from kilometers to meters use the conversion factor of $\frac{1000 \text{ m}}{1 \text{ km}}$. Thus,

$$18 \frac{\text{km}}{\text{hr}} = 18 \frac{\text{km}}{\text{hr}} \times \frac{1000 \text{ m}}{1 \text{ km}} \times ?$$

The numerator of the conversion factor for time must contain hours so that it cancels the hours in the denominator of $18 \frac{\text{km}}{\text{hr}}$. Hence, use the conversion factor $\frac{1 \text{ hr}}{60 \text{ min}}$. This gives:

$$18 \frac{\text{km}}{\text{hr}} = 18 \frac{\text{km}}{\text{hr}} \times \frac{1000 \text{ m}}{1 \text{ km}} \times \frac{1 \text{ hr}}{60 \text{ min}}$$

$$= \frac{18 \times 1000 \times \text{m}}{60 \times \text{min}}$$

$$= \frac{18,000 \text{ m}}{60 \text{ min}}$$

$$= 300 \frac{\text{m}}{\text{min}}$$

Therefore, a speed of $\textbf{300} \frac{\textbf{m}}{\textbf{min}}$ is equivalent to a speed of $18 \frac{\text{km}}{\text{hr}}$.

Exercise Set 7.3

1. Given 1 meter = 3.28 feet and 5280 feet = 1 mile.
(a) Determine the conversion factor from meters to miles.
(b) Use the conversion factor found in part **(a)** to convert $300 \frac{\text{meters}}{\text{min}}$ to $\frac{\text{miles}}{\text{hour}}$.

2–3. Convert $45 \frac{\text{meters}}{\text{hr}}$ into the stated units and express the result using scientific notation.

2. $\frac{\text{feet}}{\text{sec}}$ [Use 1 meter = 3.28 feet]   **3.** $\frac{\text{km}}{\text{min}}$

4. Convert $60 \frac{\text{miles}}{\text{hr}}$ to $\frac{\text{feet}}{\text{sec}}$.

5. Convert $66 \frac{\text{feet}}{\text{sec}}$ to $\frac{\text{miles}}{\text{hr}}$.

6. Convert $45 \frac{\text{miles}}{\text{hr}}$ to $\frac{\text{km}}{\text{min}}$. [Use 1 mile = $\frac{8}{5}$ km].

7.4 ADDING AND SUBTRACTING FRACTIONS WITH LIKE DENOMINATORS

$$\triangle \atop \text{KEY IDEAS} \atop \triangle\!\triangle$$

To **add** (or **subtract**) fractions having like denominators, write the sum (*or* difference) of the numerators over the common denominator.

Addition	Subtraction
$\dfrac{5}{9} + \dfrac{2}{9} = \dfrac{5+2}{9} = \dfrac{\mathbf{7}}{\mathbf{9}}$	$\dfrac{7}{11} - \dfrac{4}{11} = \dfrac{7-4}{11} = \dfrac{\mathbf{3}}{\mathbf{11}}$

Combining Fractions

In the methods used to add or subtract algebraic fractions are patterned after those used in arithmetic.

Examples

1. Write the sum in lowest terms: $\dfrac{5y}{9x^2} + \dfrac{y}{9x^2}$

Solution: Write the sum of the numerators over the common denominator and simplify.

$$\frac{5y}{9x^2} + \frac{y}{9x^2} = \frac{5y+y}{9x^2}$$

$$= \frac{6y}{9x^2}$$

$$= \frac{\overset{2}{\cancel{6}y}}{\underset{3}{\cancel{9}x^2}} = \frac{\mathbf{2y}}{\mathbf{3x^2}}$$

2. Write the difference in lowest terms: $\dfrac{5a+b}{10ab} - \dfrac{3a-b}{10ab}$

Solution: Write the difference of the numerators over the common denominator and simplify.

$$\frac{5a+b}{10ab} - \frac{3a-b}{10ab} = \frac{5a+b-(3a-b)}{10ab}$$

$$= \frac{5a+b-3a+b}{10ab}$$

$$= \frac{(5a-3a)+(b+b)}{10ab}$$

$$= \frac{2a + 2b}{10ab}$$

$$= \frac{\overset{1}{\cancel{2}}(a + b)}{\underset{5}{\cancel{10}ab}} = \frac{a + b}{5ab}$$

Negative Signs and Fractions

Notice that $\frac{-1}{1}$ and $\frac{1}{-1}$ are each equivalent to -1. Similarly, all of the following algebraic fractions are equivalent since they have a coefficient of -1:

$$\frac{-x}{y}, \quad \frac{x}{-y}, \quad \text{and} \quad -\frac{x}{y}.$$

An equivalent fraction results if the signs (that is, the factor of -1) are exchanged between any two of the following: the numerator, the denominator, and the sign of the fraction.

Example

3. Express as a single fraction in lowest terms: $\dfrac{4x}{x^2 - 4} + \dfrac{x + 6}{4 - x^2}$

Solution: Rewrite $4 - x^2$ as $-x^2 + 4$. Then factor out -1 so that

$$-x^2 + 4 = -(x^2 - 4).$$

$$\frac{4x}{x^2 - 4} + \frac{x + 6}{4 - x^2} = \frac{4x}{x^2 - 4} + \frac{x + 6}{-(x^2 - 4)}$$

Write the negative sign in front of the second fraction: $\quad = \dfrac{4x}{x^2 - 4} - \dfrac{x + 6}{x^2 - 4}$

Write the numerators over the common denominator: $\quad = \dfrac{4x - (x + 6)}{x^2 - 4}$

Simplify the numerator: $\quad = \dfrac{4x - x - 6}{x^2 - 4}$

Write the fraction in lowest terms: $\quad = \dfrac{3x - 6}{x^2 - 4}$

Factor, and cancel common factors: $\quad = \dfrac{3\overset{1}{\cancel{(x - 2)}}}{\underset{1}{\cancel{(x - 2)}}(x + 2)}$

$$= \frac{3}{x + 2}$$

Exercise Set 7.4

1–12. Write the sum or difference in lowest terms.

1. $\dfrac{2a}{14} + \dfrac{5a}{14}$

2. $\dfrac{5x+2}{3} - \dfrac{x+1}{3}$

3. $\dfrac{4}{a+b} + \dfrac{1}{-a-b}$

4. $\dfrac{2r+s}{r+2s} + \dfrac{r+5s}{r+2s}$

5. $\dfrac{3(n^2-2n)}{10n} + \dfrac{2n^2+n}{10n}$

6. $\dfrac{3(2b-1)}{4b} - \dfrac{5b+3}{4b}$

7. $\dfrac{7(x+y)}{12xy} - \dfrac{3x-y}{12xy}$

8. $\dfrac{p}{p^2-9} + \dfrac{3}{9-p^2}$

9. $\dfrac{5a-b+c}{2a+b} + \dfrac{a+4b-c}{2a+b}$

10. $\dfrac{x+y}{11} - \dfrac{2x+4y}{22}$

11. $\dfrac{a^2-5}{a-b} - \dfrac{b^2-5}{a-b}$

12. $\dfrac{7t+3}{8rs} - \dfrac{3t+7}{8rs} + \dfrac{2(t+5)}{8rs}$

7.5 ADDING AND SUBTRACTING FRACTIONS WITH UNLIKE DENOMINATORS

$$\bigwedge$$
KEY IDEAS

The *Lowest Common Denominator* or LCD of $\frac{1}{8}$ and $\frac{7}{12}$ is 24, since 24 is the smallest positive integer that is evenly divisible by 8 and by 12. To add $\frac{1}{8}$ and $\frac{7}{12}$, we first need to change each fraction into an equivalent fraction that has the LCD as its denominator. Thus,

$$\frac{1}{8} + \frac{7}{12} = \frac{1}{8} \cdot \left(\frac{3}{3}\right) + \frac{7}{12} \cdot \left(\frac{2}{2}\right) = \frac{3}{24} + \frac{14}{24} = \frac{17}{24}$$

Fractions that contain variables are handled in a similar way.

Fractions with Monomial Denominators

The LCD of $\frac{1}{2x}$ and $\frac{1}{3x^2}$ is $6x^2$. Notice that 6 is the smallest positive number that is evenly divisible by both 2 and 3, and x^2 is the greatest power of x that appears in either denominator. The LCD of fractions that have monomial denominators is a monomial whose numerical coefficient is the smallest positive integer that is evenly divisible by the numerical coefficients of each denominator. The variable factors of the LCD are the different variables that appear in the denominators, each written with the greatest exponent it has in any denominator.

Example

1. Find the LCD of $\dfrac{1}{15a^2b}$ and $\dfrac{1}{6ab^3}$

Solution:

Step 1: First find the numerical factor of the LCD. The numerical factor of the LCD is 30 since 30 is the smallest positive integer that is evenly divisible by both 15 and 6.

Step 2: Compare the denominators to determine the greatest power of each variable factor. The greatest power of a that is contained in any one denominator is a^2 and the greatest power of b in any one denominator is b^3.

Step 3. Write the LCD as the product of the factors found in steps 1 and 2. Thus, the LCD of $\dfrac{1}{15a^2b}$ and $\dfrac{1}{6ab^3}$ is $\mathbf{30a^2b^3}$.

Combining Fractions with Unlike Denominators

To add or subtract fractions in arithmetic, the denominators must be the same. If the denominators are *different,* each fraction must be changed into an equivalent fraction having the LCD as its denominator. This allows the fractions to be combined by writing the sum (or difference) of the numerators over the LCD. Algebraic fractions are combined in much the same way.

Examples

2. Add: $\dfrac{2x+1}{6} + \dfrac{3x-5}{8}$

Solution: The LCD of 6 and 8 is 24 since 24 is the smallest positive integer into which 6 and 8 divide evenly. Since 6 times *4* is 24, multiplying the first fraction by 1 in the form of $\frac{4}{4}$ produces an equivalent fraction with the LCD as its denominator. Since 8 times *3* is 24, multiplying the second fraction by 1 in the form of $\frac{3}{3}$ produces an equivalent fraction with the LCD as its denominator.

$$\frac{2x+1}{6} + \frac{3x-5}{8} = \left(\frac{2x+1}{6}\right)\frac{4}{4} + \left(\frac{3x-5}{8}\right)\frac{3}{3}$$

Write the numerators over the LCD:
$$= \frac{4(2x+1) + 3(3x-5)}{24}$$

Combine like terms in the numerator:
$$= \frac{8x+4+9x-15}{24}$$

$$= \frac{17x-11}{24}$$

3. Write the difference in lowest terms: $\dfrac{3}{4x} - \dfrac{1}{5x}$

Solution: The LCD of $4x$ and $5x$ is $20x$. Multiplying the first fraction by $\frac{5}{5}$ and the second fraction by $\frac{4}{4}$ produces equivalent fractions with the LCD as their denominators.

$$\frac{3}{4x} - \frac{1}{5x} = \frac{5}{5}\left(\frac{3}{4x}\right) - \frac{4}{4}\left(\frac{1}{5x}\right)$$

$$= \frac{15 - 4}{20x}$$

$$= \frac{11}{20x}$$

4. Write the sum in lowest terms: $\dfrac{a-1}{3ab} + \dfrac{2}{9b}$

Solution: The smallest positive integer that is evenly divisible by both 3 and 9 is 9. The greatest power of each of the different variable factors of the two denominators is a and b. Thus, the LCD is $9ab$. Multiplying the first fraction by $\frac{3}{3}$ and the second fraction by $\frac{a}{a}$ produces equivalent fractions that have $9ab$ as their denominators. Thus,

$$\frac{3}{3}\left(\frac{a-1}{3ab}\right) + \frac{a}{a}\left(\frac{2}{9b}\right) = \frac{3(a-1)}{9ab} + \frac{2a}{9ab}$$

$$= \frac{3a - 3 + 2a}{9ab}$$

$$= \frac{5a - 3}{9ab}$$

Combining Fractions with Polynomial Denominators (Optional Topic)

If fractions have polynomial denominators, then their LCD can be determined by first factoring the denominators, where possible, into their *prime factors*. Prime factors are expressions that cannot be factored further. The LCD is the product of all the different prime factors of the denominators with each raised to the greatest power to which that factor appears in any of the denominators.

Examples

5. Find the LCD of $\dfrac{4x}{(x+3)^2}$ and $\dfrac{5}{4x^2 + 12x}$.

Solution: The first denominator is the square of a prime expression. Factoring the second denominator gives

$$4x^2 + 12x = 4x(x + 3)$$

Since the greatest power of $(x + 3)$ that appears in any denominator is $(x + 3)^2$, the LCD is $\mathbf{4x(x + 3)^2}$.

6. Express the difference in simplest form: $\dfrac{a^2 + 1}{a^2 - 1} - \dfrac{a}{a + 1}$

Solution: Factor the first denominator.

$$a^2 - 1 = (a + 1)(a - 1)$$

The second denominator cannot be factored. Since no denominator contains a prime factor that is raised to a power, the LCD is the product of the different prime factors. Hence, LCD is $(a + 1)(a - 1)$. The first denominator already has the LCD as its denominator. Multiplying the second fraction by $\frac{a-1}{a-1}$ changes it into an equivalent fraction that has the LCD as its denominator.

$$\frac{a^2 + 1}{a^2 - 1} - \frac{a}{a + 1} = \frac{a^2 + 1}{(a + 1)(a - 1)} - \frac{a}{a + 1} \cdot \left(\frac{a - 1}{a - 1}\right)$$

$$= \frac{a^2 + 1 - a(a - 1)}{(a + 1)(a - 1)}$$

$$= \frac{a^2 + 1 - a^2 + a}{(a + 1)(a - 1)}$$

$$= \frac{\overset{1}{\cancel{a + 1}}}{\cancel{(a + 1)}(a - 1)}$$

$$= \frac{1}{a - 1}$$

7. Write the difference in simplest form: $\dfrac{3}{10xy} - \dfrac{10x - y}{5xy^2}$

Solution: The LCD is $10xy^2$. Multiply the numerator and denominator of each fraction by the factors contained in the LCD that are missing from the denominator of that fraction.

$$\left(\frac{3}{10xy}\right)\frac{y}{y} - \left(\frac{10x - y}{5xy^2}\right)\frac{2}{2} = \frac{3y}{10xy^2} - \frac{2(10x - y)}{10xy^2}$$

$$= \frac{3y - 20x + 2y}{10xy^2}$$

$$= \frac{5y - 20x}{10xy^2}$$

$$= \frac{\overset{1}{\cancel{5}}(y - 4x)}{\underset{2}{\cancel{10}xy^2}}$$

$$= \frac{y - 4x}{2xy^2}$$

Exercise Set 7.5

1–24. Write each of the following sums or differences as a single fraction in simplest form.

1. $\dfrac{2x}{3} + \dfrac{5x}{6}$

2. $\dfrac{7}{12x} - \dfrac{1}{3x}$

3. $\dfrac{4b}{5x} - \dfrac{3b}{10x}$

4. $\dfrac{4a + 1}{2} - \dfrac{a}{3}$

5. $\dfrac{3y - 5}{5xy} - \dfrac{1}{10x}$

6. $\dfrac{1}{a} + \dfrac{1}{b}$

7. $\dfrac{a}{b} - \dfrac{b}{a}$

8. $\dfrac{3x - 1}{7x} + \dfrac{x + 9}{14x}$

9. $\dfrac{3x + 1}{5x} - \dfrac{x + 2}{10x}$

10. $\dfrac{y - x}{x^2 y} + \dfrac{y + x}{xy^2}$

11. $\dfrac{y - 4}{8y} + \dfrac{y + 6}{12y}$

12. $\dfrac{p - 3}{4p} + \dfrac{3p + 1}{6p}$

13. $\dfrac{b - 5}{10b} + \dfrac{b + 10}{15b}$

14. $\dfrac{a - 2}{2a} + \dfrac{3a + 5}{5a}$

15. $\dfrac{10x + 1}{14x} - \dfrac{x + 5}{21x}$

16. $\dfrac{3t - 7}{4t} + \dfrac{14 - 5t}{8t}$

17. $\dfrac{2c + 5}{6c} + \dfrac{5c - 7}{21c}$

18. $\dfrac{a + b}{10c} + \dfrac{6a - 2b}{4c}$

19. $\dfrac{3}{ab} - \dfrac{4a + 1}{a^2 b} + \dfrac{b + 2}{ab^2}$

20. $\dfrac{10xy + 1}{5x^2 y} - \dfrac{x + 6}{3x}$

21. $\dfrac{x}{3} + 1$

22. $\dfrac{3}{x} - 1$

23. $y + \dfrac{3}{y}$

24. $\dfrac{3}{x - 1} - \dfrac{2}{x}$

25–30. Combine the fractions and write the result in simplest form.

25. $\dfrac{a}{a - b} + \dfrac{b}{a + b}$

26. $\dfrac{3a + 5}{a^2 + 2a} - \dfrac{1}{2a + 4}$

27. $\dfrac{x^2}{(x+1)^2} - \dfrac{x-1}{x+1}$

28. $\dfrac{5}{x^2-1} + \dfrac{3}{(x-1)^2}$

29. $\dfrac{y}{y^2-9} - \dfrac{1}{2y+6}$

30. $\dfrac{3}{x^2-4} + \dfrac{2}{x^2+5x+6}$

7.6 SOLVING EQUATIONS INVOLVING FRACTIONS

KEY IDEAS

To solve an equation that contains fractions, transform the equation into an equivalent that does not include any fractions.

Solving Equations with Fractions

To solve an equation that contains arithmetic or algebraic fractions, clear the equation of its fractions by multiplying each member of *both* sides of the equation by the LCD of the denominators.

Examples

1. Solve for x and check: $\dfrac{x+1}{4} - \dfrac{2}{3} = \dfrac{1}{12}$

Solution: The LCD is 12.

Multiply each term by 12:
$$\overset{3}{\cancel{12}}\left(\frac{x+1}{\cancel{4}}\right) - \overset{4}{\cancel{12}}\left(\frac{2}{\cancel{3}}\right) = \overset{1}{\cancel{12}}\left(\frac{1}{\cancel{12}}\right)$$

$$3(x+1) - 4(2) = 1$$
$$3x + 3 - 8 = 1$$
$$3x - 5 = 1$$
$$3x = 5 + 1$$
$$\frac{3x}{3} = \frac{6}{3}$$
$$x = 2$$

Check:

$$\frac{x+1}{4} - \frac{2}{3} = \frac{1}{12}$$

$$\frac{2+1}{4} - \frac{2}{3}$$

$$\frac{3}{4} - \frac{2}{3}$$

$$\frac{9}{12} - \frac{8}{12}$$

$$\frac{1}{12} = \frac{1}{12}$$

2. Solve for *y:* $\dfrac{2}{y} - \dfrac{9}{10} = \dfrac{1}{5y}.$

Solution: The LCD of *y*, 10, and 5*y* is 10*y*.

Multiply each term by 10*y:*

$$10\overset{1}{\cancel{y}}\left(\frac{2}{\cancel{y}}\right) - \overset{1}{\cancel{10}}y\left(\frac{9}{\cancel{10}}\right) = \overset{2}{\cancel{10}}y\left(\frac{1}{\cancel{5}y}\right)$$

$$20 - 9y = 2$$
$$-9y = 2 - 20$$
$$\frac{-9y}{-9} = \frac{-18}{-9}$$
$$y = 2$$

The check is left for you.

Exercise Set 7.6

1–12. Solve for the variable and check.

1. $\dfrac{x}{5} - 12 = 4$

2. $\dfrac{x}{3} - \dfrac{x}{4} = 1$

3. $\dfrac{n}{2} - 3 = \dfrac{n}{5}$

4. $\dfrac{2r}{3} - \dfrac{5r}{12} = \dfrac{5}{4}$

5. $\dfrac{x-1}{2} - \dfrac{x-1}{3} = 2$

6. $\dfrac{3}{x} - \dfrac{1}{2x} = \dfrac{1}{2}$

7. $n - 4 = \dfrac{n-1}{4}$

8. $\dfrac{2x-4}{4} = 5 + \dfrac{2-x}{3}$

9. $2 + \dfrac{9}{x} = \dfrac{5}{x^2}$

10. $\dfrac{1}{x} - \dfrac{x-1}{14} = \dfrac{1}{7x}$

11. $\dfrac{x-1}{16} + \dfrac{5}{8x} = \dfrac{1}{x}$

12. $\dfrac{5}{x} + 3x = \dfrac{17}{x}$

13–16. Solve each of the following inequalities:

13. $\dfrac{3x-1}{4} > \dfrac{x+3}{2}$

14. $\dfrac{y+5}{8} < \dfrac{y-1}{4}$

15. $\dfrac{n+6}{3} - 2 \le \dfrac{n-2}{2}$

16. $\dfrac{a+1}{4} - \dfrac{3a}{8} \ge \dfrac{1}{2}$

REGENTS TUNE-UP: CHAPTER 7

Each of the questions in this section has appeared on a previous Course I Regents Examination. Here is an opportunity for you to review Chapter 7 and, at the same time, prepare for the Course I Regents Examination.

---

1. Express the sum $\dfrac{1}{3x} + \dfrac{2}{5x}$, $x \ne 0$, as a single fraction in lowest terms.

2. Solve for x: $\dfrac{x}{3} + \dfrac{x}{2} = 5$.

3. What is the sum of $\dfrac{3}{2x}$ and $\dfrac{4}{6x}$?

(1) $\dfrac{7}{8x}$ (2) $\dfrac{13}{6x}$ (3) $\dfrac{7}{6x}$ (4) $\dfrac{13}{8x}$

4. Which number represents 72 kilometers per hour as meters per hour?

(1) 7.2×10^{-2} (2) 7.2×10^{2} (3) 7.2×10^{-4} (4) 7.2×10^{4}

5. Express as a single fraction in lowest terms:

$$\dfrac{5x+2}{6} + \dfrac{2x-3}{3}$$

6. Expressed as meters per minute, $60 \dfrac{\text{km}}{\text{hr}}$ is equivalent to

(1) $3.6 \dfrac{\text{m}}{\text{min}}$ (2) $36 \dfrac{\text{m}}{\text{min}}$ (3) $100 \dfrac{\text{m}}{\text{min}}$ (4) $1000 \dfrac{\text{m}}{\text{min}}$

7. The expression $\dfrac{16y^3 + 4y^2 + 2y}{-2y}$, $y \ne 0$, is equivalent to

(1) $-8y^2 - 2y - 1$ (2) $-8y^2 - 2y$ (3) $-2y$ (4) $8y^2 - 2y - 1$

8. Which expression is the simplest form of $\dfrac{25x^4y^2 - 15x^2y}{5xy}$ if $x \ne 0$ and $y \ne 0$?

(1) $5x^3y - 15x^2y$ (2) $5x^3y - 3x$ (3) $5x^4y^2 - 3x^2y$ (4) $2x$

Section 7.1

1. $7a^3$
3. $4c$
5. $\dfrac{b-a}{2}$

7. $\dfrac{y(2y-1)}{23}$ or $\dfrac{2}{3}y^2 - \dfrac{1}{3}y$

9. $0.6x - 0.2$
11. -3

13. $\dfrac{3}{x}$
15. $\dfrac{10}{y-3}$

Section 7.2

1. $\dfrac{9ac^2}{2}$

3. $\dfrac{(y+2)(3-x)}{3}$

5. $\dfrac{4y}{3x^2}$

7. $\dfrac{2}{x-y}$

9. $\dfrac{1}{2}$

11. $\dfrac{x^4}{ab}$

Section 7.3

1. (a) $\dfrac{1\ \text{mile}}{1607.8\ \text{meters}}$

  (b) $11.2\ \dfrac{\text{miles}}{\text{hour}}$

3. $7.5 \times 10^{-4}\ \dfrac{\text{km}}{\text{min}}$

5. $45\ \dfrac{\text{miles}}{\text{hour}}$

Section 7.4

1. $\dfrac{a}{2}$

3. $\dfrac{3}{a+b}$

5. $\dfrac{n-1}{2}$

7. $\dfrac{x+2y}{3xy}$

9. 3
11. $a+b$

Section 7.5

1. $\dfrac{3x}{2}$

3. $\dfrac{b}{2x}$

5. $\dfrac{y-2}{2xy}$

7. $\dfrac{a^2-b^2}{ab}$

9. $\dfrac{1}{2}$

11. $\dfrac{5}{24}$

13. $\dfrac{b+1}{6b}$

15. $\dfrac{4x-1}{6x}$

17. $\dfrac{8c+17}{14}$

19. $\dfrac{2a-4b}{a^2b^2}$

21. $\dfrac{x+3}{3}$

23. $\dfrac{y^2+2}{y}$

25. $\dfrac{a^2+2ab-b^2}{a^2-b^2}$

27. $\dfrac{8x-2}{(x-1)^2(x+1)}$

29. $\dfrac{1}{2(y-3)}$

Section 7.6

1. 80

3. 10

5. 13

7. 5

9. $-5, \dfrac{1}{2}$

11. $-2, 3$

13. $x > 7$

15. $n \geq 6$

Regents Tune-Up: Chapter 7

1. $\dfrac{11}{15x}$

3. (2)

5. $\dfrac{9x - 4}{6}$

7. (1)

CHAPTER 8

OPERATIONS WITH IRRATIONAL NUMBERS

8.1 EVALUATING RADICALS AND CLASSIFYING REAL NUMBERS

=== ⋀ **KEY IDEAS** ===

What is the area of a square whose side is 5? To find the area of this square, we *square* 5 by writing $5 \times 5 = 25$. The area of the square is 25.

What is the length of a side of a square whose area is 25? To find the length of a side of this square, we find the *square root* of 25 by thinking, "The product of which two identical numbers is 25?" Since $25 = 5 \times 5$, the square root of 25 is 5.

Finding the square of a number and finding the square root of a number are inverse operations in the same sense that adding and subtracting are inverse operations.

Square Roots

The **square root** of a nonnegative number N is one of two identical factors whose product is N. The symbol $\sqrt{}$ is called a **radical sign** and is used to indicate that we wish to *extract* the square root of any number, called the **radicand**, that is written underneath the symbol. For example,

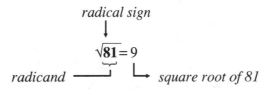

The square root of 81 is 9 since $9 \times 9 = 81$. Notice that there is another pair of identical factors whose product is 81, that is, $-9 \times -9 = 81$. However, the expression $\sqrt{N}$ will always be taken to mean the *positive* square root of N. The positive root of a number is sometimes called the *principal square root*. Compare the following notation, and observe the effect of the placement of the negative sign:

$$\sqrt{81} = 9;$$
$$-\sqrt{81} = -9;$$
$$\pm\sqrt{81} = \pm 9 \quad (\pm \text{ is read as "positive or negative."})$$

173

Example

1. Express each of the following as a rational number:

(a) $\sqrt{64}$ (b) $-\sqrt{121}$ (c) $\sqrt{\dfrac{49}{100}}$ (d) $\pm\sqrt{0.16}$

Solution:

(a) $\sqrt{64} = \mathbf{8}$

(c) $\sqrt{\dfrac{49}{100}} = \dfrac{\mathbf{7}}{\mathbf{10}}$

(b) $-\sqrt{121} = \mathbf{-11}$

(d) $\pm\sqrt{0.16} = \mathbf{\pm\,0.4}$

Perfect Squares

The numbers 1, 4, 9, and 16/25 are examples of *perfect squares*. A **perfect square** is any rational number whose square root is also a rational number. The number 8 and 5/9 are *not* perfect squares since their square roots are not rational numbers.

Negative Radicands

Notice that $\sqrt{-25}$ cannot be evaluated since we cannot find two identical numbers whose product is *negative* 25. In general $\sqrt{N}$ represents a real number only when N is greater than or equal to 0.

Example

2. What is the largest possible integer value of x for which the expression $\sqrt{26 - x}$ is a positive integer?

Solution: The value of x must be an integer less than or equal to 26; otherwise the radicand would be negative. If $x = 25$, then

$$\sqrt{26 - x} = \sqrt{26 - 25} = \sqrt{1} = 1.$$

The largest possible integer value of x is **1**.

Rational Numbers as Decimals

A decimal number like 0.25 $\left(=\frac{1}{4}\right)$ does not contain any other nonzero decimal digits after the digit 5 so that it "ends" after the 5. The decimal number 0.001 $\left(=\frac{1}{1000}\right)$ is a decimal number that ends after the digit 1. Some decimal numbers never end. Using a calculator,

$$\frac{2}{3} = 2 \div 3 = 0.6666666\ldots \text{ and } \frac{3}{11} = 3 \div 11 = 0.272727\ldots$$

The decimal number 0.6666666 . . . never ends. The digit 6 endlessly repeats. The decimal number 0.272727 . . . is also a nonending decimal in which the block of digits 27 endlessly repeats. These observations suggest the following facts:

- Every rational number can be represented by a decimal number that ends or endlessly repeats one or more digits.
- Every decimal number that ends or endlessly repeats one or more digits represents a rational number.

Irrational and Real Numbers

The square root of a number that is *not* a perfect square is *not* rational. For example, $\sqrt{7}$ is not equal to a rational number since we cannot find two identical rational numbers whose product is 7. We can, however, represent $\sqrt{7}$ by an infinite, *nonrepeating* decimal number: $\sqrt{7} = 2.6457513\ldots$. Thus, $\sqrt{7}$ is an example of an irrational number. An **irrational number** is any number that can be represented by a never-ending decimal that has no pattern of repeating digits.

The square root of any nonnegative number that is *not* a perfect square is irrational. Similarly, the cube root of any number that is not a perfect cube is irrational. However, the set of irrational numbers also includes other types of infinite, nonrepeating decimal numbers such as π (= 3.1415926 . . .).

Thus, a **real number** is a number that is either rational or irrational but not both at the same time.

Example

3. Which of the following is an irrational number?

(1) $\sqrt{49}$ (2) $\sqrt{8}$ (3) $\sqrt{\dfrac{25}{81}}$ (4) 0   (5) 0.444 . . .

Solution: Choice (2), 8, is not a perfect square. Since the square root of a number that is not a perfect square is irrational, **choice (2)** is the correct answer. Also note that choice (1), $\sqrt{49}$, = 7 = $\frac{7}{1}$; choice (3), $\sqrt{25/81}$, = 5/9; and choice (4), 0, is rational since it can be written as the numerator of any fraction having an integer denominator except 0, for example, 0/2. Choice (5) is a repeating decimal so it represents a rational number.

Exercise Set 8.1

1–10. Simplify each of the following:

1. $\sqrt{100}$

2. $-\sqrt{36}$

3. $\sqrt{0.49}$

4. $\sqrt{\dfrac{9}{25}}$

5. $\pm\sqrt{144}$

6. $\sqrt{0.09}$

7. $\sqrt{\dfrac{121}{169}}$

8. $\sqrt{\dfrac{1}{64}}$

9. $2\sqrt{\dfrac{9}{4}}$

10. $-3\sqrt{\dfrac{16}{36}}$

11. Which of the following is *not* defined?

    (1) $\sqrt{-1000}$ (2) $-\sqrt{1000}$ (3) $\dfrac{-1}{\sqrt{1000}}$ (4) $\pm\sqrt{1000}$

12. What is the *smallest* integer value of x for which the expression $\sqrt{x-16}$ is a positive integer?

13. What is the *largest* integer value of x for which the expression $\sqrt{11-x}$ is defined?

14. Which of the following expressions represents an irrational number?

    (1) $\sqrt{\dfrac{1}{16}}$ (2) $-\sqrt{\dfrac{9}{25}}$ (3) $\sqrt{-1}$ (4) $\sqrt{12}$

15–22. Evaluate each square root and then perform the indicated operation.

15. $\sqrt{9}+\sqrt{16}$

16. $(\sqrt{4})(\sqrt{100})$

17. $(\sqrt{25})^2$

18. $\dfrac{\sqrt{81}}{\sqrt{9}}$

19. $2\sqrt{64}$

20. $-6-\sqrt{36}$

21. $-4+\sqrt{9}$

22. $-1-2\sqrt{1}$

8.2 SIMPLIFYING RADICALS

A radical may be distributed over each factor of its radicand. In general,

$$\sqrt{ab} = \sqrt{a}\,\sqrt{b}.$$

For example,

$$\sqrt{12} = \sqrt{4 \cdot 3} = \sqrt{4} \cdot \sqrt{3} = 2\sqrt{3}.$$

Simplifying Radicals

The square root of a whole number is in simplest form when the number does not include any perfect square factors greater than 1. To simplify an irrational square root, factor the radicand, if possible, in such a way that one of its factors is the *largest* perfect square factor. Distribute the radical over each factor. Then evaluate the square root of the perfect square factor. For example, to simplify $\sqrt{80}$ proceed as follows:

Factor the radicand: $\sqrt{80} = \sqrt{16 \cdot 5}$
Write the radical over each factor: $= \sqrt{16} \cdot \sqrt{5}$
Evaluate the square root of the perfect square: $= 4\sqrt{5}$

If the largest perfect square factor of the radicand is not factored out initially, then the procedure must be repeated. For example,

$$\sqrt{80} = \sqrt{4 \cdot 20} = \sqrt{4} \cdot \sqrt{20} = 2 \cdot \sqrt{20}$$
$$= 2\sqrt{4}\sqrt{5} = (2 \cdot 2)\sqrt{5} = 4\sqrt{5}$$

Examples

1. Simplify: $\sqrt{45}$

Solution: $\sqrt{45} = \sqrt{9 \cdot 5} = \sqrt{9}\sqrt{5} = 3\sqrt{5}$

2. Simplify: $2\sqrt{48}$

Solution: $2\sqrt{48} = 2\sqrt{16 \cdot 3} = 2\sqrt{16} \cdot \sqrt{3} = (2 \cdot 4)\sqrt{3} = 8\sqrt{3}$

3. Simplify: $\frac{1}{3}\sqrt{108}$

Solution: $\frac{1}{3}\sqrt{108} = \frac{1}{3}\sqrt{36 \cdot 3} = \frac{1}{3}\sqrt{36} \cdot \sqrt{3}$
$$= \frac{1}{3}(6)\sqrt{3} = 2\sqrt{3}$$

4. Simplify: $\sqrt{x^5}$

Solution: Observe that $\sqrt{x^2} = x$, $\sqrt{x^4} = x^2$, $\sqrt{x^6} = x^3$ and so forth. The exponent of the greatest perfect square factor of a variable power is the largest even number that is less than or equal to the exponent of the variable radicand:

$$\sqrt{x^5} = \sqrt{x^4 \cdot x} = \sqrt{x^4} \cdot \sqrt{x} = x^2 \cdot \sqrt{x}.$$

Solving $ax^2 + c = 0$

A quadratic equation that is missing the first-degree (middle) term may be solved by writing an equivalent equation that isolates the square of the variable on one side of the equation. The variable can then be found by taking the square root of both sides of the equation.

Examples

5. Solve for x: $2x^2 - 50 = 0$

Solution:

Method 1: Extracting Roots	Method 2: Factoring
$2x^2 = 50$ $x^2 = \dfrac{50}{2}$ $x^2 = 25$ $\sqrt{x^2} = \pm\sqrt{25}$ $x = \pm 5 \ or \ \{-5, 5\}$	$2(x^2 - 25) = 0$ $2(x - 5)(x + 5) = 0$ $x - 5 = 0 \ \text{or} \ x + 5 = 0$ $x = 5 \ \text{or} \quad x = -5$ $\{-5, 5\}$

6. Solve for n: $n^2 - 17 = 0$

Solution:

Method 1: Extracting Roots	Method 2: Factoring
$n^2 - 17 = 0$ $n^2 = 17$ $\sqrt{n^2} = \pm\sqrt{17}$ $n = \pm\sqrt{17}$	Since $n^2 - 17$ cannot be factored using rational numbers, this method cannot be used.

Exercise Set 8.2

1–20. Write each of the following radicals in simplest form:

1. $\sqrt{28}$

2. $\sqrt{40}$

3. $-\sqrt{63}$

4. $-\sqrt{98}$

5. $2\sqrt{75}$

6. $\sqrt{192}$

7. $\dfrac{\sqrt{48}}{4}$

8. $\dfrac{1}{2}\sqrt{72}$

9. $4\sqrt{90}$

10. $-\sqrt{112}$

11. $\sqrt{500}$

12. $\sqrt{a^8}$

13. $\sqrt{0.16x^2}$

14. $\sqrt{0.09b^4}$

15. $\sqrt{a^7}$

16. $\sqrt{x^5}$

17. $\sqrt{a^4b^2}$

18. $\sqrt{8c^4}$

19. $\sqrt{99t^3}$

20. $\sqrt{180x^4y}$

21–26. Solve for x.

21. $3x^2 - 12 = 0$

22. $2x^2 - 28 = 0$

23. $2x^2 = x^2 + 48$

24. $3x^2 = 28 - x^2$

25. $(x - 5)^2 = 16$

26. $(x + 3)^2 = 6x + 25$

8.3 MULTIPLYING AND DIVIDING RADICALS

△ KEY IDEAS △

In general,

$$(p\sqrt{a})(q\sqrt{b}) = (pq)\sqrt{ab}\,,$$

and provided that *q* and *b* are not 0,

$$(p\sqrt{a}) \div (q\sqrt{b}) = \frac{p\sqrt{a}}{q\sqrt{b}} = \frac{p}{q}\sqrt{\frac{a}{b}}\,.$$

Multiplying and Dividing Radicals

To **multiply** (*or* **divide**) square root radicals, proceed as follows:

Step 1. Multiply (*or* divide) their rational coefficients, if any.
Step 2. Multiply (*or* divide) their radicands.
Step 3. Write the resulting product (*or* quotient) in simplest form.

Examples

1. Multiply: $(2\sqrt{6})(4\sqrt{8})$

Solution: $(2\sqrt{6})(4\sqrt{8}) = (2 \cdot 4)(\sqrt{6} \cdot \sqrt{8}) = 8\sqrt{48}$
$$= 8\sqrt{16} \cdot \sqrt{3}$$
$$= 8 \cdot 4\sqrt{3}$$
$$= \mathbf{32\sqrt{3}}$$

2. Divide: $18\sqrt{120} \div 6\sqrt{3}$

Solution: $18\sqrt{120} \div 6\sqrt{3} = \dfrac{18\sqrt{120}}{6\sqrt{3}}$
$$= 3\sqrt{40}$$
$$= 3\sqrt{4} \cdot \sqrt{10} = (3 \cdot 2)\sqrt{10}$$
$$= \mathbf{6\sqrt{10}}$$

Squaring Radicals

Notice in the following examples:

$$(\sqrt{4})^2 = \sqrt{4} \cdot \sqrt{4} = \sqrt{16} = 4$$

that the product of the square roots of two identical numbers is equal to the number (radicand) itself. In general,

$$\sqrt{N} \cdot \sqrt{N} = N.$$

Examples

3. Evaluate: $(5\sqrt{3})^2$

Solution: $(5\sqrt{3})^2 = (5\sqrt{3})(5\sqrt{3}) = (5 \cdot 5)(\sqrt{3} \cdot \sqrt{3}) = 25 \cdot 3 = \mathbf{75}$

4. Simplify by using the distributive property: $\sqrt{6}(\sqrt{15} - \sqrt{6})$.

Solution: $\sqrt{6}(\sqrt{15} - \sqrt{6}) = (\sqrt{6} \cdot \sqrt{15}) - (\sqrt{6} \cdot \sqrt{6})$

$$= \sqrt{90} - 6$$
$$= \sqrt{9} \cdot \sqrt{10} - 6$$
$$= 3\sqrt{10} - 6$$

Writing Radicals in Simplest Form

The square root of a term is in simplest form if each of the following statements is true:

1. The radical does *not* contain a perfect square factor greater than 1.

Example: $\sqrt{18}$ is not in simplest form.

$\sqrt{18} = \sqrt{9 \cdot 2} = \sqrt{9} \cdot \sqrt{2} = 3\sqrt{2}$ is in simplest form.

2. The radical does *not* contain a fraction.

Example: $\sqrt{\dfrac{3}{4}}$ is *not* in simplest form.

$\sqrt{\dfrac{3}{4}} = \dfrac{\sqrt{3}}{\sqrt{4}} = \dfrac{\sqrt{3}}{2}$ is in simplest form.

3. The radical does *not* appear in the denominator. If it does, rationalize the denominator by multiplying the numerator and the denominator by the radical denominator.

Example: $\dfrac{5}{\sqrt{7}}$ is *not* in simplest form.

$\dfrac{5}{\sqrt{7}} \cdot \left(\dfrac{\sqrt{7}}{\sqrt{7}}\right) = \dfrac{5\sqrt{7}}{7}$ is in simplest form.

Use the multiplication property of 1.

Example

5. Express $\sqrt{\dfrac{9}{40}}$ in simplest form

Solution: $\qquad\qquad \sqrt{\dfrac{9}{40}} = \dfrac{\sqrt{9}}{\sqrt{40}}$

Rationalize the denominator: $= \dfrac{3}{\sqrt{40}} \cdot \dfrac{\sqrt{40}}{\sqrt{40}}$

Simplify the numerator: $= \dfrac{3\sqrt{40}}{40} = \dfrac{3 \cdot \sqrt{4} \cdot \sqrt{10}}{40} = \dfrac{3 \cdot \overset{1}{\cancel{2}} \cdot \sqrt{10}}{\underset{20}{\cancel{40}}}$

$= \dfrac{3\sqrt{10}}{20}$

Exercise Set 8.3

1–26. Multiply, and write the product in simplest form.

1. $\sqrt{7} \cdot \sqrt{11}$

2. $\sqrt{12} \cdot \sqrt{3}$

3. $3\sqrt{5} \cdot 2\sqrt{7}$

4. $2\sqrt{6} \cdot 5\sqrt{6}$

5. $(\sqrt{13})^2$

6. $(-\sqrt{8})^2$

7. $-(\sqrt{5})^2$

8. $-(-\sqrt{7})^2$

9. $(3\sqrt{2})^2$

10. $(-4\sqrt{3})^2$

11. $-(2\sqrt{5})^2$

12. $-(-2\sqrt{3})^2$

13. $-\sqrt{15} \cdot 2\sqrt{5}$

14. $(2\sqrt{8})(-3\sqrt{12})$

15. $(5\sqrt{14}) \cdot \left(\dfrac{1}{2}\sqrt{10}\right)$

16. $\sqrt{15} \cdot \sqrt{2} \cdot \sqrt{10}$

17. $\sqrt{3} \cdot \sqrt{5} \cdot \sqrt{6}$

18. $(\sqrt{2})^3$

19. $(\sqrt{3})^4$

20. $(-\sqrt{5})^3$

21. $(-\sqrt{6})^4$

22. $2(\sqrt{8})^3$

23. $\sqrt{x} \cdot \sqrt{x^3}$

24. $\sqrt{10a} \cdot \sqrt{8a}$

25. $\sqrt{y^5} \cdot \sqrt{y^4}$

26. $(2\sqrt{5b})^2$

27–36. Divide and write the quotient in simplest form.

27. $\dfrac{\sqrt{18}}{\sqrt{2}}$

28. $\dfrac{\sqrt{60}}{\sqrt{15}}$

29. $\dfrac{12\sqrt{54}}{4\sqrt{3}}$

30. $\dfrac{21\sqrt{96}}{7\sqrt{2}}$

31. $\dfrac{\sqrt{252}}{3\sqrt{7}}$

32. $\dfrac{-5\sqrt{88}}{-2\sqrt{2}}$

33. $\dfrac{\sqrt{0.8}}{\sqrt{5}}$

34. $\dfrac{\sqrt{0.018}}{\sqrt{0.2}}$

35. $\dfrac{\sqrt{x^5}}{\sqrt{x}}$

36. $\dfrac{\sqrt{90x^4}}{\sqrt{2x^2}}$

37–40. Simplify.

37. $\dfrac{10}{\sqrt{5}}$ **38.** $\dfrac{8}{3\sqrt{2}}$ **39.** $\sqrt{\dfrac{8}{49}}$ **40.** $\dfrac{\sqrt{6}}{\sqrt{8}}$

41. Solve for x: $\dfrac{2x}{3\sqrt{2}} = \dfrac{3\sqrt{2}}{x}$.

8.4 COMBINING LIKE RADICALS

$\bigwedge$
KEY IDEAS

Like radicals are radicals that have the same index and the same radicand; $3\sqrt{5}$ and $-2\sqrt{5}$ are examples of like radicals. However, $4\sqrt{11}$ and $4\sqrt{6}$ are not like radicals since they have different radicands, and $\sqrt{7}$ and $\sqrt[3]{7}$ are not like radicals since they do not have the same index.

Adding and Subtracting Radicals

To **add** or **subtract** like radicals, combine their rational coefficients and use the same radical factor. For example,

$$2\sqrt{6} + 3\sqrt{6} = (2 + 3)\sqrt{6} = \mathbf{5\sqrt{6}}.$$

Sometimes a radical must be simplified before it can be combined with another radical. This is illustrated in Examples 1–4 below.

Examples

1. Find the sum of $4\sqrt{3}$ and $\sqrt{75}$

Solution:
$$\begin{aligned}
4\sqrt{3} + \sqrt{75} &= 4\sqrt{3} + \sqrt{25} \cdot \sqrt{3} \\
&= 4\sqrt{3} + 5\sqrt{3} \\
&= (4 + 5)\sqrt{3} = \mathbf{9\sqrt{3}}
\end{aligned}$$

2. Subtract: $\sqrt{2} - \sqrt{32}$

Solution:
$$\begin{aligned}
\sqrt{2} - \sqrt{32} &= \sqrt{2} - \sqrt{16} \cdot \sqrt{2} \\
&= \sqrt{2} - 4\sqrt{2} \\
&= (1 - 4)\sqrt{2} = \mathbf{-3\sqrt{2}}
\end{aligned}$$

3. Combine: $3\sqrt{20} - \sqrt{5} + \dfrac{1}{2}\sqrt{80}$

Solution: $3\sqrt{20} - \sqrt{5} + \dfrac{1}{2}\sqrt{80} = 3\sqrt{4} \cdot \sqrt{5} - \sqrt{5} + \dfrac{1}{2}\sqrt{16} \cdot \sqrt{5}$

$$= (3 \cdot 2)\sqrt{5} - \sqrt{5} + \dfrac{1}{2}(4)(\sqrt{5})$$

$$= 6\sqrt{5} - \sqrt{5} + 2\sqrt{5}$$

$$= 5\sqrt{5} + 2\sqrt{5} = \mathbf{7\sqrt{5}}$$

4. Simplify: $\sqrt{3}(\sqrt{6} + 2\sqrt{24})$

Solution: $\sqrt{3}(\sqrt{6} + 2\sqrt{24}) = (\sqrt{3} \cdot \sqrt{6}) + 2(\sqrt{3} \cdot \sqrt{24})$

$$= \sqrt{18} + 2\sqrt{72}$$

$$= \sqrt{9 \cdot 2} + 2\sqrt{36 \cdot 2}$$

$$= \sqrt{9} \cdot \sqrt{2} + 2\sqrt{36} \cdot \sqrt{2}$$

$$= 3\sqrt{2} + (2 \cdot 6)\sqrt{2}$$

$$= 3\sqrt{2} + 12\sqrt{2} = \mathbf{15\sqrt{2}}$$

Exercise Set 8.4

1–27. Simplify.

1. $8\sqrt{11} + \sqrt{11}$

2. $3\sqrt{5} - 7\sqrt{5}$

3. $\dfrac{2}{3}\sqrt{6} - \dfrac{5}{3}\sqrt{6}$

4. $\dfrac{\sqrt{7}}{5} + \dfrac{\sqrt{7}}{3}$

5. $\dfrac{1}{3}\sqrt{13} + \dfrac{1}{4}\sqrt{13}$

6. $\sqrt{8} + \sqrt{2}$

7. $2\sqrt{54} - 4\sqrt{6}$

8. $\sqrt{32} + \sqrt{50}$

9. $2\sqrt{48} - \sqrt{27}$

10. $\sqrt{28} + 3\sqrt{2}$

11. $3\sqrt{p} - \sqrt{p} + 4\sqrt{p}$

12. $\sqrt{16x} + \sqrt{9x}$

13. $x\sqrt{12} - x\sqrt{3}$

14. $\sqrt{18} - \sqrt{200} + \sqrt{72}$

15. $\sqrt{75} + 2\sqrt{3} - \sqrt{48}$

16. $\dfrac{3}{4}\sqrt{7} - \dfrac{1}{4}\sqrt{28}$

17. $\sqrt{128x} - \sqrt{32x}$

18. $\dfrac{1}{4}\sqrt{24} + \dfrac{1}{3}\sqrt{96}$

19. $\sqrt{75y} - 2\sqrt{3y} + \sqrt{300y}$

20. $\sqrt{16} + \dfrac{1}{\sqrt{4}}$

21. $\sqrt{8} + \dfrac{24}{\sqrt{8}}$

22. $\sqrt{2}(\sqrt{2} - \sqrt{8})$

23. $\sqrt{3}(\sqrt{15} + 2\sqrt{60})$

24. $\sqrt{5}(\sqrt{5} - \sqrt{20})$

25. $\dfrac{\sqrt{50} - \sqrt{8}}{\sqrt{4}}$

26. $\dfrac{\sqrt{27} + 4\sqrt{3}}{\sqrt{12}}$

27. $\dfrac{3 - \sqrt{2}}{\sqrt{32}}$

28. $(\sqrt{2} + \sqrt{8})^2$

8.5 USING THE PYTHAGOREAN RELATIONSHIP

$\triangle$ KEY IDEAS $\triangle$

A **right triangle** is a triangle that contains a 90-degree (right) angle. Each of the sides that form the right angle is called a **leg** of the triangle. The side opposite the right angle is called the **hypotenuse**.

The lengths of the sides of a right triangle satisfy the following relationship, which is often referred to as the *Pythagorean theorem:*

$(\text{Leg 1})^2 + (\text{Leg 2})^2 = (\text{Hypotenuse})^2,$
$\quad a^2 \quad + \quad b^2 \quad = \quad c^2$

Applying the Pythagorean Relationship

When the lengths of any two sides of a *right* triangle are known, the length of the third side may be found by the following procedure:

Step 1. Use a variable to represent the length of the unknown side.

Step 2. Substitute the given lengths into the Pythagorean relationship.

Step 3. Solve for the length of the unknown side.

Example

1. For each triangle, find the value of x.

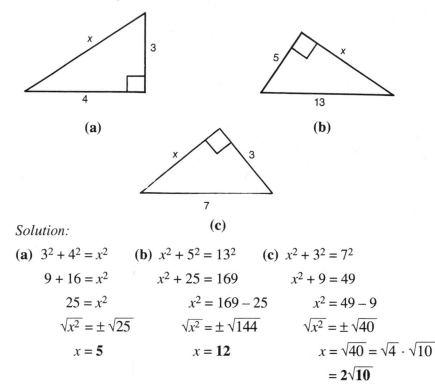

(a)

(b)

(c)

Solution:

(a) $3^2 + 4^2 = x^2$

$9 + 16 = x^2$

$25 = x^2$

$\sqrt{x^2} = \pm\sqrt{25}$

$x = 5$

(b) $x^2 + 5^2 = 13^2$

$x^2 + 25 = 169$

$x^2 = 169 - 25$

$\sqrt{x^2} = \pm\sqrt{144}$

$x = 12$

(c) $x^2 + 3^2 = 7^2$

$x^2 + 9 = 49$

$x^2 = 49 - 9$

$\sqrt{x^2} = \pm\sqrt{40}$

$x = \sqrt{40} = \sqrt{4} \cdot \sqrt{10}$

$= 2\sqrt{10}$

Note: In each of these examples the negative value of x is discarded since the length of a side of a triangle cannot be negative.

The Pythagorean relationship is useful when solving a variety of geometric problems.

Example

2. If the length of a diagonal of a square is 10, what is the length of a side of the square?

Solution: Let x = length of a side of the square.

A diagonal of a figure such as a square (or rectangle) is a line segment that connects any two nonconsecutive corners (called **vertices**) of the figure.

Apply the Pythagorean relationship in right triangle ABC.

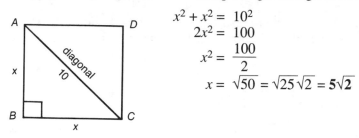

$$x^2 + x^2 = 10^2$$
$$2x^2 = 100$$
$$x^2 = \frac{100}{2}$$
$$x = \sqrt{50} = \sqrt{25}\sqrt{2} = \mathbf{5\sqrt{2}}$$

Pythagorean Triples

A **Pythagorean triple** is a set of positive integers $\{a, b, c\}$ that satisfy the equation $a^2 + b^2 = c^2$. There are many Pythagorean triples. The sets

$$\{3, 4, 5\} \quad \text{and} \quad \{5, 12, 13\} \quad \text{and} \quad \{8, 15, 17\}$$

are some commonly encountered Pythagorean triples. Observe that:

$$\frac{3^2 + 4^2 = 5^2}{9 + 16 \mid 25} \qquad \frac{5^2 + 12^2 = 13^2}{25 + 144 \mid 169} \qquad \frac{8^2 + 15^2 = 17^2}{64 + 255 \mid 289}$$
$$25 = 25\checkmark \qquad\qquad 169 = 169\checkmark \qquad\qquad 289 = 289\checkmark$$

Whole-number multiples of any Pythagorean triple also comprise a Pythagorean triple. For example, if each member of the set $\{3, 4, 5\}$ is multiplied by 2, the set $\{6, 8, 10\}$ that is obtained is also a Pythagorean triple since $6^2 + 8^2 = 10^2$ ($36 + 64 = 100$). If each member of the set $\{3, 4, 5\}$ is multiplied by 3, the set $\{9, 12, 15\}$, which is also a Pythagorean triple since $9^2 + 12^2 = 15^2$ ($81 + 144 = 225$), is obtained.

Example

3. Which of the following is *not* a Pythagorean triple?
(1) $\{9, 40, 41\}$ (3) $\{8, 12, 17\}$
(2) $\{15, 20, 25\}$ (4) $\{10, 24, 26\}$

Solution: Choice (1) represents a Pythagorean triple since

$$\frac{9^2 + 40^2 = 41^2}{81 + 1600 \mid 1681}$$
$$1681 = 1681\checkmark$$

Choice (2) is a mulitple of $\{3, 4, 5\}$ where each element is multiplied by 5. Choice (4) is a multiple of $\{5, 12, 13\}$ where each element is multiplied by 2. In choice (3) you can easily verify that $8^2 + 12^2 \neq 17^2$, so $\{8, 12, 17\}$ is *not* a Pythagorean triple. The correct answer is **choice (3)**.

Some problems become easier to solve if we recognize that a set of numbers forms a Pythagorean triple.

Examples

4. A ladder 13 feet in length rests against a vertical building. The foot of the ladder is 5 feet from the building. How far up the building does the ladder reach?

Solution: The ladder forms the hypotenuse of a right triangle. Notice that the sides of the right triangle form a 5-12-13 Pythagorean triple. The ladder reaches **12 feet** up the building.

If you did not recognize that the lengths form a Pythagorean triple, you could have solved this problem by letting $x = AB$ and proceeding as follows:

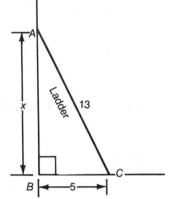

$$x^2 + 5^2 = 13^2$$
$$x^2 + 25 = 169$$
$$x^2 = 169 - 25$$
$$x = \sqrt{144} = \mathbf{12}$$

5. The perimeter of a right triangle is 60. If the length of the hypotenuse is 26, find the length of the shorter leg of the triangle.

Solution: The sum of the lengths of the legs = $60 - 26 = 34$.

Let x = length of the shorter leg.
Then $34 - x$ = length of the remaining leg.

$$x^2 + (34 - x)^2 = 26^2$$
$$x^2 + x^2 - 68x + 1156 = 676$$
$$2x^2 - 68x + 1156 - 676 = 0$$
$$2x^2 - 68x + 480 = 0$$
$$\frac{2x^2}{2} - \frac{68x}{2} + \frac{480}{2} = \frac{0}{2}$$
$$x^2 - 34x + 240 = 0$$
$$(x - 10)(x - 24) = 0$$
$$(x - 10 = 0) \text{ or } (x - 24 = 0)$$
$$x = 10 \text{ or } \qquad x = 24$$

The length of the shorter leg of the right triangle is **10**.

Converse of the Pythagorean Relationship

If the lengths of a side of a triangle satisfy the Pythagorean relationship, then the triangle is a right triangle.

Example

6. The lengths of the sides of a triangle are 2, $\sqrt{5}$, and 3. Determine whether the triangle is a right triangle.

Solution: $\dfrac{2^2 + (\sqrt{5})^2 = 3^3}{4+5 \mid 9}$
$9 = 9.$ ✓

Therefore the triangle is a right triangle.

Exercise Set 8.5

1. Find the length of the diagonal of a rectangle whose length is 15 and whose width is 20.

2. The length of a rectangle is twice its width. If the length of the diagonal of the rectangle is $\sqrt{15}$, find the width of the rectangle.

3. Find the length of a diagonal of a square if the length of a side is
   (a) 8   **(b)** 15   **(c)** $\sqrt{8}$ **(d)** 1

4. Find the length of a side of a square if the length of a diagonal is:
   (a) 12   **(b)** 9   **(c)** $16\sqrt{2}$ **(d)** $2\sqrt{6}$ **(e)** 1

5. A ladder 41 feet in length rests against a vertical building. The foot of the ladder is 9 feet from the building. How far up the building does the ladder reach?

6. The length and the width of a rectangle are in the ratio of 3 : 4. If the length of the diagonal of the rectangle is 100, what are the length and width of the rectangle?

7. The diagonal of a rectangle exceeds the length by 1. The width of the rectangle is 1 less than one-third of the length. Find the length of the diagonal.

8. The hypotenuse of a right triangle exceeds the length of one leg by 1, and is 3 less than four times the length of the other leg. Find the length of the shorter leg of the right triangle.

9. Determine whether each of the following sets of numbers can represent the lengths of the sides of a right triangle:

   (a) 15, 8, 17   (b) 2, 7, $\sqrt{53}$ (c) $\frac{1}{3}, \frac{1}{4}, \frac{1}{5}$ (d) $3\sqrt{5}, \sqrt{19}, 8$

10. The perimeter of a right triangle is 180. If the length of the hypotenuse is 82, find the length of the shorter leg of the triangle.

11. The perimeter of a right triangle is 320. If the length of the hypotenuse is 136, find the length of the shorter leg of the triangle.

REGENTS TUNE-UP: CHAPTER 8

Each of the questions in this section has appeared on a previous Course I Regents Examination. Here is an opportunity for you to review Chapter 8 and, at the same time, prepare for the Course I Regents Examination.

---

1. The legs of a right triangle have lengths of 2 and 7. Express, in radical form, the length of the hypotenuse.

2. Find $\sqrt{23}$ to the nearest tenth.

3. Simplify $\sqrt{50} + 3\sqrt{2}$.

4. What is the length of the diagonal of a rectangle whose dimensions are 5 by 7?
   (1) 5   (2) 8   (3) $\sqrt{24}$ (4) $\sqrt{74}$

5. The expression $(3\sqrt{75} - 2\sqrt{27})$ is equivalent to:
   (1) $57\sqrt{3}$ (2) 9   (3) $9\sqrt{3}$ (4) $\sqrt{48}$

6. The expression $\sqrt{125} + 2\sqrt{5}$ is equivalent to:
   (1) 35   (2) $7\sqrt{5}$ (3) $3\sqrt{5}$ (4) $3\sqrt{130}$

7. The value of π is:
   (1) rational and equal to 3.14
   (2) irrational and equal to 3.14
   (3) rational and between 3.14 and 3.15
   (4) irrational and between 3.14 and 3.15

8. The sum of $6\sqrt{6}$ and $\sqrt{54}$ is:
   (1) $3\sqrt{6}$ (2) $6\sqrt{60}$ (3) $9\sqrt{6}$ (4) $15\sqrt{6}$

9. Which is a rational number?
   (1) $\sqrt{6}$ (2) $\sqrt{2}$ (3) $\sqrt{3}$ (4) $\sqrt{4}$

10. The longer leg of a right triangle is 7 more than the shorter leg. The hypotenuse is 8 more than the shorter leg. The perimeter of the triangle is 30. Find the length of each leg.

11. Let p represent "Triangle ABC is a right triangle."
Let q represent "The square of the hypotenuse is equal to the sum of the square of the legs."
(**a**) Write in words: $p \rightarrow q$.
(**b**) Write in words the contrapositive of $p \rightarrow q$.
(**c**) If the hypotenuse is 19 and the legs are 13 and 14, respectively, is q a true statement?
(**d**) If p is false, what is the truth value of $p \rightarrow q$?
(**e**) Using your answer to part (**d**), determine the truth value of the contrapositive of $p \rightarrow q$.

MIDYEAR REGENTS TUNE-UP

Problems included in this section have appeared on a previous Course I Regents Examination.

---

1. Solve for x: $\frac{2}{3}x + 1 = 13$.

2. Factor: $x^2 - x - 12$.

3. If $xy^2 = 18$, find x when $y = -3$.

4. Solve for x: $0.03x + 7.2 = 8.34$.

5. Solve for y: $6(y + 3) = 2y - 2$.

6. From $2x^2 - 3x - 5$ subtract $x^2 - x - 6$.

7. Solve for x: $7x < 4x + 18$.

8. Solve for a: $\frac{a + 2}{12} = \frac{5}{3}$.

9. The perimeter of a square is represented by $8x - 8$. Express the length of one side of the square in terms of x.

10. Express the product $(2x - 7)(x + 3)$ as a trinomial.

11. What percent of 25 is 10?

12. Solve for p in terms of r, s, and t: $rp + s = t$.

13. The inverse of a statement is $p \rightarrow \sim q$. What is the statement?

14. Factor: $4x^2 - 9$.

15. Thirty percent of what number is 12?

191

16. Find the positive root of the equation $x^2 - x - 6 = 0$.

17. Let x represent the smaller of two integers whose sum is greater than 40. The larger integer is 7 times the smaller. Find the *smallest* possible value of x.

18. One positive number is 4 more than another. The sum of their squares is 40. Find the numbers.

19. The area of a rectangle is represented by $x^2 + 2x - 3$. If the width of the rectangle is represented by $x - 1$, the length may be represented by:
(1) $x - 3$ (2) $x - 2$ (3) $x + 3$ (4) $x + 4$

20. Which statement is false when p is false and q is false?
(1) $p \wedge q$ (2) $p \rightarrow q$ (3) $\sim p \rightarrow \sim q$ (4) $p \leftrightarrow q$

21. The quotient of $\dfrac{-4a^6b^2}{2a^2b}$ is:

(1) $2a^3b^2$ (2) $-2a^4b^2$ (3) $-2a^4b$ (4) $-6a^3b$

22. Which number is *not* a member of the solution set of $3x \le 6$?

(1) 0   (2) -1 (3) 3   (4) $\dfrac{1}{2}$

23. Which statement is the converse of "If it rains today, I'll stay home"?
(1) If I stay home, then it will rain today.
(2) If it rains today, then I'll stay home.
(3) If I don't stay home, then it won't rain today.
(4) If it doesn't rain today, then I won't stay home.

24. Let p represent the statement "$x \ge 5$," and let q represent the statement "$2x = 4$." Which is true if $x = 6$?
(1) $p \wedge q$ (2) $p \vee q$ (3) $p \rightarrow q$ (4) $p \leftrightarrow q$

25. The expression $\sqrt{48} + \sqrt{27}$ is equivalent to:
(1) $7\sqrt{3}$ (2) $\sqrt{75}$ (3) $6\sqrt{3}$ (4) $4\sqrt{6}$

26. The solution set of the equation $x^2 - 5x - 6 = 0$ is:
(1) $\{6, -1\}$ (2) $\{3, -2\}$ (3) $\{3, 2\}$ $\{-6, 1\}$

27. Find three positive consecutive odd integers such that the largest decreased by twice the second is equal to 10 less than the smallest.

28. Given the replacement set $\{5, 6, 7, 8\}$, which member of the replacement set will make the statement $(x < 6) \vee (x < 8)$ *false*?

29. The length of a side of a square is 1 more than twice the length of a side of another square. The perimeters of the two squares differ by 24 centimeters. Find, in centimeters, the length of a side of the *smaller* square.

30. Find two consecutive positive integers such that the square of the smaller is 1 more than four times the larger.

31. The domain (replacement set) for each open sentence below is $\{-3, -2, -1, 0, 1, 2, 3\}$. Write the solution set for *each* open sentence.

(a) $0.4x + 3.6 = 2.2x$

(b) $\dfrac{x-3}{2} = \dfrac{4}{5}$

(c) $-4x - 3 > 5$

(d) $2x^2 - x - 1 = 0$

(e) $x^2 - 3x = 0$

32. Below are three statements symbolized by p, q, and r:

Let p represent: 7 is an even number.

Let q represent: 9 is a prime number.

Let r represent: 25 is a perfect square.

(a) Write in words and tell whether the statement is *true* or *false*:

(1) $p \vee r$ (2) $r \rightarrow q$ (3) $\sim p \wedge \sim q$

(b) Write in symbolic form and give the truth values of the following statements:

(1) 25 is a perfect square if and only if 7 is an even number.

(2) 9 is not a prime number or 7 is not an even number.

33. (a) Each part in the box consists of a set of three statements. Write the numerals 1 through 3 and next to each numeral write the truth value (TRUE or FALSE) for the third statement in each part, based on the values given for the first two statements. If the truth value cannot be determined from the information given, write "CANNOT BE DETERMINED."

(1) Sue is 16 or John is 20.	T
John is not 20.	T
Sue is 16.	?
(2) It snows and it is cold.	F
It is cold.	F
It snows.	?
(3) If it is July, then it is warm.	T
It is not warm.	T
It is not July.	?

(b) Let p represent "John passes math" and let q represent "John studies." Write *each* of the following statements in symbolic form.

(1) It is not true that John studies and does not pass math.

(2) If John studies, then John passes math.

193

Answers to Odd-Numbered Exercises: Chapter 8

Section 8.1

1. 10	**9.** 3	**17.** 25
3. 0.7	**11.** (1)	**19.** 16
5. ±12	**13.** 11	**21.** −1
7. $\dfrac{11}{13}$	**15.** 7	

Section 8.2

1. $2\sqrt{7}$	**11.** $10\sqrt{5}$	**21.** −2, 2
3. $-3\sqrt{7}$	**13.** $0.4x$	**23.** $\pm4\sqrt{3}$
5. $10\sqrt{3}$	**15.** $a^3\sqrt{a}$	**25.** 1, 9
7. $\sqrt{3}$	**17.** a^2b	
9. $12\sqrt{10}$	**19.** $3t\sqrt{11t}$	

Section 8.3

1. $\sqrt{77}$	**11.** −20	**21.** 36	**31.** 2
3. $6\sqrt{35}$	**13.** $-10\sqrt{3}$	**23.** x^2	**33.** 0.4
5. 13	**15.** $5\sqrt{35}$	**25.** y^3	**35.** x^2
7. −5	**17.** $3\sqrt{10}$	**27.** 3	**37.** $2\sqrt{5}$
9. 18	**19.** 9	**29.** $9\sqrt{2}$	**39.** $\dfrac{2\sqrt{2}}{7}$
			41. ±3

Section 8.4

1. $9\sqrt{11}$	**11.** $6\sqrt{p}$	**21.** $\dfrac{9}{2}$
3. $-\sqrt{6}$	**13.** $x\sqrt{3}$	**23.** −2
5. $\dfrac{7\sqrt{13}}{12}$	**15.** $-\sqrt{2}$	**25.** −5
7. $2\sqrt{6}$	**17.** $\dfrac{\sqrt{7}}{4}$	**27.** $\dfrac{7}{2}$
9. $5\sqrt{3}$	**19.** $\dfrac{11\sqrt{6}}{6}$	

Section 8.5

1. 25	**7.** 25
3. (a) $8\sqrt{2}$	**9. (a)** Yes
(b) $15\sqrt{2}$	**(b)** Yes
(c) 4	**(c)** No
(d) $\sqrt{2}$	**(d)** Yes
5. 40 feet	**11.** 64

Regents Tune-Up: Chapter 8

1. $\sqrt{53}$ **2.** $8\sqrt{2}$ **5.** (3)   **7.** (4)   **9.** (4)

11. (a) If triangle ABC is a right triangle, then the square of the hypotenuse is equal to the sum of the squares of the legs.

    (b) If the square of the hypotenuse is not equal to the sum of the squares of the legs, then triangle ABC is not a right triangle.

    (c) No

    (d) True

    (e) True

ANSWERS TO MIDYEAR REGENTS TUNE-UP EXERCISES

1. 18	**16.** 3
2. $(x - 4)(x + 3)$	**17.** 6
3. 2	**18.** 2 and 6
4. 38	**19.** (3)
5. −5	**20.** (1)
6. $x^2 - 2x + 1$	**21.** (3)
7. $x < 6$	**22.** (3)
8. 18	**23.** (1)
9. $2x - 2$	**24.** (2)
10. $2x^2 - x - 21$	**25.** (1)
11. 40	**26.** (1)
12. $\dfrac{t - s}{r}$	**27.** 5, 7, and 9
13. $\sim p \rightarrow q$	**28.** 8
14. $(2x + 3)(2x - 3)$	**29.** 5
15. 40	**30.** 5 and 6

31. (a) 2   **(b)** { }   **(c)** −3   **(d)** 1   **(e)** 0, 3

32. (a) (1) 7 is an even number or 25 is a perfect square. (True)

        (2) If 25 is a perfect square, then 9 is a prime number. (False)

        (3) 7 is not an even number and 9 is not a prime number. (True)

33. (a) (1) True   **(2)** CANNOT BE DETERMINED   **(3)** True

    (b) (1) $\sim(q \wedge \sim p)$ **(2)** $q \rightarrow p$

CHAPTER 9

FUNDAMENTAL IDEAS IN GEOMETRY

9.1 BUILDING A GEOMETRY VOCABULARY

KEY IDEAS

Cement and bricks are used to give a house a strong foundation. The building blocks of geometry take the form of *undefined terms, defined terms, postulates,* and *theorems.*

Undefined Terms

Some basic terms in geometry can be described but cannot be defined by using simpler terms. *Point, line,* and *plane* are undefined terms.

A **point** indicates location and has no size or dimensions. A point is represented by a dot and named by a capital letter (Figure 9.1).

A **line** is a set of continuous points that form a straight path that extends without ending in two opposite directions. A line has no width. A line is identified by naming two points on the line or by writing a lower case letter next to the line (Figure 9.2). The notation $\overleftrightarrow{AB}$ is read as "line AB" and refers to the line that contains points A and B.

A **plane** is a flat surface that has no thickness and extends without ending in all directions. A plane is represented by a "window pane" and named by writing a capital letter in one of its corners (Figure 9.3).

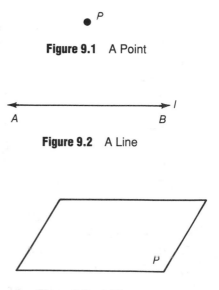

Figure 9.1 A Point

Figure 9.2 A Line

Figure 9.3 A Plane

Defined Terms

Defined terms are terms whose distinguishing characteristics can be explained by using either undefined or previously defined terms. Table 9.1 lists some basic geometric terms and their definitions.

TABLE 9.1

Term	Definition	Illustration
Line segment	A **line segment** is a part of a line consisting of two points, called endpoints, and the set of all points between them.	 Notation: $\overline{AB}$
Ray	A **ray** is a part of a line consisting of a given point, called the endpoint, and the set of all points on one side of the endpoint.	 Notation: $\overrightarrow{LM}$ A ray is always named by using two points, the first of which must be the endpoint. The arrow on top must always point to the right.
Opposite rays	**Opposite rays** are rays that have the same endpoint and that form a line.	 $\overrightarrow{KX}$ and $\overrightarrow{KB}$ are opposite rays.
Angle	An **angle** is the union of two rays having the same endpoint. The endpoint is called the vertex of the angle; the rays are called the sides of the angle.	 Vertex: K Sides: $\overrightarrow{KJ}$ and $\overrightarrow{KL}$
Collinear points	**Collinear points** are points that lie on the same line.	 Points A, B, and C are collinear. Points B, C, and D are *not* collinear.

197

Naming Angles

An angle may be named in several different ways.

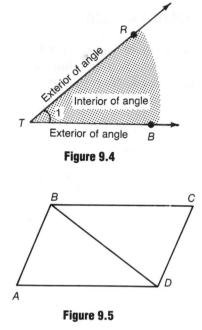

Figure 9.4

1. By using three letters, with the middle letter corresponding to the vertex of the angle and the other letters naming one point on each side of the angle. In Figure 9.4, the name of the angle may be ∠*RTB* or ∠*BTR*.

2. By placing a number at the vertex and in the interior of the angle. The angle may then be referred to by the number. In Figure 9.4, the name of the angle may be ∠1 *or* ∠*RTB* (*or* ∠*BTR*).

3. By using a *single* capital letter that corresponds to the vertex, provided that this causes no confusion. In Figure 9.5, there is no question that ∠*A* is another name for ∠*BAD*. On the other hand, ∠*D* may not be used since it is not clear whether it names ∠*ADB or* ∠*CDB or* ∠*ADC*.

Figure 9.5

Beginning Postulates

A **postulate** is a statement that is assumed to be true. For example, we accept as postulates the following statements:

Postulate 1 Exactly one line may be drawn between two given points.

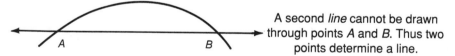

A second *line* cannot be drawn through points *A* and *B*. Thus two points determine a line.

Postulate 2 Exactly one plane contains three noncollinear points.

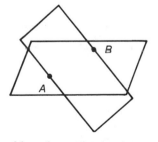

More than one plane may contain *two* points.

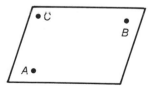

Exactly one plane contains *three* noncollinear points.

Postulate 3   If two lines intersect, they meet in exactly one point.

Intersecting lines cannot meet
in more than one point.

Theorems

A **theorem** is a generalization that can be demonstrated to be true. The Pythagorean relationship ($a^2 + b^2 = c^2$) is a theorem since it can be proved to be true for *any* right triangle.

Exercise Set 9.1

1. In the accompanying diagram:
   (**a**) Name four rays having point
      B as an endpoint.
   (**b**) Name line l in two different
      ways.
   (**c**) Name four angles that have
      the same vertex.
   (**d**) Name two pairs of opposite
      rays.

Use the following diagram for Exercises 2–5:

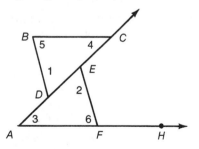

2. Name the vertex of each of the following angles:
   (**a**) 1   (**b**) 3   (**c**) 5

3. Use three letters to name each of the following angles:
   (**a**) 2   (**b**) 4   (**c**) 6

4. Name two rays whose endpoint is E.

5. Name three line segments that have F as an endpoint.

9.2 MEASURING SEGMENTS AND ANGLES

KEY IDEAS

The familiar *ruler* is used for measuring the length of a segment, and the *protractor* is used to measure an angle. Comparing the measures of segments and measures of angles leads to new terms and concepts.

Measuring Segments

The length (or measure) of a line segment is the distance between its endpoints. If the distance between points A and B is 2 inches, the length of $\overline{AB}$ is 2 inches; this may be abbreviated by writing $AB = 2$. Notice that $\overline{AB}$ represents a line segment, while AB (without the top bar) represents the *length* of $\overline{AB}$.

Measuring Angles

If you imagine the vertex as a pivot point, then the measure of an angle refers to the amount of rotation from the first side to the second side. The amount of rotation is measured by a protractor (Figure 9.6), on which the customary unit of measurement is the degree, represented by the symbol °. A ray that sweeps out one complete circle rotates 360 degrees. One degree is therefore defined to be $\frac{1}{360}$ of one complete rotation of a ray.

The measure of an angle corresponds to some number on the protractor, greater than 0 and less than *or* equal to 180. In Figure 9.6, the degree measure of angle ABC is 60 degrees. We abbreviate this by writing m $\angle ABC = 60$, read as "the measure of angle ABC is 60." It is customary to omit the degree symbol (°); we never write m $\angle ABC = 60°$ or $\angle ABC = 60$ (omitting the "m").

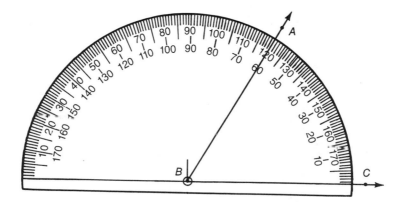

Figure 9.6   A Protractor

Example

1. In the accompanying figure, find the measures of these angles:
(**a**) ∠APZ (**b**) ∠FPZ (**c**) ∠WPB (**d**) ∠ZPB (**e**) ∠SPZ

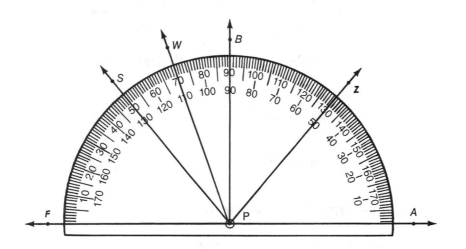

Solution:

(**a**) m ∠ *APZ* = **50** (read lower scale).
(**b**) m ∠ *FPZ* = **130** (read upper scale).
(**c**) m ∠ *WPB* = 110 – 90 = **20**.
(**d**) m ∠ *ZPB* = 90 – 50 = **40**.
(**e**) m ∠ *SPZ* = 130 – 50 = **80**.

Right Angles and Perpendiculars

An L-shaped angle is called a *right* angle. A **right angle** is an angle whose measure is 90. Lines or segments that intersect to form a right angle are said to be *perpendicular*.

    Perpendicular lines are two lines that intersect to form a right angle. If line *l* is perpendicular to line *m*, we may write *l* ⊥ *m*, where the symbol ⊥ is read as "is perpendicular to."

Classifying Angles

An angle whose sides form a straight line is called a **straight** angle and has a measure of 180. Other angles having measures between 0 and 180 may be classified according to whether their measures are less than 90 or greater than 90. Figure 9.7 shows the four types of angles.

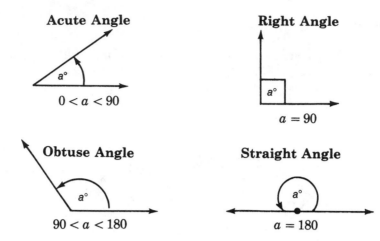

Figure 9.7 Types of Angles

Congruent Segments and Angles

Objects that have the same size and shape are said to be **congruent**. Line segments are congruent if they have the same length. Angles are congruent if they have the same degree measure. Since all right angles have 90 as their measure, *all right angles are congruent.*

Notation

The symbol $\cong$ is translated as "is congruent to." If $AB = 2$ and $RS = 2$, then $\overline{AB} \cong \overline{RS}$; this is read as "line segment AB *is congruent to* line segment RS." If m $\angle J = 60$ and m $\angle K = 60$, then $\angle J \cong \angle K$, read as "angle J *is congruent to* angle K."

When comparing diagrams, it is sometimes helpful to indicate which pairs of segments or angles, if any, are congruent. Matching bars drawn through segments are used to indicate that pairs of line segments are congruent. In Figure 9.8, $\overline{AB} \cong \overline{CD}$ and $\overline{BC} \cong \overline{AD}$. As shown in the diagram, congruent angles may also be identified by drawing matching bars. Angles 1 and 2 are marked off as being congruent. According to the diagram, angles 3 and 4 are also congruent.

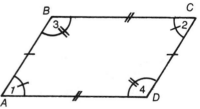

Figure 9.8 Congruent Angles and Line Segments

Bisector and Midpoint

A **bisector** of a line segment is any line, line segment, or ray that divides the segment into two shorter segments that have the same length. If line *l* *bisects* $\overline{AB}$, then the point at which line *l* intersects $\overline{AB}$ is the *midpoint* of $\overline{AB}$ (Figure 9.9). In general, a point *M* is the **midpoint** of $\overline{AB}$ if points *A, B,* and *M* are collinear and *AM = MB*. A line segment has exactly one midpoint, but an infinite number of bisectors.

An **angle bisector** is a ray that divides an angle into two congruent angles. In Figure 9.10, $\angle ABD \cong \angle CBD$ so that $\overrightarrow{BD}$ bisects $\angle ABC$. An angle has exactly one bisector.

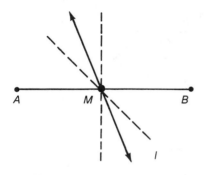

Figure 9.9   Bisection and Midpoint

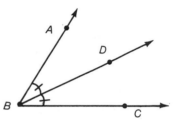

Figure 9.10   Angle Bisector

Examples

2. In the accompanying figure, point *M* is the midpoint of $\overline{RS}$. If *RM* = 18 and the length of $\overline{SM}$ is represented by $3x - 6$, find the value of *x*.

Solution: Since *M* is the midpoint of $\overline{RS}$, *SM = RM*, so

$$3x - 6 = 18$$
$$3x = 18 + 6$$
$$\frac{3x}{3} = 24$$
$$x = \mathbf{8}$$

3. For the accompanying figure, draw a conclusion, given that:
(a) $\overline{AL}$ bisects $\overline{BC}$.
(b) $\overline{BK}$ bisects $\angle ABC$.
(c) $\overline{BK}$ bisects $\overline{AL}$.
(d) $\overline{AL}$ bisects $\angle CAB$.

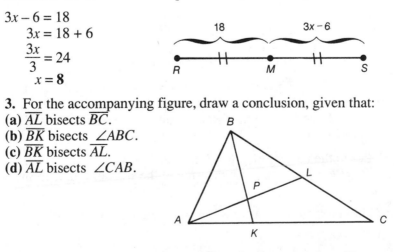

Solution: (a) $BL = CL$ (c) $AP = LP$
        (b) $\angle ABK \cong \angle CBK$ (d) $\angle CAL \cong \angle BAL$

Exercise Set 9.2

1. In the accompanying diagram, classify each of the following angles as acute, right, obtuse, or straight and approximate its measure:
   (a) $\angle TOM$
   (b) $\angle LOM$
   (c) $\angle SOM$
   (d) $\angle LOR$
   (e) $\angle ROT$
   (f) $\angle LOT$

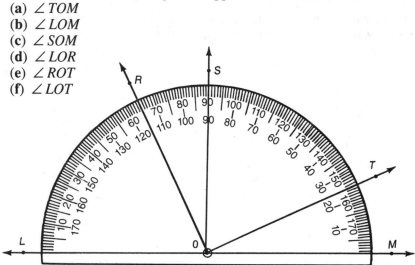

2. In the accompanying diagram, pairs of angles and segments are indicated as congruent. Using the letters in the diagram, write the appropriate congruence relationships.

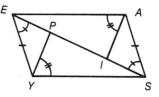

3. Line $\overleftrightarrow{HG}$ passes through point P in such a way that P lies between points H and G.
   (a) If $HP = 3$ and $PG = 5$, find HG.
   (b) If point X is the midpoint of $\overline{HG}$, find the length of $\overline{XP}$.

4. Points, $P, I,$ and Z are collinear, and $IZ = 8$, $PI = 14$, and $PZ = 6$.
   (a) Which of the three points is between the other two?
   (b) If point M is the midpoint of $\overline{PI}$ and point N is the midpoint of $\overline{IZ}$, what is the length of $\overline{MN}$?

5. $\overrightarrow{PL}$ lies in the interior of $\angle RPH$. If m $\angle RPL = x - 5$, m $\angle LPH = 2x + 18$, and m $\angle HPR = 58$, what is the measure of the smallest angle formed that has ray PL as one side?

6. $\overleftrightarrow{XY}$ bisects $\overline{RS}$ at point M. If $RM = 6.5$, find the length of $\overline{RS}$.

204

7. $\overrightarrow{PQ}$ bisects *HPJ* If m $\angle QPJ = 2x - 9$ and m $\angle QPH = x + 29$, find m $\angle HPJ$.

8. $\overrightarrow{BP}$ bisects $\angle ABC$ If m $\angle ABP = 4x + 5$ and m $\angle CBP = 3x + 15$:
(**a**) find the value of x
(**b**) classify $\angle ABC$ as acute, right, or obtuse

9. If R is the midpoint $\overline{XY}$, and $XR = 3n + 1$, and $YR = 16 - 2n$:
(**a**) find the value of n (**b**) find the length of $\overline{XY}$

In Exercises 10–12, mark off each diagram with the given information, and draw an appropriate conclusion.

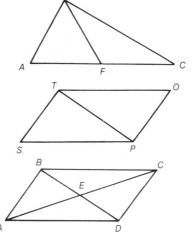

10. Given: $\overline{BF}$ bisects $\overline{AC}$.
Conclusion: ?

11. Given: $\overline{PT}$ bisects $\angle STO$.
Conclusion: ?

12. Given: $\overline{AC}$ bisects $\overline{BD}$.
$\overline{BD}$ bisects $\angle$ ADC.
Conclusion: ?

9.3 CLASSIFYING POLYGONS AND TRIANGLES

KEY IDEAS

Any figure that can be represented by stretching and/or twisting a rubber band is an example of a **closed figure**. If the rubber band does not cross over itself, then the figure is called a *simple* closed figure.

(1)   (2)   (3)   (4)

Figure (1) is *not* a simple closed curve since the curve crosses over itself. Figure (2) is *not* a simple closed curve since the "rubber band *snapped*" and the curve is "open." Figures (3) and (4) are examples of simple closed curves.

Classifying Polygons

A **polygon** is a simple closed curve that consists entirely of line segments. Each line segment is called a **side** of the polygon. Each corner of the polygon in which two sides intersect is called a **vertex** of the polygon. Polygons may be classified by the number of their sides. The names of some commonly referred to polygons are listed in Table 9.2.

TABLE 9.2 NAMES OF SOME POLYGONS

Number of Sides	Name
3	Triangle
4	Quadrilateral
5	Pentagon
6	Hexagon
8	Octagon
10	Decagon
12	Dodecagon

Regular Polygons

If each side of a polygon has the same length, the polygon is **equilateral**. If each angle of a polygon has the same measure, the polygon is **equiangular**. If a polygon is both equilateral *and* equiangular, it is called a **regular polygon**. Figure 9.11 shows some different types of quadrilaterals.

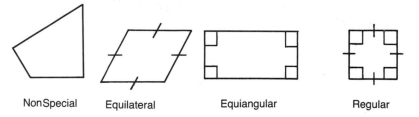

NonSpecial   Equilateral   Equiangular   Regular

Figure 9.11   Types of Quadrilaterals

Examples

1. Find the perimeter of an equilateral pentagon if the length of one of its sides is 4.6 inches.

Solution: The perimeter of a polygon is equal to the sum of the lengths of its sides. A pentagon has five sides, and, since it is regular, each side has the same length, so

$$\text{Perimeter} = 4.6 + 4.6 + 4.6 + 4.6 + 4.6$$
$$= 5(4.6)$$
$$= \textbf{23 inches}$$

2. Find the measure of each angle of a regular hexagon if the sum of the degree measures of its angles is 720.

Solution: Since a hexagon has six sides, it must have six interior angles. Each angle of a regular hexagon has the same degree measure.

Let x = measure of one interior angle.
$$6x = 720$$
$$x = \frac{720}{6} = \mathbf{120}$$

3. If the perimeter of an equilateral polygon having seven sides is represented by $14x - 35$, express the length of one of its sides in terms of x.

Solution: Let s = length of a side.
$$7s = 14x - 35$$
$$\frac{7s}{7} = \frac{14x - 35}{7}$$
$$s = \frac{7(2x - 5)}{7}$$
$$s = \mathbf{2x - 5}$$

Classifying Triangles

A triangle may be classified according to the number of congruent sides that it contains (Figure 9.12).

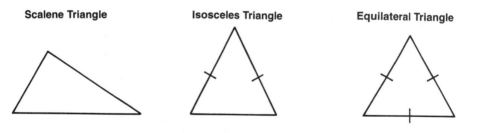

Scalene Triangle   **Isosceles Triangle**   **Equilateral Triangle**

Figure 9.12   Triangles Classified by Number of Congruent Sides

A triangle may also be classified by the measure of its greatest angle (Figure 9.13).

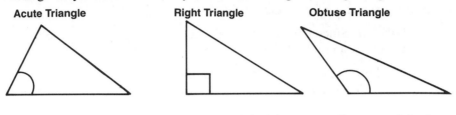

Acute Triangle   Right Triangle   Obtuse Triangle

Greatest angle is *acute*.   Greatest angle is *right*.   Greatest angle is *obtuse*.

Figure 9.13 Triangles Classified by Greatest Angle Measurement

Exercise Set 9.3

1. Find the perimeter of a regular octagon if the length of a side is 4.

2. Express in terms of x the perimeter of a regular hexagon if the length of a side is represented by $3x + 2$.

3. If the perimeter of a regular decagon is 78, find the length of a side.

4. If the sum of the measures of the angles of a regular 9-gon is 1260, find the measure of an angle of the 9-gon.

5. If the perimeter of a regular quadrilateral is represented by $24x^2 - 8x$, represent in terms of x the length of a side of the quadrilateral.

6. The perimeter of an equilateral triangle is equal to the perimeter of a certain equilateral quadrilateral. If the length of a side of the triangle exceeds the length of a side of the quadrilateral by 5, find the length of a side of the quadrilateral.

7. The perimeter of a regular pentagon is equal to the perimeter of a certain regular hexagon. If the length of a side of the pentagon exceeds the length of a side of the hexagon by 2, find the perimeter of each polygon.

8–10. Find the value of x *and classify* △ABC *as scalene, isosceles, or equilateral.*

8. Perimeter = 45; $AB = x + 7$, $BC = 2x - 1$, and $AC = 3x - 9$.

9. Perimeter = 59; $AB = 2x$, $BC = 3x - 10$, and $AC = x + 9$.

10. Perimeter = 66; $AB = x + 10$, $BC = 2x$, and $AC = 3x - 10$.

9.4 LOOKING AT SPECIAL PAIRS OF ANGLES

Key Ideas

Important geometric results often rely on whether a special relationship exists between two angles. *Supplementary, complementary*, and *vertical* angle pairs are of special importance.

Adjacent Angles

Two angles are **adjacent** if they have the same vertex and share a common side, but do not have any interior points in common (they don't overlap). In Figure 9.14, angles 1 and 2 are adjacent angles. Point A is the vertex of each angle, and $\overrightarrow{AS}$ is the common side of the two angles.

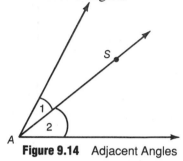

Figure 9.14 Adjacent Angles

Supplementary Angles

Two angles are **supplementary** (Figure 9.15) if the sum of their measures is 180. If m $\angle A = 60$ and m $\angle B = 120$, then $\angle A$ and $\angle B$ are supplementary, and either angle is called the *supplement* of the other angle. Observe that m $\angle A = 180 - $ m $\angle B$, and m $\angle B = 180 - $ m $\angle A$.

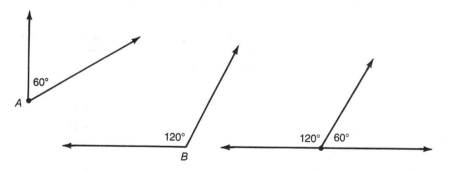

$\angle A$ is the supplement of $\angle B$, and $\angle B$ is the supplement of $\angle A$.

If two *adjacent* angles are supplementary, their noncommon (exterior) sides form a straight line.

Figure 9.15 Supplementary Angles

Complementary Angles

Two angles are **complementary** (Figure 9.16) if the sum of their measures is 90. If m $\angle A$ = 40 and m $\angle B$ = 50, then $\angle A$ and $\angle B$ are complementary, and either angle is called the *complement* of the other angle. Observe that m $\angle A$ = 90 – m $\angle B$, and m $\angle B$ = 90 – m $\angle A$.

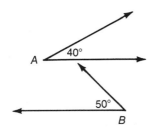

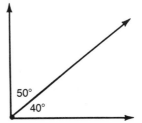

$\angle A$ is the complement of $\angle B$, and $\angle B$ is the complement of $\angle A$.

If two *adjacent* angles are complementary, their noncommon (exterior) sides are perpendicular.

Figure 9.16   Complementary Angles

Examples

1. In the accompanying diagram, find the value of x.

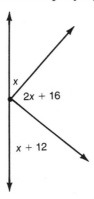

Solution: The sum of the measures of the adjacent angles having point P as their vertex is 180, so

$$x + (2x + 16) + (x + 12) = 180$$
$$4x + 28 = 180$$
$$4x = 180 - 28$$
$$4x = 152$$
$$x = \frac{152}{4} = \mathbf{38}$$

2. The measures of two complementary angles are in the ratio of 2 : 13. Find the measure of the smaller angle.

Solution: Let $2x$ = measure of the smaller angle.
Then $13x$ = measure of the larger angle.

Since the two angles are complementary, the sum of their measures is 90.

$$2x + 13x = 90$$
$$15x = 90$$
$$\frac{15x}{15} = \frac{90}{15}$$
$$x = 6$$
$$2x = 2(6) = 12.$$

The measure of the smaller angle is **12**.

3. If the measure of an angle exceeds twice its supplement by 30, find the measure of the angle.

Solution: Let x = measure of the angle.
Then $180 - x$ = measure of the angle's supplement.
$$x = 2(180 - x) + 30$$
$$x = 360 - 2x + 30$$
$$x + 2x = 390$$
$$\frac{3x}{3} = \frac{390}{3}$$
$$x = \mathbf{130}$$

Vertical Angles

If two different lines intersect, four angles are formed having the point of intersection of the lines as their vertex. Nonadjacent (opposite) pairs of these angles are called **vertical angles**. In Figure 9.17, angles 1 and 3 are vertical angles. Also, angles 2 and 4 are vertical angles.

Suppose that m $\angle 1 = 50$. Then m $\angle 2$ = $180 - $ m $\angle 1 = 180 - 50 = 130$. Also, m $\angle 3 = 180 - $ m $\angle 2 = 180 - 130 = 50$. And m $\angle 4 = 180 - $ m $\angle 1 = 130$.

Therefore m $\angle 1 = $ m $\angle 3$, and m $\angle 2$ = m $\angle 4$. This suggests the following theorem:

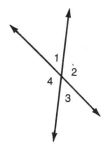

Figure 9.17 Vertical Angles

═══════ **MATH FACTS** ═══════

VERTICAL ANGLE THEOREM

Vertical angles are equal in measure and are, therefore, congruent.

Example

4. In the accompanying diagram, find the value of y:

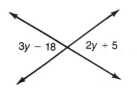

Solution: Since vertical angles are equal in measure,

$$3y - 18 = 2y + 5$$
$$3y = 2y + 5 + 18$$
$$3y - 2y = 23$$
$$y = \mathbf{23}$$

MATH FACTS

SUMMARY OF ANGLE PAIR RELATIONSHIPS

Let $m\angle A = a$, and $m\angle B = b$.

If $\angle A$ and $\angle B$ are:	*Then:*
Supplementary	$a + b = 180$
Complementary	$a + b = 190$
Vertical	$a = b$

Exercise Set 9.4

1. For the accompanying diagram, list all pairs of adjacent, supplementary, and vertical angles.

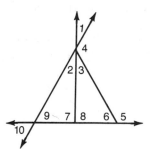

2. For each of the following, find the value of x:

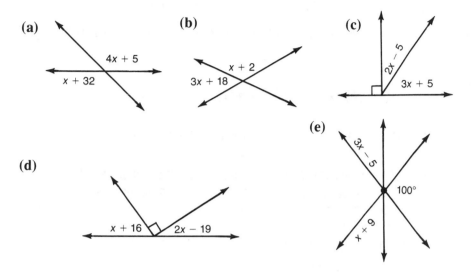

(a)

$4x + 5$

$x + 32$

(b)

$x + 2$

$3x + 18$

(c)

$2x - 5$

$3x + 5$

(d)

$x + 16$ $2x - 19$

(e)

$3x - 5$

$100°$

$x + 9$

3. If two angles are supplementary, find the measure of the smaller angle if the measures of the two angles are in the ratio of:
(**a**) $1 : 8$ (**b**) $3 : 5$

4. If two angles are complementary, find the measure of the smaller angle if the measures of the two angles are in the ratio of:
(**a**) $2 : 3$ (**b**) $3 : 1$

5. The measure of an angle exceeds three times its supplement by 4. Find the measure of the angle.

6. The measure of an angle exceeds four times the measure of its complement by 6. Find the measure of the angle.

7. The measure of an angle is 22 less than three times the measure of the complement of the angle. Find the measure of the angle.

8. The measure of the supplement of an angle is three times as great as the measure of the angle's complement. What is the measure of the angle?

9. Find the measure of an angle if it is 12 less than twice the measure of its complement.

10. The difference between the measures of an angle and its complement is 14. Find the measure of the smaller of the two angles.

11. The difference between the measures of an angle and its supplement is 22. Find the measure of the smaller of the two angles.

12. The measure of the supplement of an angle is three times as great as the measure of the complement of the same angle. What is the measure of the angle?

13. $\overleftrightarrow{XY}$ and $\overleftrightarrow{AB}$ intersect at point C. If m $\angle XCB = 4y - 9$ and m $\angle ACY = 3y + 29$, find m $\angle XCB$.

9.5 WORKING WITH PARALLEL LINES

KEY IDEAS

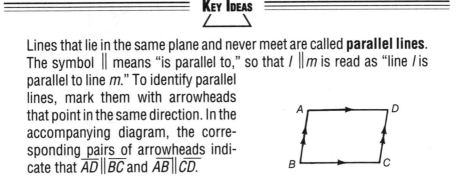

Lines that lie in the same plane and never meet are called **parallel lines**. The symbol ‖ means "is parallel to," so that $l \parallel m$ is read as "line l is parallel to line m." To identify parallel lines, mark them with arrowheads that point in the same direction. In the accompanying diagram, the corresponding pairs of arrowheads indicate that $\overline{AD} \parallel \overline{BC}$ and $\overline{AB} \parallel \overline{CD}$.

Transversals and Special Angle Pairs

A line that intersects two or more lines in different points is called a **transversal**. In Figure 9.18, line t represents a transversal since it intersects lines l and m at two different points. Angles 1, 2, 5, and 6 lie *between* lines l and m and are called **interior angles**. Angles 3, 4, 7, and 8 lie *outside* lines l and m and are called **exterior angles**.

Table 9.3 further classifies special pairs of these angles.

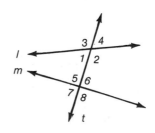

Figure 9.18 Transversal and Angle Pairs

TABLE 9.3 SPECIAL ANGLE PAIRS

Type of Angle Pair	Distinguishing Features	Examples (Figure 9.18)
Alternate interior angles	• Angles are interior angles. • Angles are on opposite sides of the transversal. • Angles do not have the same vertex.	Angles 1 and 6; angles 2 and 5.
Corresponding angles	• One angle is an interior angle; the other angle is an exterior angle. • Angles are on the same side of the transversal. • Angles do not have the same vertex.	Angles 3 and 5; angles 4 and 6; angles 1 and 7; angles 2 and 8.

In analyzing diagrams, alternate interior angle pairs may be identified by their Z shape, while corresponding angles form an F shape (Figure 9.19). The Z and F shapes, however, may be rotated so that the letter may appear reversed or upside down.

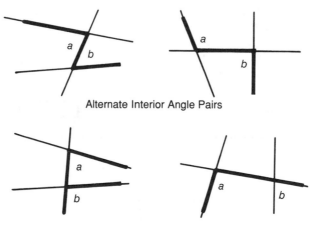

Alternate Interior Angle Pairs

Corresponding Angle Pairs

Figure 9.19 Z and F Shapes Formed by Angle Pairs

Example

1. For the accompanying diagram, name all pairs of:
(**a**) alternate interior angles
(**b**) corresponding angles

215

Solution:

(**a**) Alternate interior angle pairs: ***b*** **and** ***g;*** ***d*** **and** ***e.***
(**b**) Corresponding angle pairs: ***a*** **and** ***e;*** ***b*** **and** ***f;*** ***c*** **and** ***g;*** ***d*** **and** ***h.***

Parallel Line Angle Relationships

Any two angles formed by a transversal intersecting two parallel lines are either congruent or supplementary.

If two parallel lines are cut by a transversal, then:

- Corresponding angles have the same degree measure.
- Alternate interior angles have the same degree measure.
- Interior angles on the same side of the transversal are supplementary.

These relationships are summarized in Table 9.4.

TABLE 9.4 PARALLEL LINE ANGLE RELATIONSHIPS

Type of Angle Pair	Relationship	Diagram
Alternate interior angles	$c = e;\ \ d = f$	
Corresponding angles	$a = e;\ c = g$ $b = f;\ \ d = h$	
Interior angles on the same side of the transversal	$d + e = 180$ $c + f = 180$	

Examples

2. Given that lines *l* and *m* are parallel, find the value of *x*.

(**a**)   (**b**)

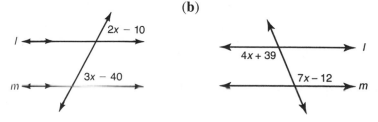

Solution: (**a**) Since corresponding angles formed by parallel lines have the same measure, we can set $3x - 40$ equal to $2x - 10$. Thus,

$$3x - 40 = 2x - 10$$
$$3x = 2x + 30$$
$$x = \mathbf{30}$$

216

(b) Since alternate interior angles formed by parallel lines have the same measure, we can set $7x - 12$ equal to $4x + 39$. Thus,

$$7x - 12 = 4x + 39$$
$$7x = 4x + 51$$
$$3x = 51$$
$$x = \frac{51}{3} = \mathbf{17}$$

3. Two parallel lines are cut by a transversal so that the measures of a pair of interior angles on the same side of the transversal are in the ratio of 5 : 13. Find the measure of the smaller of these angles.

Solution: Let $5x$ = measure of the smaller interior angle.
Then $13x$ = measure of the other interior angle.

Since the lines are parallel, interior angles on the same side of the transversal are supplementary, so the sum of their measures is 180.

$$5x + 13x = 180$$
$$18x = 180$$
$$x = \frac{180}{18} = 10$$
$$5x = 5(10) = 50$$

The measure of the smaller interior angle is **50**.

Example

4. Which of the accompanying diagrams contains a pair of parallel lines?

(1)
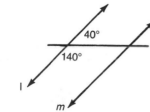

(3)
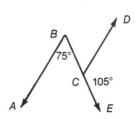

(2)
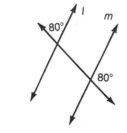

(4)
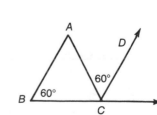

Solution: In choice (3), m $\angle BCD = 180 - 105 = 75$. Since alternate interior angles are equal in measure, $\overline{BA} \parallel \overline{CD}$. The correct answer is **choice (3)**.

Determining When Lines Are Parallel

Two lines are parallel if any one of the following statements is true:

- A pair of alternate interior angles have the same measure.
- A pair of corresponding angles have the same measure.
- A pair of interior angles on the same side of the transversal are supplementary.

Each of these statements is the *converse* of a statement that gives a property of parallel lines.

Exercise Set 9.5

1. In the accompanying diagram, $l \parallel m$. Find the value of x if:
   (a) m $\angle 1 = x$ and m $\angle 7 = 68$
   (b) m $\angle 2 = 53$ and m $\angle 7 = x$
   (c) m $\angle 3 = 64$ and m $\angle 8 = x$
   (d) m $\angle 3 = 76$ and m $\angle 5 = 3x - 5$
   (e) m $\angle 6 = 128$ and m $\angle 2 = 4x$
   (f) m $\angle 2 = 3x - 15$ and m $\angle 5 = x + 29$
   (g) m $\angle 8 = 2x - 11$ and m $\angle 5 = 3x - 47$

2-10. Given that $l \parallel m$, *find the value of* x.

2. 3.

4.

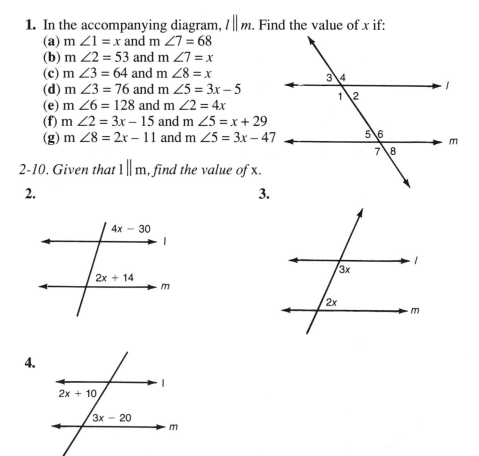

5.

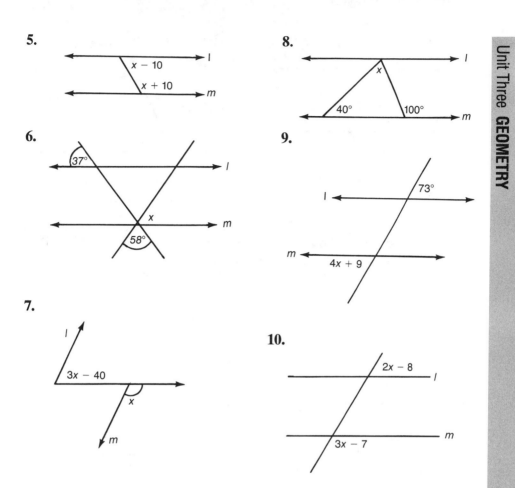

8.

6.

9.

7.

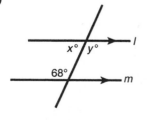

10.

11. In each of the following, given that $l \parallel m$, find the values of x and y:

(a)

(b)

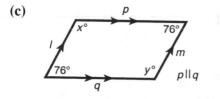

(c)

(d)

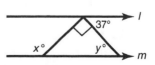

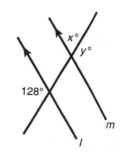

(e)

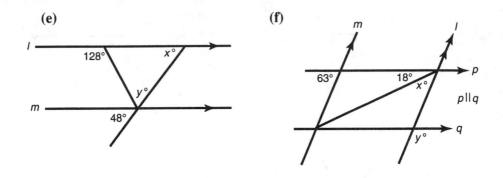

(f)

p||q

12. Two parallel lines are cut by a transversal. Find the measures of a pair of interior angles on the same side of the transversal if the angles:
(**a**) are represented by $5x - 32$ and $x + 8$
(**b**) have measures such that the measure of one angle is four times the measure of the other

13. In the accompanying diagram $\overleftrightarrow{AB} \parallel \overleftrightarrow{CD}$ and $\overrightarrow{FG}$ bisects $\angle EFD$. If m $\angle EFG = x$ and m $\angle FEG = 4x$, find x.

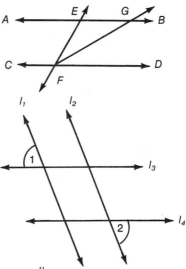

14. In the accompanying diagram, lines l_1 and l_2 are parallel and $m\angle 1 = 70$. What must $m\angle 2$ be so that lines l_3 and l_4 will be parallel?

15–18. In each of the following, determine whether l $\parallel$ m:

15.

16.

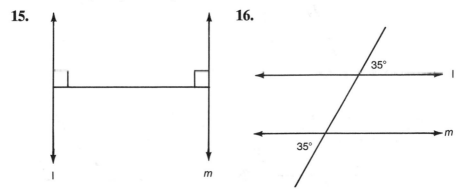

17.

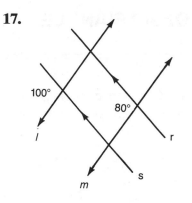

18.

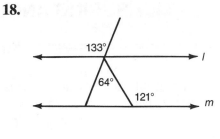

19. Two parallel lines are cut by a transversal, and the two interior angles on the same side of the transversal are bisected. What kind of angle is formed where the two angle bisectors meet?
(1) Right   (2) Obtuse   (3) Acute   (4) Straight

20. In the accompanying diagram, $\overrightarrow{AD} \parallel \overrightarrow{BC}$ and $\overrightarrow{AC}$ bisects $\angle BAD$. If m $\angle BAD = x$, express m $\angle 1$ in terms of x.

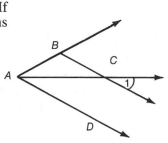

21. Given that lines p and q are parallel, determine whether line l is parallel to line m.

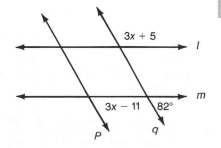

9.6 MEASURING ANGLES OF A TRIANGLE

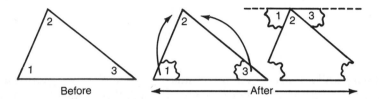

The accompanying diagrams illustrate that, after "tearing off" angles 1 and 3, their sides can be aligned with one of the sides of angle 2 so that their exterior sides form a straight line.

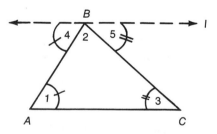

Before   After

Since the measure of a straight angle is 180, this experiment suggests that the sum of the measures of angles 1, 2, and 3 is 180 and, therefore, the sum of the measures of the angles of *any* triangle is 180.

Angles of a Triangle

In order to *prove* that the sum of the measures of the angles of a triangle is 180, a proof that does not depend on "tearing off" angles is needed. In Figure 9.20, a line has been drawn parallel to a side of the triangle and through one of its vertices. Since $l \parallel \overrightarrow{AC}$, alternate interior angles are equal in measure:

$$m \angle 1 = m \angle 4,$$

and

$$m \angle 3 = m \angle 5.$$

The sum of the measures of the angles formed at vertex B is 180:

$$m \angle 4 + m \angle 2 + m \angle 5 = 180.$$

We may therefore substitute in the above equation $m \angle 1$ for $m \angle 4$ and $m \angle 3$ for $m \angle 5$:

$$m \angle 1 + m \angle 2 + m \angle 3 = 180.$$

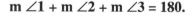

Figure 9.20   Angles of a Triangle

This analysis provides an "informal" proof of the following theorem:

====== **MATH FACTS** ======

TRIANGLE ANGLE SUM THEOREM

The sum of the measures of the angles of a triangle is 180.

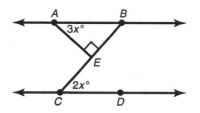

$$a + b + c = 180$$

Examples

1. In the accompanying figure, $\overleftrightarrow{AB}$ is parallel to $\overleftrightarrow{CD}$, $\overline{AE} \perp \overline{BC}$, m∠$BCD$ = 2x, m∠BAE = 3x. Find the value of *x*.

Solution: Since alternate interior angles formed by parallel lines have the same degree measure, m ∠ABE = m ∠BCD = 2x. In right triangle *AEB*, angles *BAE* and *ABE* are the acute angles and, as a result, are complementary. Hence,

m ∠BAE + m ∠ABE = 90
3x + 2x = 90
5x = 90
$$x = \frac{90}{5}$$
x = **18**

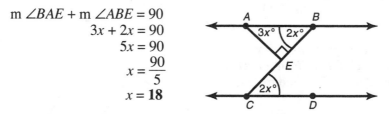

2. If a triangle is equiangular, what is the measure of each of its angles?

Solution: An equiangular triangle is a triangle in which all of the angles have the same measure. If *x* represents the measure of each angle, then

$$x + x + x = 180$$
$$3x = 180$$
$$x = \frac{180}{3} = 60$$

Each angle of an equiangular triangle has a degree measure of **60**.

3. The measures of the angles of a triangle are in the ratio of 2 : 3 : 5. What is the measure of the smallest angle of the triangle?

Solution: Let $2x$ = measure of the smallest angle of triangle.
Then $3x$ and $5x$ = measures of the remaining angles.

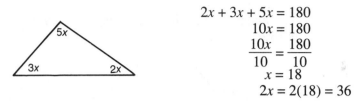

$$2x + 3x + 5x = 180$$
$$10x = 180$$
$$\frac{10x}{10} = \frac{180}{10}$$
$$x = 18$$
$$2x = 2(18) = 36$$

The measure of the smallest angle of the triangle is **36**.

4. In the accompanying diagram, $\overline{DE} \perp \overline{AEC}$. If m $\angle ADB$ = 80 and m $\angle CDE = 60$, what is m $\angle DAE$?

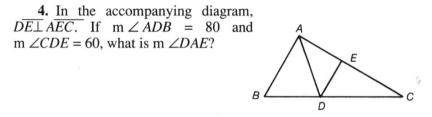

Solution: Since angles $ADB, ADE,$ and CDE form a straight angle, the sum of their measures is 180. Hence

$$80 + m\angle ADE + 60 = 180$$
$$m\angle ADE = 180 - 140 = 40$$

In $\triangle ADE$,

$$m \angle DAE + m \angle ADE + m \angle AED = 180$$
$$m \angle DAE + \quad 40 \quad + \quad 90 \quad = 180$$
$$m \angle DAE + \quad 130 \quad = 180$$
$$m \angle DAE = 180 - 130 = \mathbf{50}$$

5. In the accompanying diagram, $\overline{AD} \parallel \overline{EC}, \overline{DF} \parallel \overline{CB},$ m$\angle DAE$ = 34, and m$\angle DFE = 57$. Find m$\angle ECB$.

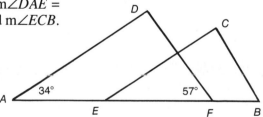

Solution: Since $\overline{AD} \parallel \overline{EC}$, transversal $\overline{AEFB}$ forms congruent corresponding angles, so that m $\angle CEB = $ m $\angle DAE = 34$. Since $\overline{DF} \parallel \overline{CB}$, transversal $\overline{AEFB}$ forms congruent corresponding angles, so that m $\angle CBE = $ m $\angle DFE = 57$. In $\triangle CEB$,

$$m \angle ECB + m \angle CEB + m \angle CBE = 180$$
$$m \angle ECB + \quad 34 \quad + \quad 57 \quad = 180$$
$$m \angle ECB + \quad\quad 91 \quad\quad = 180$$
$$m \angle ECB = 180 - 91$$
$$m \angle ECB = \mathbf{89}$$

Exterior Angles of a Triangle

At each vertex of a triangle an *exterior* angle of the triangle may be formed by extending one of the sides of the triangle (Figure 9.21). Notice that:

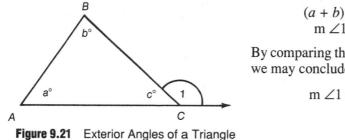

Figure 9.21   Exterior Angles of a Triangle

$$(a + b) + c = 180,$$
$$m \angle 1 + c = 180.$$

By comparing these two equations, we may conclude that:

$$m \angle 1 = a + b.$$

The angles whose measures are represented by a and b are the two interior angles of the triangle that are the most remote from $\angle 1$. With respect to $\angle 1$, these angles are nonadjacent. We may generalize as follows:

MATH FACTS

TRIANGLE EXTERIOR ANGLE THEOREM

The measure of an exterior angle of a triangle is equal to the sum of the measures of the two remote (nonadjacent) interior angles of the triangle.

Example

6. In the accompanying diagrams, find the value of x.

(a)   **(b)**

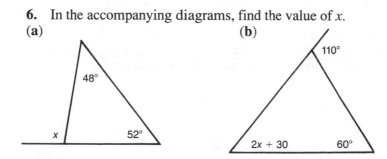

(c)

(d)

Solution:

(a) $x = 48 + 52 = \mathbf{100}$

(b) $110 = 2x + 30 + 60$
$110 = 2x + 90$
$20 = 2x$
$x = \mathbf{10}$

(c) $3x - 10 = (x + 15) + 45$
$3x - 10 = x + 60$
$3x = x + 70$
$2x = 70$
$x = \mathbf{35}$

(d) $x + 50 = 110$ so $x = \mathbf{60}$

MATH FACTS

SUMMARY OF ANGLE RELATIONSHIPS IN A TRIANGLE

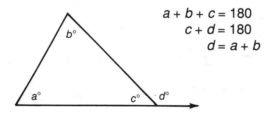

$a + b + c = 180$
$c + d = 180$
$d = a + b$

Exercise Set 9.6

1. Determine whether each of the following statements is *true* or *false*.
(a) A triangle may not contain more than one obtuse angle.
(b) The acute angles of a right triangle are supplementary.
(c) A triangle may not contain more than one right angle.
(d) If a triangle is regular, then each angle has measure 60.
(e) The sum of the measures of the exterior angles of a triangle, one formed at each of the three vertices of the triangle, is 360.

2–5. Find the measure of the smallest angle of a triangle if the measures of the three angles of the triangle are in the ratio:

2. $1:2:6$ **3.** $2:3:10$ **4.** $1:1:2$ **5.** $3:4:5$

6. In right triangle *ABC*, the measure of acute angle *A* exceeds twice the measure of $\angle B$ by 27. Find the measure of the smallest angle of the triangle.

7. When a ray bisects an angle of an equiangular triangle, what type of angle does it always form with the opposite side of the triangle?
(1) Acute (2) Right (3) Obtuse (4) Straight

8–10. For each of the following, the measures of the angles of $\triangle ABC$ are represented in terms of x. Find the value of x, and classify the triangle as acute, right, or obtuse.

8. m $\angle A = 3x + 8$
    m $\angle B = x + 10$
    m $\angle C = 5x$

9. m $\angle A = x + 24$
    m $\angle B = 4x + 17$
    m $\angle C = 2x - 15$

10. m $\angle A = 3x - 5$
    m $\angle B = x + 14$
    m $\angle C = 2x - 9$

11. In $\triangle ABC$, $\overline{BD} \perp \overline{AC}$. If m $\angle A = 72$ and m $\angle ABC = 54$, find m $\angle CBD$.

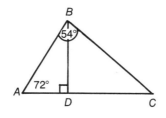

12–25. In each of the following, find the value of x:

12.

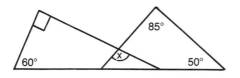

13.

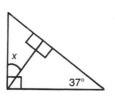

227

14.

$$\overline{AB} \parallel \overline{DE}$$

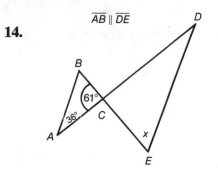

15.

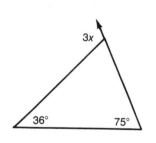

16.

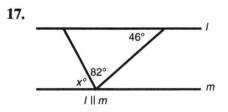

17.

l ∥ *m*

18.

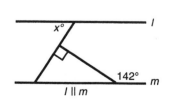

l ∥ *m*

19.

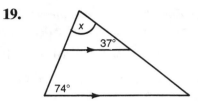

20.

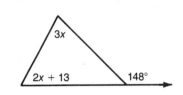

21.

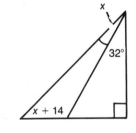

22.

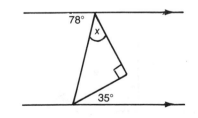

23.

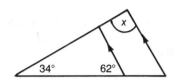

24.

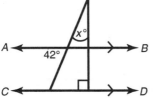

$\overline{AZ}$ and $\overline{BZ}$ are angle bisectors.

25.

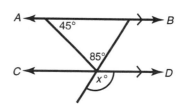

$\overline{AB} \parallel \overleftrightarrow{CD}$

26. In the accompanying figure, $\overleftrightarrow{AB}$ is parallel to $\overleftrightarrow{DE}$, m $\angle BAE = 4x$, m $\angle CDE$ = $3x$, and m $\angle ACD = 98$. Find m $\angle CEF$.

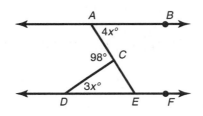

27–28. If $\overleftrightarrow{AB} \parallel \overleftrightarrow{CD}$, *find the value of* x.

27.

28.

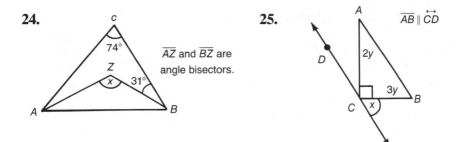

9.7 CLASSIFYING QUADRILATERALS AND THEIR ANGLE RELATIONSHIPS

KEY IDEAS

In quadrilaterals, as in triangles, special angle relationships exist.

Angles of a Quadrilateral

In Figure 9.22, diagonal $\overline{BD}$ divides quadrilateral *ABCD* into two triangles. The sum of the degree measures of the four angles of a quadrilateral can be found by combining the measures of the angles of the two triangles; this gives a sum of 360. This analysis leads to the following generalization:

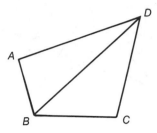

Figure 9.22   Quadrilateral Divided into Triangles

MATH FACTS

QUADRILATERAL ANGLE SUM THEOREM

The sum of the degree measures of the angles of a quadrilateral is 360.

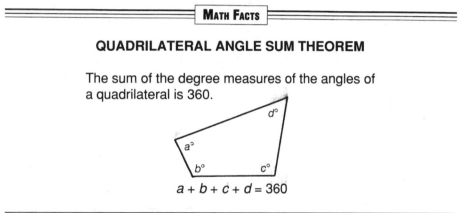

$$a + b + c + d = 360$$

Example

1. The degree measures of three angles of a quadrilateral are 72, 119, and 67. What is the degree measure of the remaining angle of the quadrilateral?

Solution:
Let x = measure of the remaining angle of the quadrilateral.
$$x + 72 + 119 + 67 = 360$$
$$x + 258 = 360$$
$$x = 360 - 258 = 102$$

The degree measure of the remaining angle of the quadrilateral is **102**.

Parallelograms

A **parallelogram** (Figure 9.23) is a quadrilateral having *two* pairs of parallel sides. The notation $\angle\!\!\!\!\angle ABCD$ is read as "parallelogram *ABCD*." The letters *A, B, C,* and *D* represent consecutive vertices of the parallelogram, and the symbol preceding these letters is a miniature parallelogram. Two angles that have a common side are called **consecutive angles**. Angles *A* and *B* are con-secutive angles, as are *B* and *C*, *C* and *D*, and *A* and *D*. Two angles that do *not* share a common side are called **opposite angles**. Angles *A* and *C* are opposite angles, as are angles *B* and *D*.

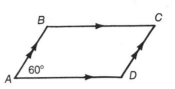

Suppose that m ∠*A* = 60. Angles *A* and *B* are interior angles on the same side of transversal $\overline{AB}$. Since $\overline{BC} \parallel \overline{AD}$, ∠*A* and ∠*B* are supplementary, so that m ∠*B* = 120. For the same reason, ∠*D* and ∠*A* are

Figure 9.23 Parallelogram

supplementary, so that m ∠*D* = 120. Since the sum of the angles of a quadrilateral is 360,

$$m \angle C = 360 - (60 + 120 + 120) = 360 - 300 = 60.$$

Notice that consecutive angles of the parallelogram are supplementary, while opposite angles of the parallelogram have the same measure.

=== **MATH FACTS** ===

SUMMARY OF RELATIONSHIPS IN A PARALLELOGRAM

1. Opposite sides are parallel:
$\overline{AB} \parallel \overline{CD}$ and $\overline{BC} \parallel \overline{AD}$.

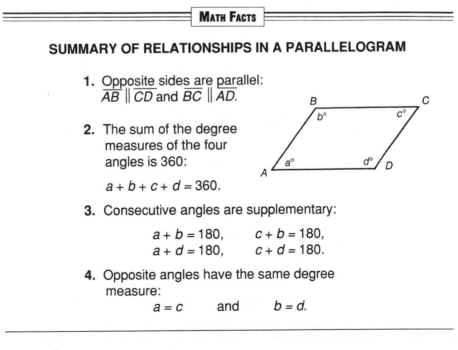

2. The sum of the degree measures of the four angles is 360:

$a + b + c + d = 360.$

3. Consecutive angles are supplementary:

$a + b = 180,$ $c + b = 180,$
$a + d = 180,$ $c + d = 180.$

4. Opposite angles have the same degree measure:

$a = c$ and $b = d.$

Examples

2. In parallelogram *JKLM*, shown in the accompanying diagram, m ∠*K* = 3*x* − 5 and m ∠*M* = 100. Find the value of *x*.

Solution: Since the opposite angles of a parallelogram have the same degree measure,

$$\text{m } \angle K = \text{m } \angle M$$
$$3x - 5 = 100$$
$$3x = 100 + 5$$
$$x = \frac{105}{3} = 35.$$

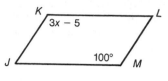

3. In parallelogram $ABCD$, shown in the accompanying diagram, the measures of $\angle A$ and $\angle B$ are in the ratio of $1 : 3$. What is the measure of $\angle A$?

Solution: Let x = measure of $\angle A$
Then $3x$ = measure of $\angle B$
Points A and B are consecutive vertices of the parallelogram, so the measures of $\angle A$ and $\angle B$ are supplementary.

$$x + 3x = 180$$
$$4x = 180$$
$$x = \frac{180}{4} = 45$$

The measure of $\angle A$ is **45**.

Special Types of Quadrilaterals

Table 9.5 classifies some special types of quadrilaterals. A *trapezoid* is the only quadrilateral appearing in the table that is *not* a parallelogram. A rhombus, a rectangle, and a square all have the special properties of a parallelogram.

TABLE 9.5 SPECIAL QUADRILATERALS

Name	Description	Figure
Rhombus	A parallelogram having four equal sides.	
Rectangle	A parallelogram having four right angles.	
Square	A rhombus with four right angles *or* a rectangle with four sides having the same length.	
Trapezoid	A quadrilateral having exactly *one* pair of parallel sides, which are called the *bases*. The nonparallel sides are called the *legs*. If the legs have the same length, then the trapezoid is isosceles.	

Exercise Set 9.7

1–4. In each of the following, find the value of x.

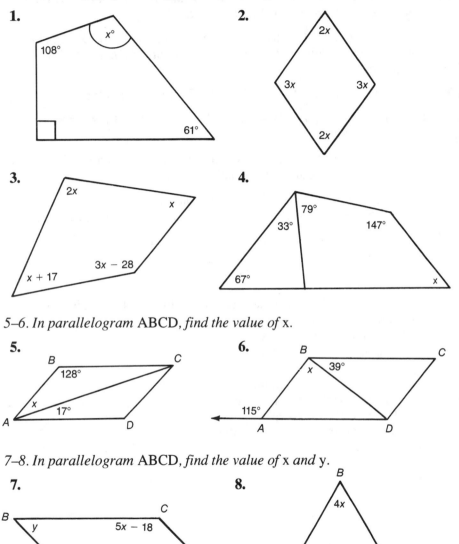

1. 108° x° 61°

2. 2x 3x 3x 2x

3. 2x x x + 17 3x − 28

4. 79° 33° 147° 67° x

5–6. In parallelogram ABCD, *find the value of* x.

5. B 128° C x 17° A D

6. B x 39° C 115° A D

7–8. In parallelogram ABCD, *find the value of* x *and* y.

7. B y C 5x − 18 142° A D

8. B 4x A 5x C y D

9. In parallelogram *ABCD*, m ∠A = 3x and m ∠B = x + 40. What is the value of *x*?

233

10. In parallelogram *MATH* the measure of ∠*T* exceeds the measure of ∠*H* by 30. Find the measure of each angle of the parallelogram.

11. In parallelogram *TRIG*, m ∠*R* = 2*a* + 9 and m ∠*G* = 4*a* − 17. Find the measure of each angle of the parallelogram.

12. In parallelogram *ABCD* the degree measure of ∠*A* is 76. Diagonal *BD* is drawn in such a way that the degree measure of ∠*CBD* is 34. What is the measure of ∠*ABD*?

13. In rhombus *ABCD*, *AB* = 3*x* + 12 and *BC* = 5*x*. What is the value of *x*?

14. In rhombus *STAR*, *ST* = 4*y* − 9 and *TA* = 2*y* + 5. What is the perimeter of rhombus *STAR*?

15. If the measures of two consecutive angles of a parallelogram differ by 20, find the degree measure of the smaller angle.

16. In parallelogram *ABCD* a line from *B* is drawn perpendicular to *AD*, intersecting *AD* at point *H*. If m ∠*D* = 119, what is the measure of ∠*ABH*?

17. Which of the following quadrilaterals is *not* a parallelogram?
(1) A rhombus   (3) A rectangle
(2) A square   (4) A trapezoid

18. In which of the following quadrilaterals are the sides in a pair of adjacent sides always congruent?
(1) A parallelogram   (3) A rectangle
(2) A rhombus   (4) A trapezoid

19. In the accompanying diagram, *E* is a point on *DC* such that *AE* bisects ∠*DAB*. If m ∠*EAB* = 30, m ∠*ABC* = 50, and m ∠*BCD* = 130, what is the m ∠*ADC*?

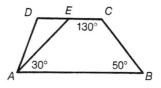

(1) 30   (3) 120
(2) 60   (4) 150

Each of the questions in this section has appeared on a previous Course I Regents Examination. Here is an opportunity for you to review Chapter 9 and, at the same time, prepare for the Course I Regents Examination.

1. The measures of the angles of a triangle are represented by x, $2x$, and $(x + 20)$. Find the number of degrees in the measure of the *smallest* angle of the triangle.

2. Two angles are supplementary and congruent. How many degrees are in the measure of each angle?

3. In the accompanying diagram $\overleftrightarrow{AB} \parallel \overleftrightarrow{CD}$ and $\overleftrightarrow{EF}$ intersects $\overleftrightarrow{AB}$ at G and $\overleftrightarrow{CD}$ at H.
If the degree measure of AGH is $(3x - 10)$ and the degree measure of $\angle GHD$ is 80, find the value of x.

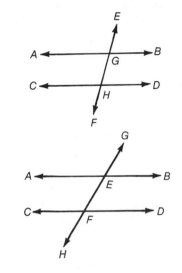

4. In the accompanying diagram, parallel lines $\overleftrightarrow{AB}$ and $\overleftrightarrow{CD}$ are intersected by transvesal $\overleftrightarrow{GH}$ at E and F, respectively. The degree measure of $\angle AEG$ is $(6x + 10)$ and of $\angle CFE$ is 130. Find x.

5. As shown in the accompanying diagram, $\overleftrightarrow{AB}$ and $\overleftrightarrow{CD}$ intersect at point E. If the degree measures of vertical angles AED and CEB are represented by $(3x + 20)$ and $(8x - 5)$, find the value of x.

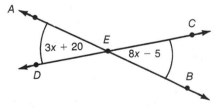

6. The measures of two complementary angles are in the ratio of 2:3. Find the measure of the *larger* angle.

7. In the accompanying diagram, m $\angle A = 70$ and m $\angle B = 30$. Find the measure of exterior angle BCD.

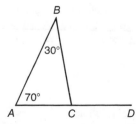

8. Two angles of a right triangle are congruent. What is the number of degrees in the measure of each of these angles?

9. The measures of two supplementary angles are in the ratio of 5 : 1. What is the measure of the *smaller* angle?

10. In $\triangle ABC$, the ratio of the measure of $\angle A$ to the measure of $\angle B$ is 3 : 5. The measure of $\angle C$ is 20 more than the sum of the measures of $\angle A$ and $\angle B$. What is the measure of each angle in $\triangle ABC$?

11. Given the true statement "If a triangle is equilateral, then it is isosceles," which statement must also be true?
(1) If a triangle is not equilateral, then it is not isosceles.
(2) If a triangle is not equilateral, then it is isosceles.
(3) If a triangle is not isosceles, then it is not equilateral.
(4) If a triangle is isosceles, then it is equilateral.

12. Let p represent "The polygon has exactly three sides," and let q represent "All angles of the polygon are right angles." What is true if the polygon is a rectangle?
(1) $p \wedge q$ (2) $p \vee q$ (3) p (4) $\sim q$

13. In the accompanying diagram, parallel lines $\overleftrightarrow{AB}$ and $\overleftrightarrow{CD}$ are intersected by transversal GH at points E and F, respectively. If m $\angle AEG$ is $(3x + 7)$ and m $\angle CFE$ is $(4x - 2)$ find x.

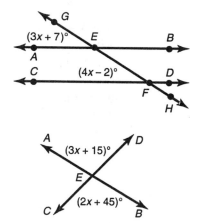

14. In the accompanying diagram, $\overleftrightarrow{AB}$ and $\overleftrightarrow{CD}$ intersect at E, and m $\angle AED = 3x + 15$. If m $\angle CEB = 2x + 45$, find the value of x.

15. In $\triangle ABC$, angle B in congruent to angle C. The measure of angle B is 20 more than twice the measure of angle A. Find the measure of *each* angle in $\triangle ABC$.

16. In $\triangle ABC$, the measure of angle B is 13 more than the measure of angle A, and the measure of angle C is 9 less than twice the measure of angle A. Find the measure of each angle in $\triangle ABC$.

Section 9.1
1. (a) *BA, BE, BC,* and *BD*
   (b) *EBD* and *DBE*
   (c) ∠*ABE,* ∠*CBE,* ∠*DBC,* and ∠*ABD*
   (d) *BA* and *BC; BD* and *BE*
3. (a) ∠*AEF* (b) ∠*BCD* (c) ∠*AFE*
5. $\overline{AF}, \overline{HF},$ and $\overline{EF}$

Section 9.2
1. (a) acute, 25   3. (a) 8   (b) 1
   (b) straight, 180   5. 10
   (c) right, 90   7. 126
   (d) acute, 65   9. (a) 3   (b) 20
   (e) right, 90   11. m ∠*STP* = m ∠*OTP*
   (f) obtuse, 155

Section 9.3
1. 32   7. 60
3. 7.8   9. *x* = 8, equilateral
5. $7x^2 - 2x$

Section 9.4
1. Adjacent angles: 1 and 4, 3 and 4, 2 and 3, 7 and 8, 5 and 6.
   Supplementary angles: 5 and 6, 7 and 8.
   Vertical angles: 1 and 2, 9 and 10.
3. (a) 20   (b) 67.5
5. 136   7. 62   9. 56   11. 79   13. 143

Section 9.5
1. (a) 78   11. (a) *x* = 122, *y* = 68
   (b) 127   (b) *x* = 52, *y* = 128
   (c) 64   (c) *x* = 104, *y* = 104
   (d) 27   (d) *x* = 127, *y* = 37
   (e) 13   (e) *x* = 48, *y* = 80
   (f) 22   (f) *x* = 45, *y* = 117
   (g) 58   13. *x* = 30
3. 36   15. parallel
5. 90   17. not parallel
7. 55   19. (1)
9. 16   21. parallel

Section 9.6

1. (a) True
   (b) False
   (c) True
   (d) True
   (e) True
3. 24

5. 54
7. (2)
9. $x = 22$, obtuse
11. 36
13. 37
15. 118

17. 52
19. 69
21. 27
23. 47
25. 54
27. 130

Section 9.7

1. 101
3. 53
5. 35
7. $x = 32, y = 38$
9. 35

11. 35, 145, 35, 145
13. 6
15. 80
17. (4)
19. (3)

Regents Tune-Up: Chapter 9

1. 40
3. 30
5. 5

7. 100
9. 30
11. (3)

13. 9
15. 28, 76, 76

CHAPTER 10

COMPARING AND MEASURING GEOMETRIC FIGURES

10.1 PROVING TRIANGLES ARE CONGRUENT

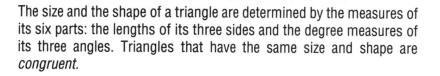

KEY IDEAS

The size and the shape of a triangle are determined by the measures of its six parts: the lengths of its three sides and the degree measures of its three angles. Triangles that have the same size and shape are *congruent*.

Congruent Triangles

In Figure 10.1, if it were possible to "slide" $\triangle ABC$ to the right, we could determine whether the two triangles can be made to coincide. If the triangles can be made to coincide, then $\triangle ABC$ is congruent to $\triangle RST$.

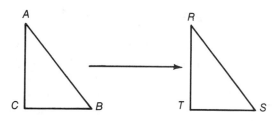

Figure 10.1   Testing Whether Triangles Coincide

Observe that, when "sliding" $\triangle ABC$, we would try to match vertex A with vertex R, vertex B with vertex S, and vertex C with vertex T. Vertices that are paired off in this way are called *corresponding* vertices. Corresponding vertices of the two triangles determine corresponding angles. Corresponding sides of the two triangles lie opposite corresponding angles.

Corresponding Angles	Corresponding Sides
$\angle A$ and $\angle R$	$\overline{BC}$ and $\overline{ST}$
$\angle B$ and $\angle S$	$\overline{AC}$ and $\overline{RT}$
$\angle C$ and $\angle T$	$\overline{AB}$ and $\overline{RS}$

Congruent triangles are triangles whose vertices can be paired off so that corresponding angles are congruent *and* corresponding sides are congruent. In naming pairs of congruent triangles, corresponding vertices are written in the same order:

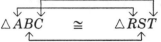

$$\triangle ABC \quad \cong \quad \triangle RST$$

Here are some other ways in which this correspondence may be written:

$$\triangle CAB \cong \triangle TRS \qquad \triangle BAC \cong \triangle SRT \qquad \triangle CBA \cong \triangle TSR.$$

Included Angles and Sides

In Figure 10.2, $\angle A$ is said to be *included* by sides $\overline{AB}$ and $\overline{AC}$ since $\angle A$ is formed by the intersection of these segments at their common endpoint, point A. Side $\overline{BC}$ is said to be included by $\angle B$ and $\angle C$ since it is a side of both angles.

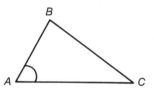

Figure 10.2 Included Angle and Side

Proving That Triangles Are Congruent

Two *polygons* are congruent if and only if their corresponding angles are congruent *and* their corresponding sides are congruent. If the polygons are triangles, then it is sufficient to show that only *three* pairs of parts are congruent, provided that they are a particular set of congruent parts.

You may conclude that two triangles are congruent if any one of the following conditions is true:

1. Three sides of one triangle are congruent to their corresponding sides in the other triangle. This is called the **SSS** (side-side-side) method.

2. Two sides and the included angle of one triangle are congruent to the corresponding parts of the other triangle. This is called the **SAS** (side-angle-side) method.

3. Two angles and the included side of one triangle are congruent to the corresponding parts of the other triangle. This is called the **ASA** (angle-side-angle) method.

4. Two angles and the not-included side of one triangle are congruent to the corresponding parts of the other triangle. This is called the **AAS** (angle-angle-side) method.

Example

Using the parts marked off in the accompanying figures as being congruent, determine in each case whether △I is congruent to △II. Give a reason for your answer.

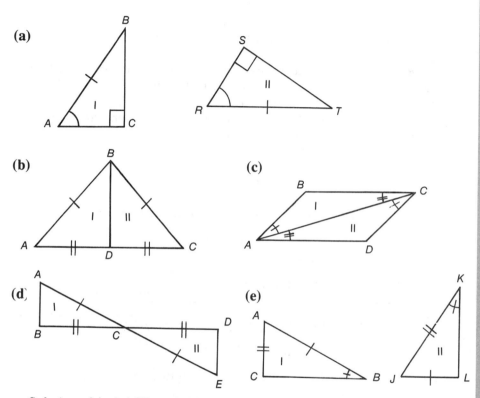

(a)

(b)

(c)

(d)

(e)

Solution: **(a)** △***ACB*** ≅ △***RST*** by applying the *AAS* method. Note that ∠*C* and ∠*S* are congruent since all right angles are congruent.

(b) △***ADB*** ≅ △***CDB***. Since $\overline{BD}$ is congruent to itself, the *SSS* method may be applied.

(c) △***ABC*** ≅ △***CDA***. Since $\overline{AC}$ is congruent to itself, the *ASA* method may be used.

(d) △***ABC*** ≅ △***EDC***. Since vertical angles are congruent, ∠*ACB* and ∠*ECD* are congruent. These angles are included by congruent pairs of corresponding sides, so that the *SAS* method may be used.

(e) △***ABC*** is *not* **congruent to** △***JKL***. The pairs of angles that are congruent are *not* included by the congruent pairs of corresponding sides.

Applying Congruent Triangles to Parallelograms

In Figure 10.3, diagonal $\overline{BD}$ divides parallelogram $ABCD$ into two congruent triangles. Since these parts are congruent:

$\angle 1 \cong \angle 2$ (Angle)
$\overline{BD} \cong \overline{BD}$ (Side)
$\angle 3 \cong \angle 4$ (Angle)

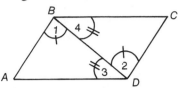

Figure 10.3   Parallelogram Divided into Two Congruent Triangles

Angles 1 and 2 are congruent since they are alternate interior angles formed by parallel lines. Angles 3 and 4 are congruent for the same reason. Triangle *BAD* is congruent to △*DCB* as proved by the ASA method. Since corresponding parts of congruent triangles are congruent, $\overline{AB} \cong \overline{BC}$ and $\overline{AB} \cong \overline{CD}$. This establishes the following important result:

MATH FACTS

PARALLELOGRAM SIDE RELATIONSHIP

Opposite sides of a parallelogram are congruent, and have, therefore, the same length.

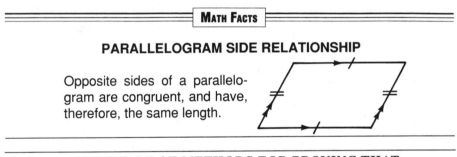

SUMMARY OF METHODS FOR PROVING THAT TRIANGLES ARE CONGRUENT

You may conclude that two triangles are congruent if it can be shown that:

1. Three sides of one triangle are congruent to the corresponding parts of the other triangle.

2. Two sides and the included angle of one triangle are congruent to the corresponding parts of the other triangle.

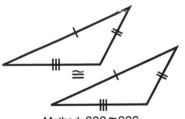

Method: SSS ≅ SSS

Method: SAS ≅ SAS

242

3. Two angles and the included side of one triangle are congruent to the corresponding parts of the other triangle.

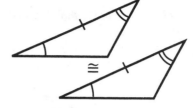

Method: ASA ≅ ASA

4. Two angles and the side opposite one of them are congruent to the corresponding parts of the other triangle.

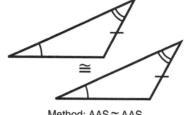

Method: AAS ≅ AAS

You may *not* conclude that two triangles are congruent when:

1. Two sides and an angle that is *not* included of one triangle are congruent to the corresponding parts of the other triangle.

SSA ≅ SSA

2. Three angles of one triangle are congruent to the corresponding parts of the other triangle.

AAA ≇ AAA

Exercise Set 10.1

1. In parallelogram $ABCD$, $BC = 5x - 17$ and $AD = 63$. What is the value of x?

2. In parallelogram $ABCD$, $AB = 4x - 17$ and $CD = 2x + 13$. What are the lengths of $\overline{AB}$ and $\overline{CD}$?

3. Which of the following is *not* a valid method for proving triangles are congruent?
   (1) ASA ≅ ASA   (3) SSA ≅ SSA
   (2) SAS ≅ SAS   (4) AAS ≅ AAS

4–11. Using the parts marked off as being congruent in the accompanying figures, determine whether △I is congruent to △II. In each case, give a reason for your answer.

4.

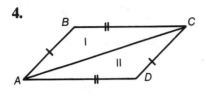

8.

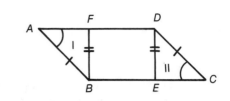

5.

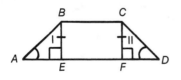

9.

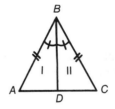

6.

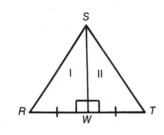

10.

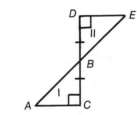

7.

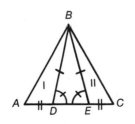

11.

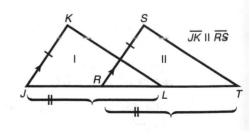

$\overline{JK} \parallel \overline{RS}$

10.2 MEASURING ANGLES OF AN ISOSCELES TRIANGLE

═══════════ △ **KEY IDEAS** △ ═══════════

In an isosceles triangle, the congruent sides are called the **legs**. The angles that are opposite the legs are called the **base angles**, and the side that they include is called the **base**. The angle opposite the base is called the **vertex angle**.

In an isosceles triangle the base angles have the same degree measure and are, therefore, congruent.

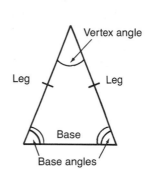

Proving That Base Angles Are Congruent

If a special line segment such as the bisector of the vertex angle of an isosceles triangle is drawn, the isosceles triangle can be "folded" along this line in such a way that the base angles exactly coincide.

This relationship may be proved using the properties of congruent triangles. In Figure 10.4, the bisector of the vertex angle is drawn, so that $\angle ABC$ is divided into two congruent angles. Comparing $\triangle ABD$ and $\triangle CBD$, we see that the following parts are congruent:

$\overline{AB} \cong \overline{BC}$ (Side)
$\angle 1 \cong \angle 2$ (Angle)
$\overline{BD} \cong \overline{BD}$ (Side)

Figure 10.4 Proving Base Angles Are Congruent

Triangles ABD and CBD are congruent as proved by the SAS method. Since the triangles are congruent, every pair of corresponding parts must be congruent. Therefore $\angle A \cong \angle C$. We have informally proved the following important relationship:

245

MATH FACTS

BASE ANGLES THEOREM

If two sides of a triangle are congruent, then the angles opposite these sides are congruent and, as a result, have the same degree measure.

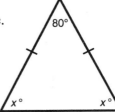

Examples

1. If the measure of the vertex angle of an isosceles triangle is 80, what is the degree measure of each base angle?

Solution: Let x = measure of each base angle.

$$x + x + 80 = 180$$
$$2x = 180 - 80$$
$$x = \frac{100}{2}$$
$$x = \mathbf{50}$$

2. If the measure of an exterior angle formed by extending the base of an isosceles triangle is 112, what is the degree measure of the vertex angle?

Solution: $m\angle A = 180 - 112 = 68$
Therefore $m\angle B$ must also equal 68

$$68 + 68 + m\angle C = 180$$
$$136 + m\angle C = 180$$
$$m\angle C = 180 - 136$$
$$m\angle C = \mathbf{44}$$

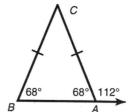

3. The degree measure of a base angle of an isosceles triangle exceeds twice the degree measure of the vertex angle by 15. Find the measure of the vertex angle.

Solution: Let x = measure of the vertex angle.
Then $2x + 15$ = measure of a base angle.

$$(2x + 15) + (2x + 15) + x = 180$$
$$5x + 30 = 180$$
$$5x = 180 - 30$$
$$x = \frac{150}{5}$$
$$x = \mathbf{30}$$

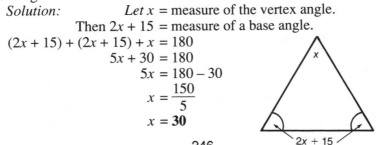

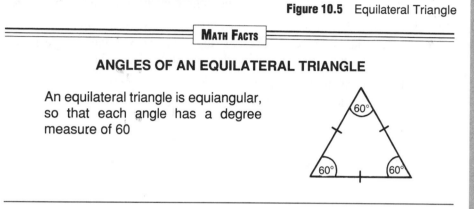

Equilateral Triangles

Since an equilateral triangle (Figure 10.5) has three equal sides, it is also an isosceles triangle, in which any pair of angles may be considered the base angles. Each angle of the triangle must, therefore, have the same measure.

Figure 10.5   Equilateral Triangle

═══════════════ **MATH FACTS** ═══════════════

ANGLES OF AN EQUILATERAL TRIANGLE

An equilateral triangle is equiangular, so that each angle has a degree measure of 60

Exercise Set 10.2

1. Find the measure of the vertex angle of an isosceles triangle if the measure of a base angle is:
   (a) 30   **(b)** 74   **(c)** 60   **(d)** x

2. Find the measure of each base angle of an isosceles triangle if the measure of the vertex angle is:
   (a) 50   **(b)** 90   **(c)** 100   **(d)** x

3. Find the measure of each angle of an isosceles triangle if the ratio of the measure of a base angle to the measure of the vertex angle is:
   (a) 2 : 1   **(b)** 3 : 2   **(c)** 1 : 4   **(d)** 2 : 5

4. The degree measure of a base angle of an isosceles triangle exceeds three times the degree measure of the vertex angle by 13. Find the measure of the vertex angle.

5. The measure of an exterior angle at the base of an isosceles triangle is 108. What is the measure of the vertex angle of the triangle?

6. The measure of the vertex angle of an isosceles triangle is 96. What is the measure of an exterior angle at the base of the triangle?

7. The degree measure of the vertex angle of an isosceles triangle is 5 less than three times the degree measure of a base angle. What is the degree measure of each base angle?

8. In the accompanying figure, *ABCD* is a parallelogram. If *EB* = *AB* and m∠*CBE* = 57, what is the value of *x*?

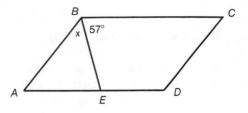

9. If the degree measure of an exterior angle formed at the vertex of an isosceles triangle is 118, what is the degree measure of each base angle?

10. The degree measure of an exterior angle formed at the vertex of an isosceles triangle exceeds one-half the measure of a base angle by 87. What is the degree measure of each base angle of the triangle?

11. In the accompanying diagram, *ABC* is an *isosceles right triangle* with *AC* = *BC*.
(**a**) What are m∠*A* and m∠*B*?
(**b**) If *AC* = 5, find *AB*.
  (*Hint:* Use the Pythagorean theorem.)
(**c**) If *AB* = 8, find *BC*.

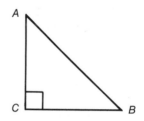

12. The length of a leg of an isosceles right triangle is 10. Find the length of the altitude drawn to the hypotenuse of the right triangle.

13–20. In each of the following, find the value of x:

13.

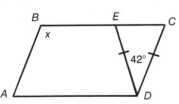

ABCD is a parallelogram.

14.

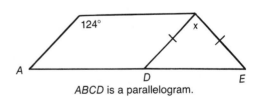

ABCD is a parallelogram.

15.

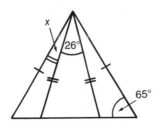

16.

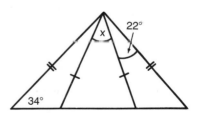

248

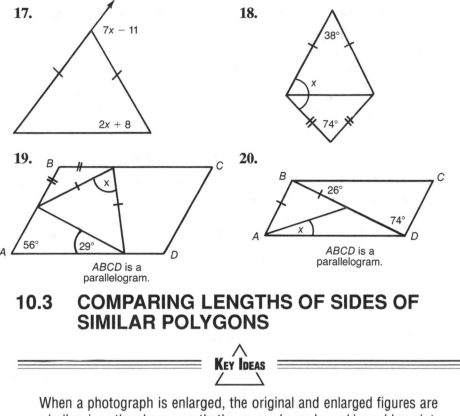

17. 7x − 11

2x + 8

18. 38°

x

74°

19. B

x

A 56° 29° D

ABCD is a
parallelogram.

20. B

26°

74°

A x D

ABCD is a
parallelogram.

10.3 COMPARING LENGTHS OF SIDES OF SIMILAR POLYGONS

△ **KEY IDEAS** △

When a photograph is enlarged, the original and enlarged figures are *similar* since they have exactly the same shape. In making a blueprint, every object must be drawn to scale so that the figures in the blueprint are in proportion and are similar to their real-life counterparts.

Similar Polygons

Congruent polygons have the same shape *and* the same size, while similar figures have the same shape, but may differ in size. The triangles in Figure 10.6 are similar.

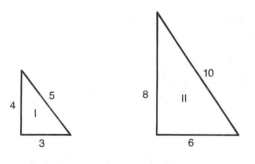

Figure 10.6 Similar Triangles

Notice that the lengths of corresponding sides of triangles I and II have the same ratio and are, therefore, in proportion:

$$\frac{\text{Side in } \triangle \text{ I}}{\text{Side in } \triangle \text{ II}} = \frac{3}{6} = \frac{4}{8} = \frac{5}{10} = \frac{1}{2} \text{ or } 1 : 2.$$

Two triangles (or any other two polygons having the same number of sides) are **similar** if both of the following conditions are met:

1. Corresponding angles have the same degree measure.

2. The lengths of corresponding sides are in proportion.

The symbol for similarity is ~. The notation $\triangle ABC \sim \triangle RST$ is read as "triangle *ABC* is *similar to* triangle *RST*."

Also observe that the perimeters of two similar triangles have the same ratio as the lengths of a pair of corresponding sides:

$$\frac{\text{Perimeter of } \triangle \text{I}}{\text{Perimeter of } \triangle \text{II}} = \frac{3+4+5}{6+8+10} = \frac{1}{2} = \text{or } 1 : 2.$$

MATH FACTS

COMPARING PERIMETERS OF SIMILAR TRIANGLES

The perimeters of two similar triangles have the same ratio as the lengths of any pair of corresponding sides.

Examples

1. Quadrilateral *ABCD* is similar to quadrilateral *RSTW*. The lengths of the side of quadrilateral *ABCD* are 6, 9, 12, and 18. If the length of the longest side of quadrilateral *RSTW* is 24, what is the length of its shortest side?

Solution: Let x = length of the shortest side of quadrilateral *RSTW*. Since the lengths of corresponding sides of similar polygons must have the same ratio, the following proportion is true:

$$\frac{\text{Shortest side of quad } ABCD}{\text{Shortest side of quad } RSTW} = \frac{\text{Longest side of quad } ABCD}{\text{Longest side of quad } RSTW}.$$

$$\frac{6}{x} = \frac{18}{24}$$

Write $\frac{18}{24}$ in lowest terms: $\quad \dfrac{6}{x} = \dfrac{3}{4}$

Cross-multiply:

$$3x = 24$$

$$x = \frac{24}{3} = 8$$

The length of the shortest side of quadrilateral *RSTW* is **8**.

2. The lengths of the three sides of a triangle are 6, 8, and 10. If the perimeter of a similar triangle is 72, what is the length of the *shortest* side of the second triangle?

Solution: Since the lengths of the three sides of the first triangle are 6, 8, and 10, the perimeter of the first triangle is $6 + 8 + 10$ or 24. The perimeter of the second triangle is given as 72.

Let $x =$ the length of the shortest side of the second triangle.

$$\frac{\text{perimeter of first triangle}}{\text{perimeter of second triangle}} = \frac{\text{side of first triangle}}{\text{side of second triangle}}$$

$$\frac{24}{72} = \frac{6}{x}$$

Write the fraction on the left side of the equation in lowest terms:

$$\frac{1}{3} = \frac{6}{x}$$

Cross-multiply:

$$1 \cdot x = 3 \cdot 6$$

$$x = 18$$

The length of the shortest side of the second triangle is **18**.

Proving That Triangles Are Similar

To prove that two *polygons* are similar, it is necessary to demonstrate that *all* pairs of corresponding angles have the same degree measure *and* the lengths of all pairs of corresponding sides have the same ratio. If two *triangles* are drawn having two pairs of congruent angles, then:

1. The remaining pair of angles is also congruent since the measures of the three angles of a triangle must always add up to 180.

2. The ratios of the lengths of the corresponding sides of the triangles are determined so that they are in proportion. Thus the two triangles are similar.

This suggests the following shortcut method for proving that two triangles are similar:

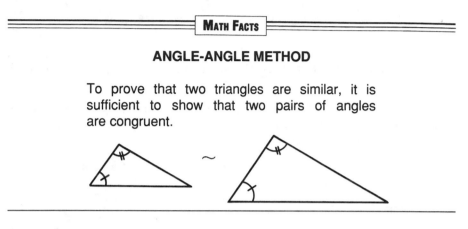

MATH FACTS

ANGLE-ANGLE METHOD

To prove that two triangles are similar, it is sufficient to show that two pairs of angles are congruent.

Examples

3. In the accompanying figure, $\overline{DE} \parallel \overline{AB}$.

(a) Is $\triangle DEC \sim \triangle ABC$? Give a reason for your answer.

(b) If $CD = 6$, $CA = 18$, and $DE = 4$, what is the length of $\overline{AB}$?

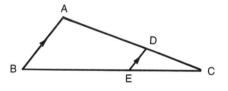

Solution: **(a)** $\overline{DE} \parallel \overline{AB}$, so transversal $\overline{AC}$ forms congruent corresponding angles, making m$\angle$A = m$\angle$CDE. Also, transversal $\overline{BC}$ forms congruent corresponding angles, making m$\angle B$ = m$\angle$CED. **Triangle $DEC \sim \triangle ABC$** since two angles of $\triangle DEC$ are equal in degree measure to two angles of $\triangle ABC$.

(b) Let x = length of $\overline{AB}$.

Since the lengths of corresponding sides of similar triangles are in proportion,

$$\frac{CD}{CA} = \frac{DE}{AB}$$
$$\frac{6}{18} = \frac{4}{x}$$
$$\frac{1}{3} = \frac{4}{x}$$
$$x = 3 \cdot 4 = 12$$

The length of $\overline{AB}$ is **12**.

4. A pole 10 feet high casts a 15-foot-long shadow on level ground. At the same time a man casts a shadow that is 9 feet in length. How tall is the man?

Solution: By assuming that the shadows are perpendicular to the pole and the man, we may use right triangles to represent these situations, where x represents the height of the man.

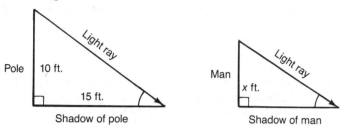

We also assume that in each triangle the light rays make angles with the ground that have the same degree measure. Since all right angles have the same degree measure, the two right triangles are similar and the lengths of their sides are in proportion.

$$\frac{\text{Height of pole}}{\text{Height of man}} = \frac{\text{Shadow of pole}}{\text{Shadow of man}}$$

$$\frac{10}{x} = \frac{15}{9}$$

$$\frac{10}{x} = \frac{5}{3}$$

$$5x = 3 \cdot 10$$

$$x = \frac{30}{5} = 6$$

The man is **6 feet** tall.

Exercise Set 10.3

1. The lengths of the corresponding sides of two similar polygons are 3 and 4, respectively. If the perimeter of the smaller polygon is 18, find the perimeter of the larger polygon.

2. Triangle ABC is similar to $\triangle XYZ$, $AB : XY = 5 : 3$, and $BC = 20$. What is the length of $\overline{YZ}$?

3. A person 6 feet tall casts a shadow 4 feet long. At the same time, a nearby tower casts a shadow 32 feet long. How many feet are in the height of the tower?

4. The lengths of three sides of a triangle are 10, 24, and 26.
(a) If the perimeter of a similar triangle is 48, what is the length of the shortest side of the similar triangle?
(b) If the length of the longest side of a similar triangle is 39, find the perimeter of this triangle.

5. The lengths of the three sides of a triangle are 8, 15, and 17.
(a) If the length of the longest side of a similar triangle is 51, find the perimeter of this triangle.
(b) If the perimeter of a similar triangle is 30, find the length of the shortest side of the similar triangle.

6. A vertical antenna 75 meters tall casts a shadow 35 meters long. At the same time, a flagpole nearby casts a shadow 14 meters long. What is the number of meters in the height of the flagpole?

7. The lengths of the sides of a triangle are 5, 6, and 9. If the length of the *shortest* side of a similar triangle is 15, find the perimeter of the larger triangle.

8. In the accompanying diagram, $\angle A \cong \angle D$ and $\angle B \cong \angle E$. If $AB = 6$, $DE = 8$, and $DF = 12$, find AC.

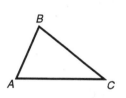
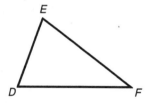

9. In the accompanying diagram, $\overline{DE} \parallel \overline{AB}$.
(a) Explain why $\triangle DCE \sim \triangle BCA$.
(b) If $DE = 6$, $AB = 10$, and $BC = 15$, find DC.
(c) If $CE = 8$, $EA = 6$, and $BC = 21$, find BD.

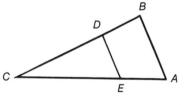

10. In the accompanying diagram, $\overline{AB} \perp \overline{BC}$ and $\overline{DC} \perp \overline{BC}$.
(a) Explain why $\triangle ABE \sim \triangle DCE$.
(b) If $CD = 9$, $AB = 6$, and $DE = 12$, find AE. $AB = 4$.
(c) If $BC = 20$ and $CD = 12$, find BE.

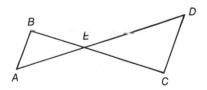

11. Given that $\triangle ABC \sim \triangle RST$, $\overline{BH}$ $\perp \overline{AC}$, and $\overline{SK} \perp \overline{RT}$.

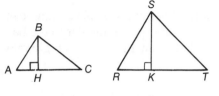

(a) Explain why $\triangle ABH \sim$ $\triangle RSK$.

(b) If $AB : RS = 4 : 9$ and $SK =$ 18, find BH.

12. In two similar triangles the ratio of the lengths of a pair of corresponding sides is 5 : 8. If the perimeter of the larger triangle exceeds the perimeter of the smaller triangle by 12, find the perimeter of the larger triangle.

10.4 FINDING AREAS OF A RECTANGLE, SQUARE, AND PARALLELOGRAM

△ KEY IDEAS

In the accompanying figure, a rectangle encloses a total of 36 square boxes, so that its area is 36 square units. When we speak of the *area* of a figure, we simply mean the total number of square boxes that the figure can enclose. If a side of each little square box measures 1 centimeter (cm), then the area of the rectangle is 36 cm^2, where cm^2 is read as "square centimeters."

Area = 36

Area of a Rectangle and Square

The formulas for the areas of a rectangle and a square should be familiar.

=== MATH FACTS ===

AREA FORMULAS

- The area of a rectangle is equal to the product of the length and the width.

$$w \quad \boxed{A = lw}$$
$$l$$

- The area of a square is equal to the product of two sides.

$$s \quad \boxed{A = s^2}$$
$$s$$

Sometimes one side of a rectangle is referred to as the **base**, and an adjacent side (which is perpendicular to the base) is called the **altitude** or **height**. The formula for the area of a rectangle may be expressed alternatively as

$$A = bh,$$

where b represents the length of the base and h represents the length of the height or altitude. The terms *base* and *altitude* are also used in connection with the area formulas for other types of geometric figures.

Examples

1. Find the area of a rectangle whose base has a length of 8 and whose diagonal has a length of 10.

Solution: Drawing diagonal $\overline{AC}$ of rectangle $ABCD$ forms right triangle ABC, in which the diagonal is the hypotenuse. The lengths of the sides of this right triangle form a 6-8-10 Pythagorean triple, where height $AB = 6$.

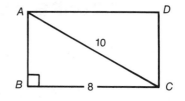

Area of rectangle $= bh = 8 \times 6 = \mathbf{48}$.

2. If the length of a side of a square is doubled, what change occurs in each of the following?
(a) The perimeter of the square   **(b)** The area of the square

Solution: In the formulas for perimeter and area, replace s by $2s$.
(a) Perimeter of original square $= 4s$
      Perimeter of new square $= 4(2s)$ or $2(4s)$
The perimeter of the square is **doubled**.
(b) Area of original square $= s^2$
      Area of new square $= (2s)^2 = 4(s^2)$
The area of the square is multiplied by **4**.

3. Find the length of a side of a square that has the same area as a rectangle whose base is 27 and whose height is 3.

Solution: Area of rectangle $= bh = (27)(3) = 81$
          Area of square $= s^2 = 81$.
Therefore the length of each side of the square is **9**.

4. A certain rectangle and square have the same area. The base of the rectangle exceeds three times its altitude by 4. If the length of a side of the square is 8, find the dimensions of the rectangle.

Solution: Let $x =$ length of altitude of rectangle.
          Then $3x + 4 =$ length of base of rectangle.
     Area of rectangle $=$ Area of square
       Base $\times$ Altitude $=$ (Side)^2
            $(3x + 4)x = 8^2$

$$3x^2 + 4x = 64$$
$$3x^2 + 4x - 64 = 0$$
$$(3x + 16)(x - 4) = 0$$
$$(3x + 16) = 0 \text{ or } (x - 4) = 0$$

$3x = -16$	$x = 4$
$x = \dfrac{-16}{3}$	Altitude $= x = 4$
Reject	Base $= 3(4) + 4 = 16$

Check: The area of the square is 64 since $8^2 = 64$. The rectangle has the same area since $4 \times 16 = 64$.

Area of a Parallelogram

Any side of a parallelogram may be considered the base. The height or altitude of the parallelogram is the length of a segment ($\overline{BH}$ in Figure 10.7) drawn perpendicular to the base from any point on the opposite side. The altitude is usually dropped from a vertex of the parallelogram. A parallelogram and a rectangle that have equal bases and altitudes have equal areas.

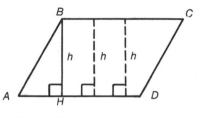

Figure 10.7 Parallelogram

MATH FACTS

AREA OF A PARALLELOGRAM

The area A of a parallelogram is equal to the product of the length of the base b and the length of the altitude h drawn to that base. Thus, $A = bh$.

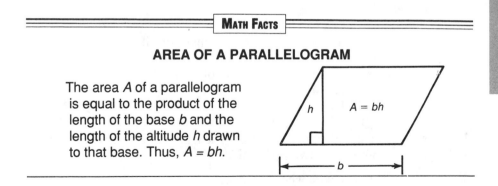

Examples

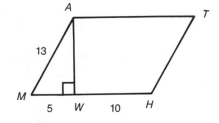

5. In the accompanying diagram, quadrilateral *MATH* is a parallelogram. If $MA = 13$, $MW = 5$, and $HW = 10$, what is the area of parallelogram *MATH*?

Solution: $\triangle MAW$ is a right triangle whose sides form a 5-12-13 Pythagorean triple, where $AW = 12$. Base $MH = MW + HW = 5 + 10 = 15$.

$$
\begin{aligned}
\text{Area of } \square MATH &= \text{Base} \times \text{Altitude} \\
&= MH \times AW \\
&= 15 \times 12 \\
&= \textbf{180 square units}
\end{aligned}
$$

6. The area of parallelogram *ABCD* is 40. $\overline{BH}$ is the altitude drawn to base $\overline{AD}$. If $BH = 5$, what is the length of $\overline{BC}$?

$$
\begin{aligned}
\textit{Solution:} \qquad \text{Area of } \square ABCD &= \text{Base} \times \text{Height} \\
40 &= AD \times BH \\
40 &= AD \times 5 \\
\frac{40}{5} &= AD \\
8 &= AD
\end{aligned}
$$

Since opposite sides of a parallelogram have the same length, $BC = AD = \textbf{8}$.

Exercise Set 10.4

1. If the area of a rectangle is 108, find its dimensions if:
(a) the length is three times the width.
(b) the width and the length have a ratio of 1 : 12.
(c) the length exceeds six times the width by 3.
(d) the sum of the length and the width is 21.

2. If the area of a parallelogram is 52 and the length of a side is 13, find the length of the altitude drawn to that side.

3. Find the area of a square whose perimeter is:
(a) 18   (b) $32x$ (c) $20x - 4$

4. Find the perimeter of a square whose area is: (a) 100   (b) $49x^2$

5. Fill in the following table, where A represents the area of a rectangle, b is the length of its base, h is the height, and d is the length of a diagonal:

	b	**h**	**d**	**A**
(a)	?	5	13	?
(b)	40	?	41	?
(c)	7	?	25	?
(d)	15	?	?	120

6. The base of a parallelogram is represented by $x + 5$, and the altitude drawn to that base is represented by $x - 5$. Express the area of the parallelogram as a binomial in terms of x.

7. If the length and the width of a rectangle are each doubled, what change occurs in each of the following?
   (a) The perimeter of the rectangle   (b) The area of the rectangle

8. If the perimeter of a square is tripled by increasing the length of each side by the same amount, what change occurs in the area of the square?

9. The length of a rectangle is 4 less than three times its width. If its area is 119, what are the dimensions of the rectangle?

10. If the length of a side of a square is doubled and the length of an adjacent side is diminshed by 3, a rectangle is formed whose area is 80. Find the original dimensions of the square.

11. In parallelogram *ABCD*, the base and the altitude drawn to that base are in the ratio of 5 : 3. If the area of the parallelogram is 60, find the lengths of the base and the altitude.

12. The length of a rectangle is three times its width. If the length of the rectangle is decreased by 7 and the width increased by 3, a square is formed whose area is 64 square units. Find the dimensions of the original rectangle.

13. An altitude is drawn from vertex *B* of rhombus *ABCD*, intersecting $\overline{AD}$ at point *H*. If *AB* = 10 and *AH* = 6, what is the area of rhombus *ABCD*?

14. The base of a parallelogram exceeds the length of the altitude drawn to that base by 5. If the area of the parallelogram is 84, find the length of the altitude.

15. The base of a rectangle exceeds three times its height by 1. If the height of the rectangle is represented by x, and the area is represented by $2x^2 + 12$, find the dimensions of the rectangle.

16. The length of a rectangle exceeds its width by 7. If the rectangle has the same area as a square whose perimeter is 48, find the dimensions of the rectangle.

17. The lengths of two adjacent sides of a parallelogram are 10 and 15. If the length of the altitude drawn to the shorter side of the parallelogram is 9, find the length of the altitude drawn to the longer side of the parallelogram.

18. Find the area of a square whose diagonal has a length of:
(a) 10   **(b)** 18   **(c)** $6\sqrt{2}$

(*Hint:* First find the length of a side of the square by using the Pythagorean theorem.)

10.5 FINDING AREAS OF A TRIANGLE AND TRAPEZOID

KEY IDEAS

The formula for the area of a triangle is also expressed in terms of the length of a base and the length of the altitude drawn to that base. Any side of a triangle may be the base. The altitude is the perpendicular dropped to the base from the opposite vertex of the triangle.

A trapezoid has *two* bases, which are the parallel sides of the trapezoid. The altitude is a segment drawn from any point on one base and perpendicular to the opposite side. The nonparallel sides are called *legs.*

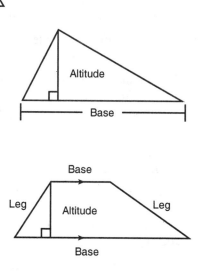

Area of a Triangle

A diagonal of a parallelogram divides the parallelogram into two congruent triangles (Figure 10.8). The area of each triangle is the same and is equal, therefore, to one-half the area of the entire parallelogram.

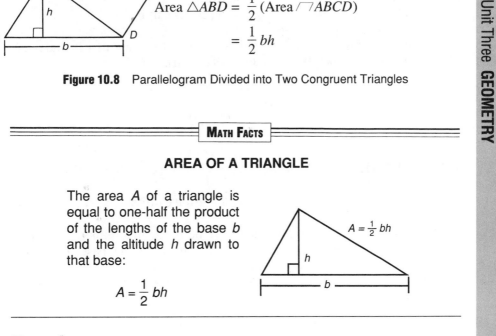

Area $\triangle ABD = \dfrac{1}{2}$ (Area $\square ABCD$)

$$= \dfrac{1}{2} bh$$

Figure 10.8   Parallelogram Divided into Two Congruent Triangles

===== **MATH FACTS** =====

AREA OF A TRIANGLE

The area A of a triangle is equal to one-half the product of the lengths of the base b and the altitude h drawn to that base:

$$A = \dfrac{1}{2} bh$$

$A = \dfrac{1}{2} bh$

Examples

1. Find the area of each of the following triangles:

(a)

(b)

(c)

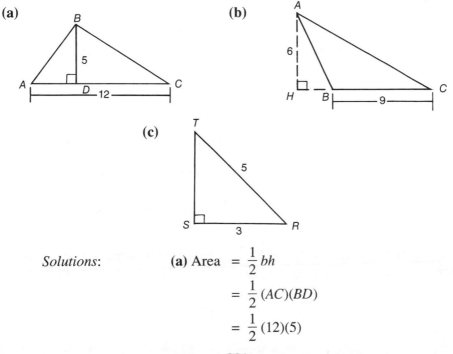

Solutions:	**(a)** Area	$= \dfrac{1}{2} bh$
		$= \dfrac{1}{2} (AC)(BD)$
		$= \dfrac{1}{2} (12)(5)$

261

$$= (6)(5)$$
$$= \textbf{30 square units}$$

(b) In an obtuse triangle, the altitude drawn to a side that includes the obtuse angle of the triangle will fall *outside* the triangle.

$$\text{Area} = \frac{1}{2} bh$$
$$= \frac{1}{2}(AH)(BC)$$
$$= \frac{1}{2}(6)(9)$$
$$= (3)(9)$$
$$= \textbf{27 square units}$$

(c) Since the legs of a right triangle are perpendicular, either leg may be considered the base, while the other leg is the altitude to that base. Therefore, the area of a *right* triangle is equal to one-half the product of the lengths of the legs of the right triangle. The lengths of the sides of right triangle RST form a 3-4-5 right triangle in which $ST = 4$.

$$\text{Area } \triangle RST = \frac{1}{2}(RS)(ST)$$
$$= \frac{1}{2}(3)(4)$$
$$= \frac{1}{2}(12)$$
$$= \textbf{6 square units}$$

2. If the lengths of the base and the altitude of a triangle are each doubled, what is the change in the area of the triangle?

Solution: In the formula for the area of a triangle, replace b by $2b$ and h by $2h$.

$$\text{Area of original } \triangle = \frac{1}{2} bh$$
$$\text{Area of new } \triangle = \frac{1}{2}(2b)(2h)$$
$$= \frac{1}{2}(4bh) \text{ or } 4\left(\frac{1}{2} bh\right)$$

The area of the new triangle is the area of the original triangle multiplied by **4**.

Areas of Similar Triangles

For any two triangles I and II (Figure 10.9) the ratio of their areas is as follows:

$$\frac{\text{Area } \Delta \text{ I}}{\text{Area } \Delta \text{ II}} = \frac{\frac{1}{2}(b_1)(h_1)}{\frac{1}{2}(b_2)(h_2)}$$

$$= \frac{b_1 h_1}{b_2 h_2}$$

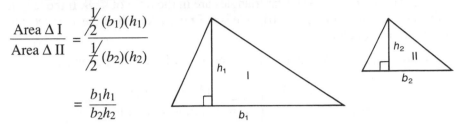

Figure 10.9   Areas of Similar Triangles

If $\triangle$I is similar to $\triangle$II, then the altitudes drawn to corresponding sides of these triangles are in proportion to the lengths of these sides. Hence we can replace the ratio of the altitudes, $\frac{h_1}{h_2}$, with the ratio of the bases.

$$\frac{\text{Area } \Delta \text{ I}}{\text{Area } \Delta \text{ II}} = \frac{b_1}{b_2} \cdot \frac{h_1}{h_2} = \frac{b_1}{b_2} \cdot \frac{b_1}{b_2} = \left(\frac{b_1}{b_2}\right)^2.$$

Since the triangles are similar, the ratio of the lengths of the bases can be replaced by the ratio of the lengths of any other pair of corresponding sides.

MATH FACTS

COMPARING AREAS

The ratio of the areas of two similar triangles is equal to the square of the ratio of the lengths of any pair of corresponding sides or altitudes.

$$\frac{\text{Area } \Delta_I}{\text{Area } \Delta_{II}} = \left(\frac{\text{Side I}}{\text{Side II}}\right)^2$$

provided that $\triangle$I ~ $\triangle$II.

Examples

3. If the length of the longest side of a triangle is 10 and the length of the longest side of a similar triangle is 20, how many times greater is the area of the larger triangle than the area of the smaller triangle?

Solution: The ratio of the area of the larger triangle to the area of the smaller triangle is equal to the square of the ratio of 20 to 10.

$$\frac{\text{Area of larger } \Delta}{\text{Area of smaller } \Delta} = \left(\frac{20}{10}\right)^2 = \left(\frac{2}{1}\right)^2 = \frac{4}{1}$$

The area of the larger triangle is **four** times as great as the area of the smaller triangle.

4. The areas of two similar triangles are in the ratio of $4 : 9$. If the length of the shortest side of a similar triangle is 6, find the length of the shortest side of the larger triangle.

Solution: Let x = length of the shortest side of the larger triangle.

$$\frac{\text{Area of smaller } \Delta}{\text{Area of larger } \Delta} = \left(\frac{\text{Side of smaller } \Delta}{\text{Corresponding side of larger } \Delta} \right)^2$$

$$\frac{4}{9} = \left(\frac{6}{x} \right)^2 = \frac{36}{x^2}$$

Cross-multiply: $\quad 4x^2 = 9 \cdot 36$

$$4x^2 = 324$$

$$x^2 = \frac{324}{4} = 81$$

$$x = \sqrt{81} = 9$$

The length of the shortest side of the larger triangle is **9**.

Area of a Trapezoid

To find a formula for the area of a trapezoid, add the areas of the two triangles that are formed when a diagonal of the trapezoid is drawn. In Figure 10.10, the altitude drawn from B to lower base AD (b_1) is labeled with the letter h. The altitude drawn from D to upper base BC extended (b_2) is labeled with the same letter, h. The two altitudes have the same length since they are drawn between two parallel lines and parallel lines always remain the same distance apart. We obtain the following formula:

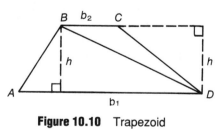

Figure 10.10 Trapezoid

$$\text{Area trap } ABCD = \text{Area } \triangle ABD + \text{Area } \triangle BCD$$

$$= \frac{1}{2} b_1 h + \frac{1}{2} b_2 h.$$

Since $\frac{1}{2}$ and h are factors of each monomial term, they can be factored out:

$$\text{Area trap } ABCD = \frac{1}{2} h(b_1 + b_2).$$

The formula states the following:

MATH FACTS

AREA OF A TRAPEZOID

The area A of a trapezoid is equal to the product of one-half the length of the altitude h and the sum of the lengths of the bases $b_1 + b_2$. Thus,

$$A = \frac{1}{2} h (b_1 + b_2).$$

Examples

5. Find the area of trapezoid *ABCD*.

(a) **(b)**

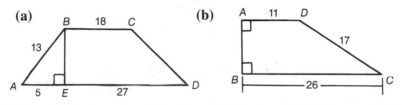

Solution: **(a)** The lengths of the sides of right triangle *AEB* form a 5-12-13 Pythagorean triple, where altitude $BE = 12$. The length of lower base $AD = AE + ED = 5 + 27 = 32$.

$$\begin{aligned}
\text{Area trap } ABCD &= \frac{1}{2} h(b_1 + b_2) \\
&= \frac{1}{2} (BE)(AD + BC) \\
&= \frac{1}{2} (12)(32 + 18) \\
&= 6(50) \\
&= \textbf{300 square units}
\end{aligned}$$

(b) Drawing altitude $\overline{DE}$ forms rectangle *ABED*. Since opposite sides of a rectangle have the same length, $EB = 11$; therefore $EC = 26 - 11$ or 15. The lengths of the sides of right triangle *DEC* form a 8-15-17 Pythagorean triple in which the length of altitude $DE = 8$.

$$\begin{aligned}
\text{Area trap } ABCD &= \frac{1}{2} h(b_1 + b_2) \\
&= \frac{1}{2} (DE)(BC + AD) \\
&= \frac{1}{2} (8)(26 + 11)
\end{aligned}$$

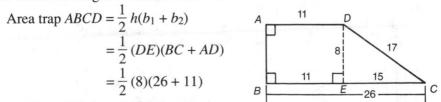

$$= 4(37)$$
$$= \textbf{148 square units}$$

6. In trapezoid *RTSW* the length of base $\overline{RW}$ is twice the length of base $\overline{ST}$. If the area of the trapezoid is 42 square inches and the length of the altitude is 4, find the length of the shorter base.

Solution: Let x = length of base $\overline{ST}$.
Then $2x$ = length of base $\overline{RW}$.
Area of trap $RTSW = \dfrac{1}{2}\, h\,(x + 2x)$

$$42 = \frac{1}{2}(4)(3x)$$

$$42 = 6x$$

$$\frac{42}{6} = \frac{6x}{6}$$

$$7 = x$$

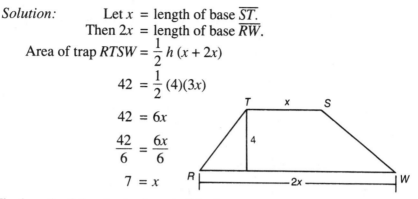

The length of the shorter base is **7 inches**.

SUMMARY OF AREA FORMULAS

Figure	Area Formula
Rectangle	$A = \textit{length} \times \textit{width}$ *or* $= \textit{base} \times \textit{height}$
Square	$A = (\textit{side})^2$
Parallelogram	$A = \textit{base} \times \textit{height}$
Triangle	$A = \dfrac{1}{2} \times \textit{base} \times \textit{height}$
Trapezoid	$A = \dfrac{1}{2} \times \textit{height} \times (\textit{base}_1 + \textit{base}_2)$

Exercise Set 10.5

1–8. In each of the following, find the area of either $\triangle ABC$ or trapezoid ABCD:

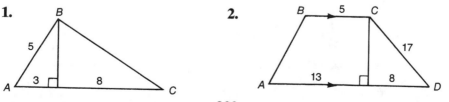

1.

2.

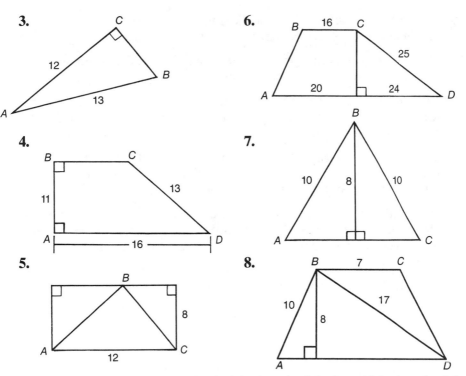

3.

6.

4.

7.

5.

8.

9. The area of a triangle is 28. Find the length of the base if the length of the altitude drawn to that base is 7.

10. The base and the altitude of a triangle are in the ratio of 1 : 2. If the area of the triangle is 36, find the length of the base.

11. Find the length of an altitude of a trapezoid if:
 (**a**) its area is 72, and the sum of the lengths of its bases is 36.
 (**b**) the ratio of the lengths of the altitude, the lower base, and the upper base is 1 : 1 : 5, and the area of the trapezoid is 147.
 (**c**) the sum of the lengths of the bases is numerically equal to one third of the area of the trapezoid.

12. The lengths of the bases of a trapezoid are in the ratio of 1 : 3. If the area of the trapezoid is 140 and the length of an altitude is 5, what is the length of the shorter base?

13. The length of the base of a triangle is 1 less than twice the length of an altitude drawn to it. If the area of the triangle is 33, what are the lengths of the base and the altitude?

14. Triangle $JKL \sim \triangle RST$. If $JL = 20$ and $RT = 15$, find the ratio of the areas of the two triangles.

15. The area of two similar triangles are 25 and 81. What is the ratio of the lengths of a pair of corresponding sides?

16. The area of a trapezoid is 50, and the length of an altitude is 4. If the length of one of the bases is 7, find the length of the other base.

17. The area of two similar triangles are 81 and 121. If the perimeter of the smaller triangle is 45, find the perimeter of the larger triangle.

18. The lengths of a pair of corresponding sides of a pair of similar triangles are in the ratio of 5 : 8. If the area of the smaller triangle is 75, find the area of the larger triangle.

19. The length of the longer base of a trapezoid exceeds twice the length of the shorter base by 3. If the length of an altitude is 9 and the area of the trapezoid is 81, find the length of the shorter base.

20. In a certain trapezoid the lengths of the shorter base and the altitude are the same. The length of the longer base is 11, and the area of the trapezoid is 63. What are the lengths of the shorter base and the altitude?

21. The sum of the lengths of the legs of a right triangle is 31. If the area of the triangle is 84, find the lengths of the legs and the hypotenuse of the triangle.

22. As shown in the accompanying diagram, $ABCD$ is a rectangle and a line segment drawn from B intersects $\overline{CD}$ at E.

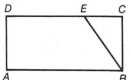

(a) If the measure of $\overline{AB}$ is 9 and the area of the rectangle is 36, find AD.
(b) Point E separates $\overline{DC}$ into two segments such that $DE : EC = 2 : 1$. Find DE and EC.
(c) Find BE.
(d) Find the area of trapezoid $ABED$.

10.6 FINDING CIRCUMFERENCE AND AREA OF A CIRCLE

$\bigwedge$ **KEY IDEAS** $\bigwedge$

If you were able to locate and connect all points that were a distance of 5 inches from a given point, a *circle* having a *radius* of 5 inches would be formed. The distance around the circle is called the *circumference* of the circle. The region enclosed by the circle represents the *area* of the circle.

The size of a circle depends on the length of its radius. The longer the radius, the greater the circumference and the area of the circle.

Some Parts of a Circle

A **circle** (Figure 10.11) is the set of all points at a given distance from a fixed point. The fixed point is called the **center** of the circle. The center of a circle is named by a single capital letter. If the center of a circle is named by the letter O, then the circle is referred to as *circle O*. Any segment drawn from the center of the circle to a point on the circle is called a **radius** (plural: *radii*) of the circle. All radii of the same circle have the same length. A **diameter** is a line segment that passes through the center of a circle and whose endpoints are points on the circle. The length of a diameter is twice the length of a radius.

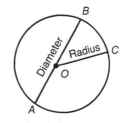

Figure 10.11 Circle

Any curved section of the circle is called an **arc** of the circle. A diameter divides a circle into two congruent arcs, each of which is called a **semicircle**. In Figure 10.11, AB (read as "arc AB") is a semicircle. Also, ACB (read as "arc ACB") is a semicircle.

Circumference of a Circle

When a circle's **circumference** (that is, the distance around the circle), is divided by its diameter, the number obtained is always the same regardless of the size of the circle. This constant value is represented by the Greek letter π and is read as "pi."

$$\frac{\text{Circumference}}{\text{Diameter}} = \frac{C}{D} = 3.1415926... = \pi \text{ (pi)}$$

Since the length of a diameter D is twice the length of the radius r, the circumference C may also be expressed in the following equivalent ways:

$$C = \pi D \quad \text{and} \quad C = 2\pi r.$$

The number p is an irrational number that takes the form of a never-ending, nonrepeating decimal number. When an estimate of the value of the circumference is needed, π may be replaced by a rational approximation, such as 3.14 or $\frac{22}{7}$. An exact value of the circumference is represented whenever the circumference is written in terms of π.

Examples

1. For a circle having a diameter of 5, find each of the following:
   (a) an approximation for the circumference of the circle
      (Use $\pi = 3.14$.)
   (b) the exact value of the circumference of the circle

Solution: (a) $C = \pi D = 3.14 \times 5 = \textbf{15.7}$
(b) $C = \pi D = \pi(5) = \textbf{5}\boldsymbol{\pi}$

2. Find an approximation for the length of a semicircle having 14 as its radius. (Use $\pi = \frac{22}{7}$.)

Solution: First find the circumference of the *entire* circle. The length of the semicircle will be one half of this amount.

$$C = 2\pi r = 2\pi(14) = 28\pi = 28 \times \frac{22}{7} = 88$$

$$\text{Circumference of semicircle} = \frac{1}{2}(88) = \textbf{44}$$

Area of a Circle

To find the area, A, of a circle, multiply π by the square of the radius, r:

$$A = \pi r^2.$$

Examples

3. Find the exact area of a circle whose radius is 6.

Solution: $A = \pi r^2 = \pi(6)^2 = \textbf{36}\boldsymbol{\pi}$

4. Express in terms of π the circumference of a circle whose area is 49π.

Solution:
$$A = \pi r^2$$
$$49\pi = \pi r^2$$
$$\frac{49\pi}{\pi} = \frac{\pi r^2}{\pi}$$
$$49 = r^2$$
$$r = \sqrt{49} = 7.$$
$$C = 2\pi r = 2\pi(7) = \textbf{14}\boldsymbol{\pi}$$

5. In the accompanying figure, $\overline{AB}$ is a semicircle of circle O. Find the area of the semicircle.

Solution: Side $\overline{AB}$ of $\triangle ABC$ is a diameter of the circle and of right $\triangle ABC$. The lengths of the sides of right $\triangle ABC$ form a 6-8-10 Pythagorean triple, in which $AB = 10$.

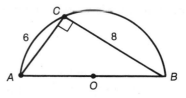

The radius of circle O is $5(10 \div 2 = 5)$. To find the area, A, of circle O use the formula

$$A = \pi r^2 = \pi(5^2) = 25\pi$$

Since the area of the *entire* circle is 25π, the area of the semicircle must be one half of 25π. The area of semicircle O is $\frac{25}{2}\pi$ or **12.5π**.

Comparing Circles

If the radius of a circle is represented by r_1 and the radius of a second circle is represented by r_2, then

$$\frac{C_1}{C_2} = \frac{2\pi r_1}{2\pi r_2} = \frac{r_1}{r_2} \quad and \quad \frac{A_1}{A_2} = \frac{\pi(r_1)^2}{\pi(r_2)^2} = \frac{(r_1)^2}{(r_2)^2} = \left(\frac{r_1}{r_2}\right)^2.$$

In words, the circumferences of two circles have the same ratio as their radii, and the areas of two circles have the same ratio as the *square* of their radii. To illustrate, if the length of the radius of one circle is twice the length of the radius of another circle, then we may conclude that:

1. The *circumference* of the larger circle is twice the circumference of the other circle since

$$\frac{\text{Circumference of larger circle}}{\text{Circumference of smaller circle}} = \frac{\text{Radius of larger}}{\text{Radius of smaller}} = \frac{2}{1}.$$

2. The *area* of the larger circle is *four* times the area of the other circle since

$$\frac{\text{Area of larger circle}}{\text{Area of smaller circle}} = \left(\frac{\text{Radius of larger}}{\text{Radius of smaller}}\right)^2 = \left(\frac{2}{1}\right)^2 = \frac{4}{1}.$$

Example

6. The ratio of the areas of two circles of $9 : 16$. If the radius of the larger circle is 20 units, what is the length of the radius of the smaller circle?

Solution: Let x = radius of the smaller circle.

$$\frac{9}{16} = \left(\frac{x}{20}\right)^2$$

Since $\frac{9}{16}$ is a perfect square, solve the proportion by taking the square root of each side of the equation:

$$\sqrt{\frac{9}{16}} = \sqrt{\left(\frac{x}{20}\right)^2}$$

$$\frac{3}{4} = \frac{x}{20}$$

Cross-multiply: $4x = 60$

$$x = \frac{60}{4} = 15$$

The radius of the smaller circle is **15 units**.

Exercise Set 10.6

1. The ratio of the areas of two circles is 1 : 9. If the radius of the smaller circle is 5, find the length of the radius of the larger circle.

2. The ratio of the lengths of the radii of two circles is 4 : 25. If the area of the smaller circle is 8 cm², find the area of the larger circle.

3. Find the circumference and the area of the semicircle whose diameter is 16.

4. If the length of the radius of a circle is doubled, what is the change in each of the following?
   (a) The circumference of the circle   **(b)** The area of the circle

5. In the accompanying figure, find the area of $\triangle ABC$.

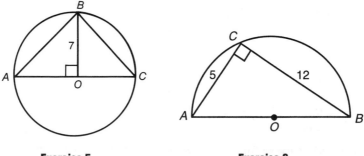

Exercise 5   **Exercise 6**

6. In the accompanying figure, find the area of the semicircle.

7. Find the area of a circle whose circumference is 10π.

8. Find the radius of a circle whose area is numerically equal to twice its circumference.

272

10.7 FINDING AREAS OF MORE COMPLICATED REGIONS

To find the area of a region whose boundaries are formed by intersecting geometric shapes, it may be necessary to apply more than one area formula.

Finding Areas Using Addition

In Figure 10.12, arcs AB, BC, and CD are semicircles with diameters $\overline{AB}$, $\overline{BC}$, and $\overline{CD}$, respectively. $ABCD$ is a rectangle, $BC = 20$, and $AB = 12$. To find the area of the entire region, we must calculate the areas of the rectangle and the three semicircles. The sum of these areas equals the area of the entire region. When finding the area of a semicircle, remember to multiply the area of the circle by $\frac{1}{2}$.

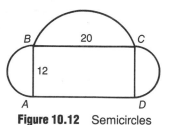

Figure 10.12 Semicircles

Area of rectangle $ABCD$:

$$\text{Area} = (BC)(AB) = (20)(12) = 240$$

Area of semicircle BC:

$$\text{Radius } r = \frac{1}{2}(20) = 10$$

$$\text{Area of semicircle } BC = \frac{1}{2}(\pi r^2)$$

$$= \frac{1}{2}(\pi)(10^2)$$

$$= \frac{1}{2}(100\pi) = 50\pi$$

Area of semicircle AB:

$$\text{Radius } r = \frac{1}{2}(12) = 6$$

$$\text{Area of semicircle } AB = \frac{1}{2}(\pi r^2)$$

$$= \frac{1}{2}(\pi)(6^2)$$

$$= \frac{1}{2}(36\pi) = 18\pi$$

Area of semicircle *CD*: Since opposite sides of a rectangle have the same length, the diameter and, therefore, the radii of semicircles *AB* and *CD* are equal. As a result, they must also have equal areas.

$$
\begin{aligned}
\text{Areas of semicircle } CD &= \text{Area of semicircle } AB = 18\pi \\
\text{Area of entire region} &= 240 + 50\pi + 18\pi + 18\pi \\
&= \mathbf{240 + 86\pi}
\end{aligned}
$$

Finding Areas Using Subtraction

When one geometric figure lies within another, the difference in their area represents the area of the region that they do not have in common. For example, in Figure 10.13, the radius of the inner circle is 5 and the radius of the outer circle is 8. The two circles are **concentric circles** since they have the same center, but have radii of different lengths. The shaded region represents the area which the concentric circles do *not* have in common. This area may be found by *subtracting* the area of the inner circle from the area of the outer circle.

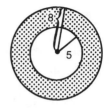

Figure 10.13   Concentric Circles

$$
\begin{array}{r l}
\text{Area of outer circle} & = \pi(8^2) = 64\pi \\
- \text{ Area of inner circle} & = \pi(5^2) = 25\pi \\
\hline
\text{Area of shaded region} & = 39\pi
\end{array}
$$

Examples

1. In the accompanying diagram, *ABCD* is a square. Find the area of the shaded region.

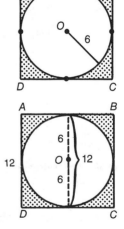

Solution: The diameter of the inscribed circle is 12, so the length of a side of the circumscribed square is also 12.

$$
\begin{aligned}
\text{Area of shaded region} &= \text{Area of square} - \text{Area of circle} \\
&= s^2 \qquad\qquad\quad\; - \pi r^2 \\
&= (12)^2 \qquad\qquad\; - \pi(6)^2 \\
&= \mathbf{144} \qquad\qquad\quad - 36\pi
\end{aligned}
$$

2. In the accompanying diagram, $\triangle BCE$ is inscribed in square $ABCD$. If the length of side $\overline{BC}$ is 4 centimeters, what is the area, in square centimeters, of the shaded portion of the diagram?

Solution: Draw an altitude of $\triangle BCE$ from E to $\overline{BC}$. Since parallel lines ($\overline{AD}$ and $\overline{BC}$) are always the same distance apart, $EH = AB = 4$.

Area of shaded region = Area of square $ABCD$ – Area of $\triangle BCE$

$$= 4^2 \qquad\qquad -\frac{1}{2}(4)(4)$$

$$= 16 \qquad\qquad -\frac{1}{2}(16)$$

$$= 16 \qquad\qquad -8$$

$$= 8\text{cm}^2$$

Exercise Set 10.7

1. In the accompanying diagram, find the area of the region bounded by square $ABCD$ and semicircle BC.

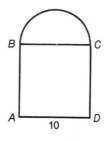

2. In the accompanying diagram, find the area of the region bounded by isosceles right triangle ABC and semicircle BC.

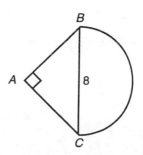

3. In the accompanying diagram, find the area of the region that is inside the circle O and outside circle O'.

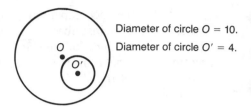

Diameter of circle O = 10.

Diameter of circle O' = 4.

4. In the accompanying diagram find the area between the two concentric circles.

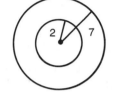

5. The area of square $ABCD$ as shown in the accompanying diagram is 121. $\overline{EKF}$ is drawn parallel to $\overline{AD}$, and $\overline{GKH}$ is drawn parallel to $\overline{AB}$, so as to form the smaller squares $AGKE$ and $FCHK$. If the area of $AGKE$ is 36, find the area of $FCHK$.

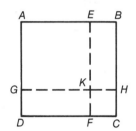

6–8. *In the accompanying diagrams* ABCD *is a square. Find the area of the shaded regions.*

6.

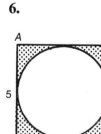

7.

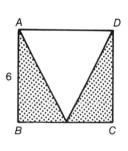

8.

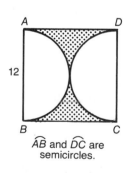

$\overset{\frown}{AB}$ and $\overset{\frown}{DC}$ are semicircles.

9–12. Find the area of the shaded regions.

9.

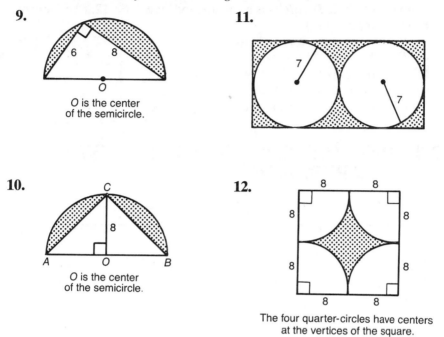

6   8

O

O is the center
of the semicircle.

11.

7

7

10.

C

8

A   *O*   *B*

O is the center
of the semicircle.

12.

8   8

8   8

8   8

8   8

The four quarter-circles have centers
at the vertices of the square.

13. In the accompanying diagram, △*RST* is inscribed in circle *O* with diameter $\overline{RT}$. Radius $\overline{OS}$ is an altitude of △*RST*, and *OS* = 4.

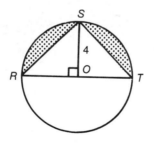

S

4

R   *O*   *T*

(**a**) Find *RT*.
(**b**) Express, in terms of π, the area of circle *O*.
(**c**) Find the area of △*RST*.
(**d**) Express, in terms of π, the area of the shaded region.
(**e**) Find *RS*. (Answer may be left in radical form.)

14. In the accompanying diagram, $\overline{AD}$ is a diameter of circle O, $\overline{BC}$ is a diameter of circle P, $ABCD$ is a square, and side $AB = 12$. (Answers may be left in terms of π.)

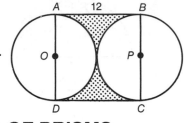

(**a**) Find the circumference of circle O.

(**b**) Find the area of circle O.

(**c**) Find the area of a semicircle of circle P.

(**d**) Find the area of square $ABCD$.

(**e**) Find the area of the shaded portion.

10.8 MEASURING VOLUMES OF PRISMS AND PYRAMIDS

KEY IDEAS

The amount of space that a solid encloses is represented by its volume. The volume of a solid, expressed in cubic units, is the number of cubes having an edge length of 1 unit that the solid can enclose.

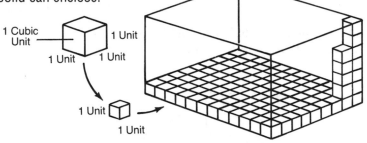

Special Prisms

A polygon is "flat" since all its sides lie in the same plane. The figure that is the counterpart of a polygon in space is called a **polyhedron**.

A **prism** (Figure 10.14) is a special type of polyhedron. The sides of a prism are called **faces**. All prisms have these two properties:

1. Two of the faces, called **bases**, are congruent polygons lying in parallel planes.

2. The faces that are not bases, called **lateral faces**, are parallelograms.

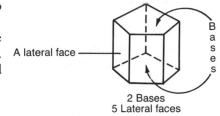

Figure 10.14 A Prism

A **rectangular solid** (Figure 10.15) is a prism whose two bases and four lateral faces are rectangles. If the bases and lateral faces are squares, then the rectangular solid is called a **cube**.

Figure 10.15 A Rectangular Solid

Volume of Rectangular Solids

The volume, V, of *any* prism is equal to the product of the area of its base, B, and its height, h:

Volume of a prism: $V = Bh$.

By using this relationship, special formulas can be derived for determining the volumes of a cube and a rectangular solid:

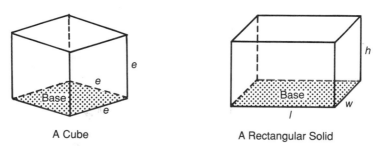

A Cube

A Rectangular Solid

Volume of a cube:

$$V = Bh = (e^2)e$$
$$V \text{ cube} = e^3$$

Volume of a rectangular solid:

$$V = Bh = (l \times w)h$$
$$V \text{ rect solid} = lwh$$

Examples

1. What is the volume of a rectangular solid having a length, width, and height of 5 cm, 4 cm, and 3 cm, respectively?

Solution: $V = lwh = (5)(4)(3) = \textbf{60 cm}^3$

2. A rectangular solid has the same volume as a cube whose edge length is 6. What is the height of the rectangular solid if its length is 8 and its width is 3?

Solution:

$$\text{Volume of rectangular solid} = \text{Volume of cube}$$
$$lwh = e^3$$
$$(8)(3)h = 6^3$$
$$24h = 216$$
$$h = \frac{216}{24} = 9$$

The height of the rectangular solid is **9**.

3. What is the change in volume of a cube whose edge length is doubled?

Solution: In the formula $V = e^3$, replace e with $2e$, so that the new volume is $V = (2e)^3 = 8e^3$.

The new volume is **8** times as great as the original volume.

Volume of a Pyramid

A **pyramid** (Figure 10.16) is a solid figure formed by connecting each vertex of a polygon base to a given point that lies in a different plane. A prism has two bases, whereas a pyramid has *one* base. The vertex of the pyramid is the point at which the triangular faces intersect. The *altitude* of the pyramid is the perpendicular segment dropped from the vertex to the plane that contains the base. The length of this segment is the *height* of the pyramid. The volume, *V*, of a pyramid is *one-third* the volume of a prism having the same base area (*B*) and height (*h*):

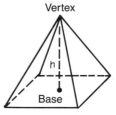

Figure 10.16 Pyramid

$$V_{\text{pyramid}} = \frac{1}{3}Bh.$$

Examples

4. A pyramid 12 cm high has a square base that has a perimeter of 20 cm. What is the volume of the pyramid?

Solution:

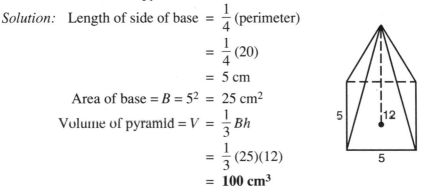

$$\text{Length of side of base} = \frac{1}{4}(\text{perimeter})$$

$$= \frac{1}{4}(20)$$

$$= 5 \text{ cm}$$

$$\text{Area of base} = B = 5^2 = 25 \text{ cm}^2$$

$$\text{Volume of pyramid} = V = \frac{1}{3}Bh$$

$$= \frac{1}{3}(25)(12)$$

$$= \textbf{100 cm}^3$$

5. The volume of a pyramid having an isosceles right triangle as its base is 48 cm^3. If the length of a leg of the right triangle is 6 cm, what is the height of the pyramid?

Solution: Area of base $= B = \dfrac{1}{2}(6)(6) = 18$ cm^2

Volume of pyramid $= V = \dfrac{1}{3}Bh$

$$48 = \frac{1}{3}(18)h$$
$$48 = 6h$$
$$\frac{48}{6} = h$$
$$8 = h$$

The height of the pyramid is **8 cm**.

Comparing Volumes of Similar Solids

If two solids are similar, then their volumes have the same ratio as the cube of the ratio of any pair of corresponding dimensions. For example, the volumes V_1 and V_2 of two rectangular solids have the same ratio as the cube of the ratio of their lengths, widths, or heights:

$$\frac{V_1}{V_2} = \left(\frac{l_1}{l_2}\right)^3 = \left(\frac{w_1}{w_2}\right)^3 = \left(\frac{h_1}{h_2}\right)^3.$$

Example

6. The volumes of two rectangular solids are 8 cm^3 and 27 cm^3. If the height of the smaller solid is 2 cm, find the height of the larger solid.

Solution:

$$\frac{V_1}{V_2} = \left(\frac{h_1}{h_2}\right)^3$$

$$\frac{8}{27} = \left(\frac{2}{h}\right)^3 = \frac{8}{h^3}$$

Cross-multiply: $8h^3 = (8)(27)$

$$h^3 = \frac{(8)(27)}{8} = 27$$

$$h = \sqrt[3]{27} = \textbf{3 cm}$$

Exercise Set 10.8

1. Find the volume of a cube whose edge length is 9.

2. Find the volume of a rectangular solid if its dimensions are
$l = 1\dfrac{1}{2}$, $w = \dfrac{4}{9}$, $h = 0.75$.

3. Find the height of a rectangular solid if its other dimensions and its volume are $l = 10$, $w = 2.5$, $V = 75$.

4. What is the edge length of a cube whose volume is 125?

5. A rectangular solid has the same volume as a cube whose edge length is 4. What is the height of the rectangular solid if its length is 2 and its width is 8?

6. A cube has the same volume as a pyramid whose base has an area of 25 and whose height is 15. What is the edge length of the cube?

7. If the edge length of a cube is tripled, by what number is the volume multiplied?

8. If the length and the height of a rectangular solid are each doubled, by what number is the volume multiplied?

9. The length of a rectangular box is 8 and its width is 5. If the height of the box is equal to the edge length of a cube whose volume is 27, what is the volume of the rectangular box?

10. A certain prism has the same length and the same volume as a pyramid. If the area of the base of the pyramid is 42, what is the area of the base of the prism?

11. A pyramid has a height of 6 cm and a right triangle as a base. If the length of the hypotenuse of the right triangle is 10 cm and the length of one of the legs is 8 cm, what is the volume of the pyramid?

12. The height of a rectangular solid is 3, and its length exceeds twice its width by 1. If the volume of the solid is 30, find the length and the width.

13. A pyramid having a square base has a volume of 98 cm^3. If a side of the square is 7 cm, what is the height of the pyramid?

14. The volumes of two cubes are 27 and 216. If the length of an edge of the smaller cube is 4.5, what is the length of an edge of the larger cube?

15. The smaller of two boxes of breakfast cereal has a volume of 12, while the larger box has a volume of 96. If the height of the smaller box is 8, what is the height of the larger box?

16. A pyramid having a rectangular base has a volume of 182 cm^3 and a height of 6 cm. If the length of the base is 1 less than twice the width, find the dimensions of the base of the pyramid.

10.9 MEASURING VOLUMES OF CIRCULAR SOLIDS

△ **KEY IDEAS** △

When the term *cylinder* is used, we will always mean a right circular cylinder, which can be visualized as a tin can. Similarly, when discussing a *cone*, we will be referring to a right circular cone, whose shape may be thought of as being the same as that of an ice cream cone.

Volume of a Cylinder

A cylinder (Figure 10.17) has *two* bases that are congruent circles lying in parallel planes. Like a prism that also has two congruent bases, the volume of a cylinder is also given by the formula $V = Bh$. In this formula, B represents the area of one of its two identical bases. The area of a cylinder's circular base is πr^2.

$$V_{\text{cylinder}} = \pi r^2 h$$

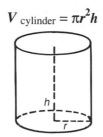

Figure 10.17 Cylinder

Volume of a Cone

A **cone** (Figure 10.18), like a pyramid, has one base and a volume given by the formula $V = \frac{1}{3}Bh$. Here B represents the area of its circular base.

$$V_{\text{cone}} = \frac{1}{3}\pi r^2 h$$

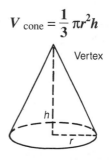

Figure 10.18 Cone

283

Examples

1. The radius of a base of a cylinder is 5 and its height is 6. What is its volume?

Solution: $V = \pi r^2 h$
$$= \pi(5^2)6 = \pi(25)(6) = \mathbf{150\pi}$$

2. The *slant height, l,* of a cone is a segment that connects the vertex to a point on the circular base. What is the volume of a cone having a radius of 6 and a slant height of 10?

Solution: The slant height is the hypotenuse of a right triangle whose legs are a radius and the altitude of the cone. The lengths of this right triangle form a 6-8-10 right triangle, where 8 represents the height of the cone. Therefore

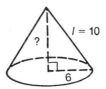

$$V = \frac{1}{3}\pi r^2 h$$

$$= \frac{1}{3}\pi(6)^2 \cdot 8$$

$$= \frac{1}{\cancel{3}}\pi(\cancel{36})^{12} \cdot 8$$

$$= \mathbf{96\pi}$$

3. The circumference of the base of a cone is 8π cm. If the volume of the cone is 16π cm^3, what is the height?

Solution: Since $C = \pi D = 8\pi$, the diameter is 8 cm; this means that the radius of the base is 4 cm.

$$V = \frac{1}{3}\pi r^2 h$$

$$16\pi = \frac{1}{3}\pi(4^2)h$$

$$48\pi = 16\pi h$$

$$\frac{48\pi}{16\pi} = h$$

$$3 = h$$

The height of the cone is **3 cm.**

Volume of a Sphere

A sphere (Figure 10.19) is a solid that represents the set of all points in space that are at a given distance from a fixed point. Its volume, V, can be determined using the formula

$$V_{sphere} = \frac{4}{3}\pi r^3$$

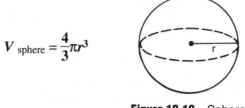

Figure 10.19   Sphere

Examples

4. Find the volume of a sphere whose diameter is 6 cm.

Solution: Radius $= \frac{1}{2}(6) = 3$

$$V = \frac{4}{3}\pi(3)^3$$

$$= \frac{4}{3}\pi(27)$$

$$= 4\pi(9)$$

$$= \mathbf{36\pi \ cm^3}$$

5. The radius of a sphere is twice as long as the radius of a smaller sphere. How many times greater is the volume of the larger sphere than the volume of the smaller sphere?

Solution: If the radii of the two spheres are represented by r_1 and r_2, then their volumes, V_1 and V_2, have the ratio

$$\frac{V_1}{V_2} = \frac{\frac{4}{3}\pi(r_1)^3}{\frac{4}{3}\pi(r_2)^3} = \frac{r_1{}^3}{r_2{}^3} = \left(\frac{r_1}{r_2}\right)^3.$$

Since the radii are in the ratio of 2 : 1,

$$\frac{V_1}{V_2} = \left(\frac{2}{1}\right)^3 = \frac{8}{1}.$$

The volume of the larger sphere is **8** times the volume of the smaller sphere.

SUMMARY OF VOLUME FORMULAS

Figure	Volume Formula
Rectangular Box	$V = \text{length} \times \text{width} \times \text{height}$ $= lwh$
Cube	$V = \text{edge} \times \text{edge} \times \text{edge}$ $= (\text{edge})^3$
Pyramid	$V = \dfrac{1}{3} \times \text{area of base} \times \text{height}$
Cylinder	$V = \pi \times (\text{radius})^2 \times \text{height}$ $= \pi r^2 h$
Cone	$V = \dfrac{1}{3} \times \pi \times (\text{radius})^2 \times \text{height}$ $= \dfrac{1}{3} \pi r^2 h$
Sphere	$V = \dfrac{4}{3} \times \pi \times (\text{radius})^3$ $= \dfrac{4}{3} \pi r^3$

Exercise Set 10.9

1. Find the volume of each cone.

  (a)   **(b)**   **(c)**

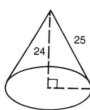 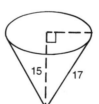 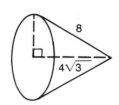

2. Find the volume of a sphere if the length of its radius is π.

3. Fill in the missing volumes in the following table:

	r	h	V_{cylinder}	V_{cone}
(a)	7	3	?	?
(b)	4.5	10	?	?
(c)	?	5	45π	?
(d)	4	?	?	96π

4. If the radius of the base of a cone is multiplied by 3 while its height remains the same, by what number is its volume multiplied?

5. If the radius of a sphere is tripled, by what number is its volume multiplied?

6. The circumference of a cylinder is 10π cm, and its height is 3.5 cm. What is the volume of the cylinder?

7. A cylinder and a cone have congruent bases and equal heights. How many times greater is the volume of the cylinder than the volume of the cone?

8. The radius of a sphere is 6, and the radius of a cylinder is 4. If these solids have the same volume, what is the height of the cylinder?

9. The radii of two spheres are 30 and 50. If the volume of the smaller sphere is 270π, what is the volume of the larger sphere?

10. The volumes of two spheres are 343π and 729π. If the radius of the larger sphere is 18, what is the radius of the smaller sphere?

11. A right triangle whose legs are 8 cm and 15 cm is revolved in space about the shorter leg. What is the volume of the resulting cone?

12. A container having the shape of a cylinder with a radius of 2 cm and a height of 9 cm is filled with sand. What is the radius of a container having the shape of a sphere that can hold the same amount of sand?

13. A sphere is inscribed in a cube having an edge length of 8 so that each side of the cube touches the sphere in one point. What is the volume of the sphere?

14. A sphere having a volume $32/3\pi$ cm^3 is inscribed in a cube. What is the volume of the cube?

REGENTS TUNE-UP: CHAPTER 10

Each of the questions in this section has appeared on a previous Course I Regents Examination. Here is an opportunity for you to review Chapter 10 and, at the same time, prepare for the Course I Regents Examination.

---

1. If the base of a triangle is represented by $2x$ and the altitude drawn to that base is equal to 10, express the area of the triangle in terms of x.

2. A building casts a shadow 15 feet long at the same time that a woman 6 feet tall casts a shadow 5 feet long. Find the number of feet in the height of the building.

3. If the diameter of a circle is 14, find the area of the circle.

4. The sides of a triangle are, in centimeters, 24, 20, and 8. The longest side of a similar triangle is 12 centimeters. Find the number of centimeters in the *shortest* side of this triangle.

5. Find the radius of a circle if its circumference is 48π.

6. The area of a triangle is 48. If the base of this triangle is 12, what is the length of the altitude to this base?

7. In the accompanying figure, $\triangle ABC$ is congruent to $\triangle RST$, $\overline{AB} \cong \overline{RS}, \overline{BC} \cong \overline{ST}$, and $\overline{AC} \cong \overline{RT}$. What angle in $\triangle RST$ is congruent to $\angle BAC$?

8. In the accompanying diagram, $\triangle ABC$ is similar to $\triangle XYZ$, with $\angle A \cong \angle X$, $\angle B \cong \angle Y$, and $\angle C \cong \angle Z$. If $AB = 35$, $XY = 7$, $BC = 10$, and $YZ = 2$, how many times larger is side $\overline{AC}$ than side $\overline{XZ}$?

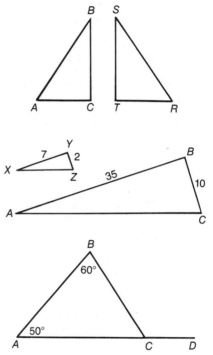

9. In the accompanying figure, $\angle BCD$ is an exterior angle to $\triangle ABC$ at vertex C. If m $\angle A = 50$ and m $\angle B = 60$, find the number of degrees in the measure of $\angle BCD$.

10. If the ratio of the corresponding sides of two similar triangles is $3 : 5$, the ratio of the areas of these triangles is:
(1) $3 : 5$ (3) $9 : 25$
(2) $\sqrt{3} : \sqrt{5}$ (4) $27 : 125$

11. If the edge of a cube is $3x$, the volume of the cube is:
(1) $27x^3$ (2) $9x^3$ (3) $3x^3$ (4) $9x^2$

12. If the radius of a circle is tripled, then the area of the circle is multiplied by:
(1) 27 (2) 9 (3) 3 (4) 6

13. The volume of a cylinder is found by using the formula $V = \pi r^2 h$. If h is doubled, then the volume is:
(1) doubled (3) increased by 4
(2) increased by 2 (4) multiplied by 4

14. In the accompanying figure, quadrilateral *ABCD* is a trapezoid with $\overleftrightarrow{AB}$ $\parallel$ $\overleftrightarrow{DC}$, $\overleftrightarrow{AD} \perp \overleftrightarrow{AB}$, $\overleftrightarrow{DB} \perp \overleftrightarrow{BC}$; *AB* = *AD* = 4, and *DC* = 8.

(a) Find *DB* in radical form.
(b) Find the area of $\triangle ABD$.
(c) Find the area of trapezoid *ABCD*.
(d) Find the area of $\triangle DBC$.

15. In the accompanying figure, rectangle *ABCD* is inscribed in circle *O* and $\overline{DB}$ is a diameter. The radius of the circle is 5. (Answers may be left in terms of π.)

(a) Find the area of the circle.
(b) If *CD* = 8, find *BC*.
(c) Find the area of $\triangle BCD$.
(d) Find the area of rectangle *ABCD*.
(e) Find the area of the shaded portion of the figure.

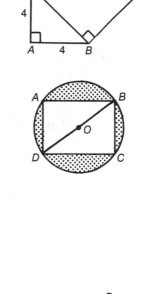

16. In the accompanying figure, $\triangle ABC$ is an isosceles triangle, *AB* = *CB*, the measure of $\angle BCA$ is 2*x*, and the measure of the exterior angle *DAB* is 3*x*.

(a) Express the measure of $\angle BAC$ in terms of *x*.
(b) What is the value of *x*?
(c) How many degrees are in the measure of $\angle BAC$?
(d) How many degrees are in the measure of $\angle BAD$?
(e) How many degrees are in the measure of $\angle ABC$?

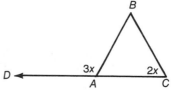

17. In the diagram below, $\triangle ABC$ is a right triangle with the right angle at *C*. Segment $\overline{DE}$ is perpendicular to $\overline{AC}$ at *E*, *BC* = 8, *AC* = 6, and *DE* : *BC* = 1 : 2.

(a) Find *DE*.
(b) Find *AB*.
(c) Find *AE*.
(d) Find the area of $\triangle ABC$.
(e) Find the area of trapezoid *ECBD*.

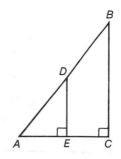

18. In the diagram of parallelogram $ABCD$, $\overline{DE}$ is perpendicular to $\overline{AB}$, $AD = 15$, and $AE = 9$.

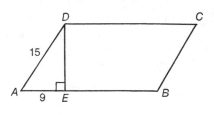

(a) Find DE.
(b) If BE is 2 less than twice AE, find BE.
(c) Find AB.
(d) Find the area of $\triangle AED$.
(e) Find the area of $ABCD$.
(f) Find the area of trapezoid $EBCD$.

19. In the accompanying diagram, $ABCD$ is a rectangle and AGE is an isosceles triangle with $AG = EG$, $\overline{GF} \perp \overline{AD}$, E is the midpoint of $\overline{AD}$, $AF = FE$, $AB = 8$, and $AD = 24$.

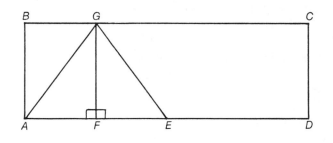

(a) What is the length of $\overline{AE}$?
(b) What is the length of $\overline{AF}$?
(c) What is the length of $\overline{AG}$?

(d) What is the area of $\triangle AGE$?
(e) What is the area of trapezoid $ADCG$?

20. In the accompanying diagram, $ABCD$ is a rectangle, E is a point on $\overline{AB}$, $DE = 10$, $AE = 6$, and $DC = 15$. Circle O has a radius of 3.

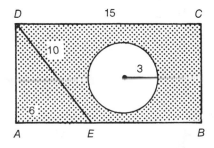

(a) Find AD.
(b) Find the area of $\triangle ADE$.
(c) Find the area of circle O.

(d) Find the area of trapezoid $EBCD$.
(e) Find the area of the shaded portion.

21. In the accompanying diagram, *ABCD* is an isosceles trapezoid with bases $\overline{AB}$ and $\overline{DC}$, *DA* = 13, *CDEF* is a square, and circle *O* has a diameter of 12. The length of a side of the square is equal to the diameter of the circle.

(a) Find the measure of
  (1) $\overline{AE}$ (2) $\overline{AB}$

(b) Find the area of trapezoid *ABCD*.

(c) Find the area of the circle in terms of π.

(d) Using π = 3.14, find, to the *nearest integer*, the area of the shaded region.

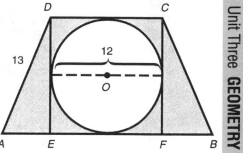

22. In the accompanying diagram, points *A, B, C,* and *D* are on circle *O* such that *ABCD* forms a square. Diagonals $\overline{AC}$ and $\overline{BD}$ intersect at *O* and *AC* = 6.

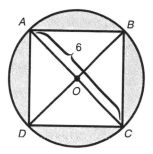

(a) Express the length of $\overline{AD}$ in simplest radical form.

(b) Find the number of square units in the area of square *ABCD*.

(c) Express the area of circle *O* in terms of π.

(d) Find, to the *nearest tenth,* the number of square units in the area of the shaded portion. [Use π = 3.14]

ANSWERS TO ODD-NUMBERED EXERCISES: CHAPTER 10

Section 10.1

1. 16

3. (3)

5. Since $\overline{RW} \cong \overline{TW}$, Right $\angle SWR \cong$ Right $\angle SWT$, and $\overline{SW} \cong \overline{SW}$, the two triangles are congruent by SAS $\cong$ SAS.

7. Since $\overline{BD} \cong \overline{BE}$, $\angle BDA \cong \angle BEC$ (supplements of congruent angles are congruent), and $\overline{AD} \cong \overline{CE}$, the two triangles are congruent by SAS $\cong$ SAS.

9. Since $\angle C \cong \angle D$, $\overline{BC} \cong \overline{BD}$, and $\angle ABC \cong \angle EBD$, the two triangles are congruent by ASA $\cong$ ASA.

11. Since $\overline{TR} \cong \overline{RS}$, $\angle J \cong \angle R$ (corresponding angles formed by parallel lines are congruent), and $\overline{JL} \cong \overline{RT}$, the two triangles are congruent by SAS $\cong$ SAS.

Section 10.2

1. (a) 120
   (b) 32
   (c) 60
   (d) $180 - 2x$
3. (a) 72, 72, 36
   (b) 67.5, 67.5, 45
   (c) 30, 30, 120
   (d) 40, 40, 100
5. 36

7. 37
9. 59
11. (a) 45, 45
    (b) $5\sqrt{2}$
    (c) $4\sqrt{2}$
13. 111
15. 12
17. 9
19. 66

Section 10.3

1. 24   3. 48   5. (a) 120   (b) 6   7. 60
9. (a) Since corresponding angles formed by parallel lines are congruent, $\angle B \cong \angle CDE$ and $\angle A \cong \angle CED$. Triangles DCE and BCA are similar since two angles of one triangle are congruent to two angles of the other triangle.
   (b) 9   (c) 9
11. (a) Since corresponding angles of similar triangles are congruent, $\angle A \cong \angle R$. Also, right angles BHA and SKR are congruent. Hence, triangles ABH and RSK are similar.
    (b) 8

Section 10.4

1. (a) 6, 18   (b) 3, 36   (c) 4, 27   (d) 9, 12

3. (a) $\dfrac{81}{4}$ (b) $64x^2$ (c) $25x^2 - 10x + 1$

5. (a) $b = 12, A = 60$ (c) $h = 24, A = 175$
   (b) $h = 9, A = 360$ (d) $h = 8, d = 17$

7. (a) doubled   (b) multiplied by 4
9. 7, 17
11. $b = 10, h = 6$
13. 80   15. 3, 10   17. 6

Section 10.5

1. 22   9. 8   13. 6

3. 30   11. (a) 4   15. $\dfrac{5}{9}$

5. 48   (b) 7   17. 55

7. 48   (c) 6   19. 5   21. 7, 24, 25

Section 10.6
1. 15   **5.** 49
3. $C = 8\pi, A = 32\pi$ **7.** 25π

Section 10.7
1. 112.5π **5.** 25   **9.** $\dfrac{25}{2}\pi - 24$

3. 21π **7.** 18   **11.** $392 - 98\pi$
13. (a) 8   **(b)** 16π **(c)** 16   **(d)** $8\pi - 16$ **(e)** $4\sqrt{2}$

Section 10.8
1. 729   **5.** 4   **9.** 120   **13.** 6
3. 3   **7.** 27   **11.** 48   **15.** 64

Section 10.9
1. (a) 392π **(b)** 320π **(c)** $\dfrac{64\pi\sqrt{3}}{3}$

3. (a) Volume of cylinder $= 147\pi$
       Volume of cone $= 49\pi$
   (b) Volume of cylinder $= \dfrac{405}{2}\pi$
       Volume of cone $= \dfrac{405}{6}\pi$
   (c) $r = 3$, Volume of cone $= 15\pi$
   (d) $h = 18$, Volume of cylinder $= 288\pi$

5. 27   **7.** 3   **9.** 1250π **11.** 600π **13.** $\dfrac{256}{3}\pi$

Regents Tune-Up: Chapter 10
1. $10x$ **15. (a)** 25π **19. (a)** 12
3. 49π **(b)** 6   **(b)** 16
5. 12   **(c)** 24   **(c)** 25
7. $\angle SRT$ **(d)** 48   **(d)** 300
9. 110   **(e)** $25\pi - 48$ **(e)** 246
11. (1)   **17. (a)** 4   **21. (a) (1)** 5   **(2)** 22
13. (1)   **(b)** 10   **(b)** 204
              (c) 3   **(c)** 36π
              (d) 24   **(d)** $204 - 36\pi$
              (e) 18

ELEMENTS OF COORDINATE GEOMETRY AND TRANSFORMATIONS

CHAPTER 11

EQUATIONS OF LINES; INTRODUCTION TO TRANSFORMATIONS

11.1 GRAPHING POINTS IN THE COORDINATE PLANE

KEY IDEAS

A **coordinate plane** may be created by drawing a horizontal number line called the **x-axis** and a vertical number line called the **y-axis**. The x-axis and the y-axis are sometimes referred to as the **coordinate axes**, and their point of intersection is called the **origin**. The process of locating a point or a series of points in the coordinate plane is called **graphing**.

Graphing Ordered Pairs

Each point in the coordinate plane is located (see Figure 11.1) using an ordered pair of numbers of the form (x, y), in which the first number of the pair is the x-coordinate, and the second number is the y-coordinate. The x-coordinate, sometimes called the **abscissa**, tells the number of units the point is located to the right $(x > 0)$ or to the left $(x < 0)$ of the origin. The y-coordinate, sometimes called the **ordinate**, gives the number of units the point is located above $(y > 0)$ or below $(y < 0)$ the origin. For example, to graph the point $(3, 5)$, start at the origin, move 3 units to the right, and then 5 units up.

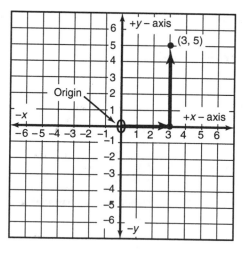

Figure 11.1 Graphing Ordered Pairs

The Four Quadrants

The coordinate axes divide the plane into four regions called **quadrants**. As shown in Figure 11.2, the quadrants are numbered in counterclockwise order, beginning at the upper right and using Roman numerals. Notice that the points $A(2, 3)$, $B(-4, 5)$, $C(-3, -6)$, and $D(3, -3)$ lie in different quadrants. The signs of the x- and y-coordinates of a point determine the quadrant in which a point lies.

Coordinates	Location of Point
(+, +)	Quadrant I
(−, +)	Quadrant II
(−, −)	Quadrant III
(+, −)	Quadrant IV

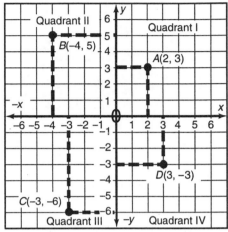

Figure 11.2   The Four Quadrants

Lengths of Horizontal and Vertical Segments

In Figure 11.3, the line segment joining $L(2, 5)$ and $M(6, 5)$ is parallel to the x-axis. The length of $\overline{LM}$ may be obtained in one of two ways:

1. By counting the number of boxes between points L and M. There are four square boxes between points L and M, so the length of $\overline{LM}$ is 4 units.

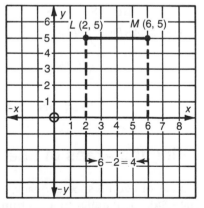

Figure 11.3   Lengths of Line Segments

2. By subtracting to find the difference between the corresponding coordinates of the two points that are different. Since the y-coordinates of points L and M are the same, subtract the x-coordinates of these points: $LM = 6 - 2 = 4$. Keep in mind that, since length must be a positive number, you must always subtract the smaller coordinate from the larger coordinate.

Examples

1. Find the length of the line segment joining points $A(3, -2)$ and $B(3, 5)$.

Solution: $\overline{AB}$ is parallel to the y-axis, and its length is equal to the difference in the y-coordinates: $5 - (-2) = 5 + 2 = \textbf{7 units}$.

2. Find the area of the triangle whose vertices are $A(0, 8)$, $B(6, 0)$, and $C(0, 0)$.

Solution: The vertices determine a right triangle, whose right angle is at C. First find the lengths of the legs of the right triangle.

$$AC = 6 - 0 = 6$$
$$BC = 8 - 0 = 8$$
$$\text{Area} \triangle ABD = \frac{1}{2} bh$$
$$= \frac{1}{2}(6)(8)$$
$$= 3(8)$$
$$= \textbf{24 square units}$$

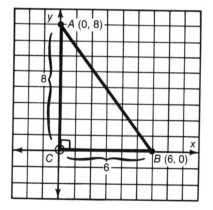

Finding the Areas of Special Quadrilaterals

The areas of certain special quadrilaterals can be easily found if at least one of their sides is parallel to a coordinate axis. The altitude drawn to that base will be parallel to the other coordinate axis, so that the lengths of the base and the altitude can then be determined by counting boxes or by subtracting coordinates.

Examples

3. Graph a parallelogram whose vertices are $A(2, 2), B(5, 6),$ $C(13, 6),$ and $D(10, 2),$ and then find its area.

Solution: In the accompanying graph, altitude $\overline{BH}$ has been drawn to base $\overline{AD}.$ Count boxes to find the lengths of these segments.

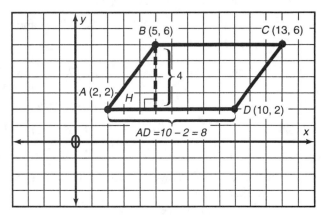

Area of $\square ABCD = bh$
$$= (AD)(BH)$$
$$= (8)(4)$$
$$= \textbf{32 square units}$$

4. Graph a trapezoid whose vertices are $A(-4, 0), B(-4, 3), C(0, 6)$ and $D(0, 0),$ and then find its area.

Solution: In the accompanying graph, $\overline{AB}$ and $\overline{CD}$ are the bases of trapezoid $ABCD.$ $\overline{AD}$ may be considered an altitude since it is perpendicular to both bases.

Area of trap $ABCD = \dfrac{1}{2}h(b_1 + b_2)$

$$= \frac{1}{2}(AD)(AB + CD)$$

$$= \frac{1}{2}(4)(3 + 6)$$

$$= 2(9)$$

$$= \textbf{18 square units}$$

Exercise Set 11.1

1. Starting at the origin, move up 3 units and move to the right 2 units. Then move to the left 5 units and down 1 unit. Label this point A. What are the coordinates of point A?

2. In what quadrant is point $P(x, y)$ located if $xy < 0$ and $y > x$?

3–8. Graph the line segment determined by each of the following pairs of points, and then determine its length:

3. (3, 8) and (7, 8)   **6.** (–2, 5) and (–6, 5)

4. (9, 4) and (1, 4)   **7.** (–3, –8) and (–3, 2)

5. (0, –1) and (0, 5)   **8.** (7, –1) and (–7, –1)

9–12. Find the area of the triangle whose vertices are:

9. $A(0, 5), B(6, 0), C(0, 0)$ **11.** $A(2, 2), B(2, 7), C(5, 2)$

10. $A(–4, 0), B(0, 0), C(0, –9)$ **12.** $A(–3, 0), B(0, 8), C(3, 0)$

13. The rectangle whose vertices are $A(0, 0), B(0, 5), C(h, k)$, and $D(8, 0)$ lies in the first quadrant.
(a) What are the values of h and k?
(b) What is the area of rectangle $ABCD$?

14–15. Find the area of the parallelogram whose vertices are:

14. $A(2, 3), B(5, 9), C(13, 9) D(10, 3)$

15. $A(–4, –2), B(–2, 6), C(10, 6), D(8, –2)$

16–18. Find the area of the trapezoid whose vertices are:

16. $A(0, 0), B(0, 5), C(7, 11), D(7, 0)$

17. $A(0, 0), B(–2, –6), C(9, –6), D(7, 0)$

18. $A(–4, –4), B(–1, 5) C(6, 5), D(9, –4)$

11.2   FINDING THE SLOPE OF A LINE

KEY IDEAS

If you imagine that a line represents a hill, you realize that some lines will be more difficult to "walk up" than other lines. In the accompanying figure, line *l* will be more difficult to climb than line *k* since line *l* is *steeper* than line *k*. Another name for steepness is *slope*. If, as you move along a line from left to right, you are "walking *uphill*," then the line has a *positive* slope; if you are "walking *downhill*," then the line has a *negative* slope.

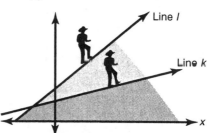

Defining Slope

The **slope** of a line is a measure of its steepness and may be expressed as a number. To do this, select any two different points on the line. In traveling from one point to the other, find the ratio of the change in the vertical distance (difference in the y-coordinates) to the change in the horizontal distance (difference in the x-coordinates). Sometimes the notation Δy (read as "delta y") is used to represent the change in the vertical distance, and Δx (read as ("delta x") to represent the change in the horizontal distance.

In general, the slope, m, of a nonvertical line that passes through points $A(x_1, y_1)$ and $B(x_2, y_2)$ (Figure 11.4) is given by the ratio of the change in the values of their y-coordinates to the change in the value of their x-coordinates:

$$\text{Slope} = m = \frac{\Delta y}{\Delta x} = \frac{y_2 - y_1}{x_2 - x_1}$$

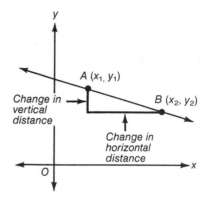

Figure 11.4 Slope of a Line

Example

1. Determine the slope of the line that contains each of the following pairs of points:

    (a) $A(4, 2)$ and $B(6, 5)$ **(c)** $J(4, 3)$ and $K(8, 3)$
    (b) $P(0, 3)$ and $Q(5, -1)$ **(d)** $W(2, 1)$ and $C(2, 7)$

Solution: **(a)** See Figure (a).

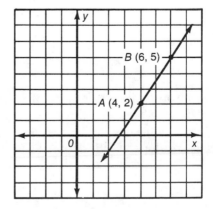

299

To find the slope of $\overleftrightarrow{AB}$, observe that:

$$\Delta y = 5 - 2 = 3$$

$A(4, 2)$ $B(6, 5)$

$$\Delta x = 6 - 4 = 2$$

$$m = \frac{\Delta y}{\Delta x} = \frac{y_2 - y_1}{x_2 - x_1} = \frac{5 - 2}{6 - 4} = \frac{3}{2}$$

The slope of $\overleftrightarrow{AB}$ is $\frac{3}{2}$. The slope of $\overleftrightarrow{AB}$ was calculated assuming that point B was the second point. When using the slope formula, either of the two points may be considered the second point. For example, let's repeat the slope calculation considering point A to the second point (that is, $x_2 = 4$ and $y_2 = 2$):

$$m = \frac{\Delta y}{\Delta x} = \frac{y_2 - y_1}{x_2 - x_1} = \frac{2 - 5}{4 - 6} = \frac{-3}{-2} = \frac{3}{2}$$

The same value, $\frac{3}{2}$, is obtained for the slope of $\overleftrightarrow{AB}$, regardless of whether A is taken as the first or the second point.

(b) See Figure (b).

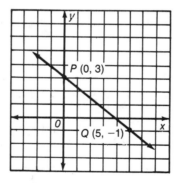

$$m = \frac{\Delta y}{\Delta x} = \frac{y_2 - y_1}{x_2 - x_1} = \frac{-1 - 3}{5 - 0} = \frac{-4}{5}$$

The slope of $\overleftrightarrow{PQ}$ is $\frac{-4}{5}$. The slope of $\overleftrightarrow{AB}$ in part **(a)** was *positive* in value, and the slope of $\overleftrightarrow{PQ}$ has a *negative* value. Compare the directions of these lines. Notice that, if a line has a positive slope, it climbs up *and* to the right; if a line has a negative slope, it falls down *and* to the right.

(c) See Figure (c).

Since the line $\overleftrightarrow{JK}$ is horizontal, there is no change in y, so that $\Delta y = 0$.

$$m = \frac{\Delta y}{\Delta x} = \frac{0}{\Delta x} = 0$$

The slope of $\overleftrightarrow{JK}$ is **0**.

(d) See Figure (d).

Since the line is vertical, there is no change in x, so that $\Delta x = 0$. The slope of $\overleftrightarrow{WC}$ in **undefined** since division by 0 is undefined.

Three Generalizations Concerning Slope

1. An **oblique** line is a line that is *not* parallel to a coordinate axis. The slope of an oblique line may be either positive or negative.

If, as x increases, the line *rises,* then the slope m of the line is *positive.*

If, as x increases, the line *falls,* then the slope m of the line is *negative.*

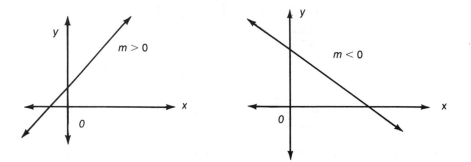

2. The slope m of a horizontal line is 0.

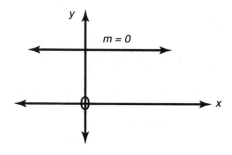

3. The slope m of a vertical line is not defined.

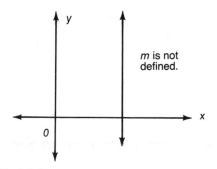

Slopes of Parallel Lines

In Figure 11.5, the slope of the line containing $A(1, 3)$ and $B(6, 8)$ is the same as the slope of the line containing $C(3, 0)$ and $D(7, 4)$. The graphs of the two lines determined by these points show that the lines are parallel.

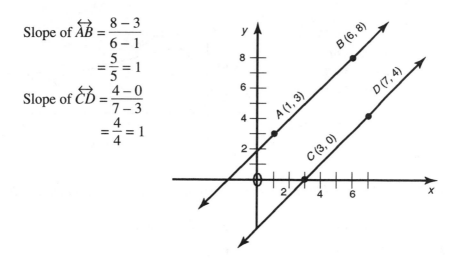

$$\text{Slope of } \overleftrightarrow{AB} = \frac{8-3}{6-1}$$

$$= \frac{5}{5} = 1$$

$$\text{Slope of } \overleftrightarrow{CD} = \frac{4-0}{7-3}$$

$$= \frac{4}{4} = 1$$

Figure 11.5   Slopes of Parallel Lines

Here are two generalizations about parallel lines and slope (Figure 11.6):
- If two nonvertical lines are parallel, then their slopes are equal.
- If two nonvertical lines have the same slope, then they are parallel.

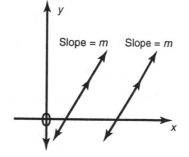

Figure 11.6   Parallel Lines and Slope

Example

2. The coordinates of the vertices of quadrilateral *MATH* are $M(-3, 2)$, $A(4, 8)$, $T(15, 5)$, and $H(8, y)$. If *MATH* is a parallelogram, find the value of y.

Solution: The opposite sides of a parallelogram are parallel and have, therefore, the same slope.

$$\text{Slope } \overline{MA} = \text{Slope } \overline{HT}$$

$$\frac{8-2}{4-(-3)} = \frac{y-5}{8-15}$$

303

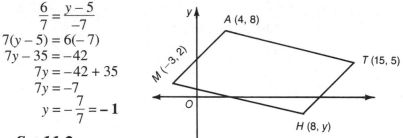

$$\frac{6}{7} = \frac{y-5}{-7}$$
$$7(y-5) = 6(-7)$$
$$7y - 35 = -42$$
$$7y = -42 + 35$$
$$7y = -7$$
$$y = -\frac{7}{7} = -1$$

Exercise Set 11.2

1–8. Find the slope of the line that passes through each of the following pairs of points:

1. $(2, -7)$ and $(5, -7)$ **5.** $(-3, -4)$ and $(3, 8)$

2. $(4, 3)$ and $(4, -1)$ **6.** $(5, -6)$ and $(0, -6)$

3. $(-3, 5)$, and $(4, 7)$ **7.** $(0, -7)$ and $(-2, -3)$

4. $(5, 4)$ and $(-7, 8)$ **8.** $(-4, 9)$ and $(-4, -9)$

9–11. In each of the following cases, determine whether $\overleftrightarrow{AB}$ is parallel to $\overleftrightarrow{CD}$:

9. $A(1, 5)$, $B(-1, 9)$; $C(2, 6)$, $D(1, 8)$

10. $A(-2, 7)$, $B(1, 4)$; $C(-8, 3)$, $D(-7, 4)$

11. $A(1, -5,)$ $B(-4, 5)$; $C(0, -7)$, $D(4, 9)$

12. The slope of $\overleftrightarrow{AB}$ is $\frac{3}{5}$, and the slope of $\overleftrightarrow{CD}$ is $\frac{9}{k}$. If $\overleftrightarrow{AB}$ is parallel to $\overleftrightarrow{CD}$, what is the value of k?

13. Points $A(2, k)$ and $B(6, 10)$ determine a line whose slope is $\frac{1}{2}$. Find k.

14. The coordinates of the vertices of parallelogram $ABCD$ are $A(0, 0)$, $B(5, 0)$, $C(8, 1)$, and $D(x, 1)$. The value of x is:
(1) 1   (2) 2   (3) 3   (4) 4

15. The coordinates of the vertices of parallelogram $ABCD$ are $A(x, y)$, $B(4, 10)$, $C(12, 10)$, and $D(9, 4)$. Find the value of x and y.

16. The line joining $A(-2, 0)$ and $B(10, 3)$ is parallel to the line joining $C(5, 7)$ and $D(1, k)$. Find the value of k.

17. The coordinates of the vertices of trapezoid $ABCD$ are $A(1, 5)$, $B(7, k)$, $C(2, -4)$, and $D(-7, -1)$. If $\overline{AB}$ and $\overline{DC}$ are the bases of the trapezoid, find the value of k.

18. In quadrilateral $ABCD$, the coordinates of the vertices are $A(3, 10)$, $B(11, 10)$, $C(3, -2)$, and $D(-5, -2)$.
(a) Using slopes, show that $ABCD$ is a parallelogram.
(b) Find the area of $ABCD$.

19. In trapezoid $ABCD$ with bases $\overline{AD}$ and $\overline{BC}$, the coordinates of the vertices are $A(3, 1)$, $B(1, 7)$, $C(4, \underline{9})$, and $D(k, 5)$.
(a) What is the slope of $\overline{BC}$?
(b) Using your answer in part (a), find k.

20. The coordinates of the vertices of quadrilateral $ABCD$ are $A(2, 0)$, $B(10, 2)$, $C(6, 7)$, and $D(2, 6)$.
(a) Using slopes, show that $\overline{AB}$ is parallel to $\overline{CD}$, and state a reason for your conclusion.
(b) Using slopes, show that quadrilateral $ABCD$ is *not* a parallelogram, and state a reason for your conclusion.

21. The vertices of parallelogram $STWU$ are $S(1, 1)$, $T(-2, 3)$, $W(0, b)$, and $U(3, -5)$.
(a) Find the slope of $\overline{ST}$.   (b) Find the value of b.

22. Parallelogram $ABCD$ has vertices $A(2, -1)$, $B(8, 1)$, and $D(4, k)$. The slope of $\overline{AD}$ is equal to 2.
(a) Find k.   (b) Find the coordinates of C.

23. The vertices of a triangle are $P(1, 2)$, $Q(-3, 6)$, and $R(4, 8)$.
(a) Find the slope of $\overline{PR}$.
(b) A line through point Q is parallel to $\overleftrightarrow{PR}$. If this line contains point $(x, 14)$, find the value of x.

24–27. Three points lie on the same line if the slope of the line determined by the first two points is the same as the slope of the line determined by the first and the last point. Determine whether the points in each of the following sets are collinear:

24. $(-4, -5)$, $(0, -2)$, and $(8, 4)$ **26.** $(1, 2)$, $(5, 8)$ and $(-3, -4)$

25. $(-3, 2)$, $(4, 2)$, and $(-5, 2)$ **27.** $(2, 1)$, $(10, 7)$, and $(-4, -6)$

28. If $D(7, k)$ is a point on the line joining $A(1, 1)$ and $B(10, 4)$, what is the value of k?

29. If $E(h, 6)$ is a point on the line joining $A(0, 1)$ and $B(-2, -1)$, what is the value of h?

30. The coordinates of the vertices of trapezoid $ABCD$ are $A(3, 0)$, $B(7, 0)$, $C(7, 11)$, and $D(3, 8)$.
(a) Find the slope of diagonal $\overline{BD}$.
(b) Find the area of the trapezoid.
(c) Find the perimeter of the trapezoid.

31. In the accompanying diagram, find:

    (a) the slope of $\overleftrightarrow{BD}$

    (b) the slope of $\overleftrightarrow{CD}$

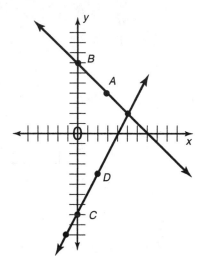

11.3 USING THE SLOPE-INTERCEPT FORM OF A LINEAR EQUATION

KEY IDEAS

In a *linear equation* the highest power of any variable is 1. The equation $x + y = 5$ is an example of a linear equation. A linear equation may also have a single variable. The equations $x = 2$ and $y = -1$ are also examples of linear equations.

    In general, a **linear equation** is any equation that can be written in the form

$$Ax + By + C = 0,$$

where A, B and C are constants and A and B are not both equal to 0. The graph of a linear equation is a straight line.

Equations of Lines

The solution set of the equation $x + 2 = 5$ has exactly *one* member, which is 3. The solution set of the equation $x + y = 5$ contains *all* ordered pairs of numbers such that the sum of the x- and y-coordinates is 5. Thus, there are an *infinite* number of ordered pairs in the solution set. Some of these are:

(1, 4) since 1 + 4 = 5,   (−3, 8) since −3 + 8 = 5,
(2, 3) since 2 + 3 = 5,   (2.5, 2.5) since 2.5 + 2.5 = 5.

The graph of each of these ordered pairs of numbers (Figure 11.7) is a different point on the same line. The line represents the set of *all* ordered pairs of numbers that satisfy the equation $x + y = 5$.

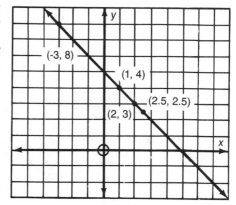

Figure 11.7   Graph of Ordered Pairs of Numbers

Examples

1. The graph of $3x - y = 6$ contains which of the following points?
(1) (0, 2)   (2) (5, 9)   (3) (4, −6)   (4) (1, 9)

Solution: Through trial and error determine the point that satisfies the equation. Substituting the values in choice (2) gives

$$\begin{array}{r} 3x - y = 6 \\ \hline 3(5) - (9) \\ 15 - 9 \\ 6 = 6 \checkmark \end{array}$$

The correct answer is choice (**2**).

2. The line whose equation is $y = 3x - 1$ contains point $A(2, k)$. What is the value of $k?$

Solution: Since the point lies on the line, its coordinates must satisfy the equation of the line. To find k, replace x by 2 and y by k in the original equation.

$$\begin{aligned} k &= 3x - 1 \\ &= 3(2) - 1 \\ &= 6 - 1 = \mathbf{5} \end{aligned}$$

3. The line whose equation is $Ax + 3y = 13$ passes through point (8, −1). What is the value of $A?$

Solution: Since the line passes through the point, the coordinates of the point must satisfy the equation. To find the value of A, replace x by 8 and y by − 1.

$$\begin{aligned}
Ax + 3y &= 13 \\
8A + 3(-1) &= 13 \\
8A - 3 &= 13 \\
8A &= 13 + 3 \\
\frac{8A}{8} &= \frac{16}{8} \\
A &= \mathbf{2}
\end{aligned}$$

Slope-Intercept Form

The y-coordinate of the point at which a nonvertical line intersects the y-axis is called the **y-intercept** of the line. Whenever a linear equation is written in the form $y = mx + b$, the equation is said to be in *slope-intercept* form since the graph of this equation (Figure 11.8) is a nonvertical line that has a slope of m and intersects the y-axis at (o, b). For example,

$$y = 2x + 4$$

slope of line ⎯⎯

y-intercept of line ⎯⎯

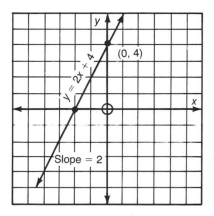

Figure 11.8 Graph of an Equation in Slope-Intercept Form

Finding the Slope and the *y*-Intercept of a Linear Equation

To determine the slope and the y-intercept of a line from its equation, write the equation in the form $y = mx + b$. In order to do this, it may be necessary to solve the equation for y in terms of x.

Examples

4. Find the slope and the *y*-intercept of a line whose equation is:

(a) $y = x + 2$ (c) $y - 5x = 0$

(b) $y = \dfrac{2x}{5} - 3$ (d) $y + 2x - 7 = 0$

Solution: (a) $y = x + 2 = 1 \cdot x + 2$. The slope is **1**, and the *y*-intercept is **2**.

(b) The slope is $\dfrac{2}{5}$, and the *y*-intercept is **–3**.

(c) $y - 5x = 0$ may be rewritten as $y = 5x + 0$. The slope is **5**, and the *y*-intercept is **0**. Note that, if the *y*-intercept of a line is 0, then the line passes through the origin.

(d) Solve for *y* in terms of *x*:

$$y + 2x - 7 = 0$$

Subtract 2*x* from each side: $y - 7 = -2x$
Add 7 to each side: $y = -2x + 7$

The slope of the line is **–2**, and the *y*-intercept is **7**.

5. Find the slope of a line that is parallel to the line whose equation is $2x + 3y = 12$.

Solution: First determine the slope of the line whose equation is obtained by solving for *y* in terms of *x*:

$$2x + 3y = 12$$

Subtract 2*x* from each side: $3y = -2x + 12$

Divide each side by 3: $\dfrac{\overset{1}{\cancel{3}y}}{\cancel{3}} = \dfrac{-2x}{3} + \dfrac{\overset{4}{\cancel{12}}}{\underset{1}{\cancel{3}}}$

$$y = \dfrac{-2x}{3} + 4$$

The slope of the given line is $\tfrac{-2}{3}$. Since parallel lines have the same slope, the slope of the line parallel to the line whose equation is $2x + 3y = 12$ is also $\tfrac{-2}{3}$.

6. Determine whether the line whose equation is $2y - 4x = 9$ is parallel to the line whose equation is $y + 4 = 2x$.

Solution: Compare the slopes of the two lines after expressing each equation in the form $y = mx + b$:

$$2y - 4x = 9 \qquad\qquad\qquad y + 4 = 2x$$
$$2y = 4x + 9 \qquad\qquad\qquad y = 2x - 4$$
$$\frac{2y}{2} = \frac{4x}{2} + \frac{9}{2}$$
$$y = 2x + \frac{9}{2}$$

Since the slope of each line is 2, **the lines are parallel**.

Writing an Equation of a Line

Knowing certain information about a line allows us to determine the values of m and b in the equation $y = mx + b$.

Examples

7. Write an equation of a line whose slope is -2 and which passes through point $(1, 4)$.

Solution: Step 1. Find the value of b.

$$y = mx + b$$
Replace m by -2: $\qquad\qquad y = (-2)x + b$
Replace x by 1 and y by 4: $\qquad 4 = (-2)(1) + b$
Simplify: $\qquad\qquad\qquad\qquad 4 = -2 + b$
Solve for b: $\qquad\qquad\qquad 4 + 2 = b$
$$6 = b$$

Step 2. Replace m and b with their numerical values. Since $m = -2$ and $b = 6$, the equation of the line is **$y = -2x + 6$**.

8. Write an equation of the line that contains points $A(6, 0)$ and $B(2, -6)$.

Solution: Step 1: Find the slope of the line.

$$m = \text{slope of } \overleftrightarrow{AB} = \frac{-6 - 0}{2 - 6}$$

$$= \frac{-6}{-4} = \frac{3}{2}$$

Step 2. Using the x and the y value of either point, find the value of b.

$$y = mx + b$$
Replace m by $\dfrac{3}{2}$: $\qquad\qquad y = \dfrac{3}{2}x + b$
Replace x by 6 and y by 0: $\qquad 0 = \dfrac{3}{2}(6) + b$
Simplify: $\qquad\qquad\qquad\qquad 0 = 9 + b$
Solve for b: $\qquad\qquad\qquad -9 = b$

310

Step 3. Replace *m* and *b* with their numerical values.

Since $m = \dfrac{3}{2}$ and $b = -9$ an equation of the line is $y = \dfrac{3}{2}x - 9$.

9. Write an equation of the horizontal line whose *y*-intercept is –7.

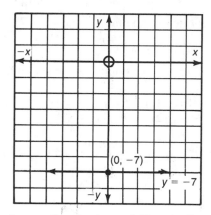

Solution: The slope of a horizontal line is 0, so $y = mx + b$ becomes $y = 0x - 7 = \mathbf{-7}$.

10. Write an equation of the line that is parallel to the *y*-axis and 4 units to the right of the origin.

Solution: The desired line is a vertical line, so the *x*-coordinate of each point is always equal to the same number. Since the line is 4 units to the right of the origin, *x* is always equal to 4.

An equation of the line is $x = 4$.

The preceding examples suggest the generalizations in the table that follows:

General Form of Equation	Description
$x = a$	• Vertical line (parallel to *y*-axis). • Each point on the line has *a* as its *x*-coordinate.
$y = b$	• Horizontal line (parallel to *x*-axis). • Each point on the line has *b* as its *y*-coordinate.
$y = mx + b$	• Nonvertical line whose slope is equal to *m* and whose *y*-intercept is *b*.

Exercise Set 11.3

1. Which point lies on the graph of $2x + y = 10$?
   (1) $(0, 8)$ (2) $(10, 0)$ (3) $(3, 4)$ (4) $(4, 3)$

2. Which point does *not* lie on the graph of $3x - y = 7$?
   (1) $(2, -1)$ (2) $(3, 2)$ (3) $(-1, 4)$ (4) $(1, -4)$

3. Which of the following pairs of points determines a line that is parallel to the x-axis?
   (1) $(3, 5)$ and $(5, 3)$ (3) $(3, 5)$ and $(3, -2)$
   (2) $(3, 5)$ and $(-3, -5)$ (4) $(3, 5)$ and $(-2, 5)$

4. Which of the following pairs of points determines a line that is parallel to the y-axis?
   (1) $(4, 7)$ and $(7, 4)$ (3) $(4, 7)$ and $(-4, 7)$
   (2) $(4, 7)$ and $(4, -7)$ (4) $(4, 7)$ and $(-4, -7)$

5. Which of the following is an equation of a line that is parallel to the y-axis and 2 units to the right of it?
   (1) $x = 2$ (2) $x = -2$ (3) $y = 2$ (4) $y = -2$

6. Which of the following is an equation of a line that is parallel to the x-axis and 3 units below it?
   (1) $x = 3$ (2) $x = -3$ (3) $y = 3$ (4) $y = -3$

7. The line whose equation is $3x + Ay = 17$ passes through point $(3, 2)$. What is the value of A?

8. The line whose equation is $2y + x = 9$ contains point $P(-1, k.)$ What is the value of k?

9–14. Find the slope and the y-intercept of the graphs of each of the following equations:

9. $y - 3x = 1$ 12. $3y + x = 6$

10. $y + 4x - 2 = 0$ 13. $2y - x = 10$

11. $y + 1 = \dfrac{2}{5} x$ 14. $4y - 3x = 12$

15–18. If m represents the slope of a line and b its y-intercept, write an equation of the line when:

15. $m = -2$ and $b = 3$ 17. $m = \dfrac{1}{3}$ and $b = -1$

16. $m = 4$ and $b = 0$ 18. $m = 0$ and $b = -2$

19. Write an equation of a line that is parallel to the x-axis and passes through $(4, -2)$.

20. Write an equation of a line that is parallel to the y-axis and passes through $(-3, 1)$.

21–26. Find an equation of the line that has slope m *and contains the given point.*

21. $m = 1$; $(2, 2)$

22. $m = 3$; $(-1, 4)$

23. $m = -2$; $(5, -3)$

24. $m = \dfrac{1}{2}$; $(-4, -5)$

25. $m = 0$; $(3, 1)$

26. m is undefined; $(-2, 7)$

27. Write an equation of a horizontal line that:
(a) is 4 units above the x-axis
(b) has a y-intercept of -1
(c) contains point $(-2, 3)$
(d) has the same y-intercept as the line whose equation is $y + 2 = 3x$.

28. Write an equation of a vertical line that:
(a) is 5 units to the right of the y-axis
(b) has an x-intercept of -3
(c) contains point $(-8, -7)$
(d) has the same x-intercept as the line whose equation is $2x - 3y = 5$.

29. (a) Find the slope of the line parallel to the line whose equation is $2y = x + 3$.
(b) Write an equation of the line that contains point $(4, -1)$ and is parallel to the line whose equation is $2y = x + 3$.

30. (a) Find the area of the triangle whose vertices are $A(0, 6)$, $B(0, 0)$, and $C(-8, 0)$.
(b) Write and equation of the line that passes throught B and is parallel to $\overline{AC}$.

31. Determine whether the lines whose equations are given are parallel.
(a) $2y - x = 9$
$y = \dfrac{1}{2}x + 5$
(b) $2x - 4y = 8$
$2y - x = 9$
(c) $3x - y = 6$
$2y + 6x = 12$
(d) $3y - 2x = 10$
$6x - 4 = 9y$

32–37. Write an equation of the line that is determined by each pair of points.

32. $(1, 5), (2, 1)$ **35.** $(5, 7), (3, 6)$

33. $(-2, 9), (1, -6)$ **36.** $(-8, -3), (-1, 4)$

34. $(2, -4), (-3, -4)$ **37.** $\left(\dfrac{1}{2}, 5\right), \left(\dfrac{3}{2}, 2\right)$

38. Write an equation of one of the lines that are parallel to the line whose equation is $3y - 12x = 6$ and which intersects the line whose equation is $y + 2x = 10$.

39. **(a)** For the accompanying graph, determine and then write an equation of line *l* in slope-intercept form.
(b) Determine and then write an equation of line *m* in slope-intercept form.

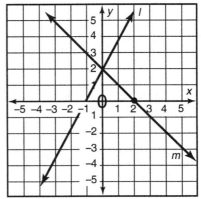

11.4   GRAPHING LINEAR EQUATIONS

KEY IDEAS

Here are some general guidelines for graphing linear equations:
- Draw the coordinate axes on graph paper, using a straightedge.
- Label the horizontal axis *x*, and the vertical axis *y*.
- Label the origin, and number the boxes along each axis sequentially.
- Choose one of the methods illustrated in this section.
- Graph the line, and label it with its equation.

Three-Point Method

To draw the graph of a linear equation, we must obtain the coordinates of at least two points that satisfy the equation. Graphing a third point and verifying that this point also lies on the line serves as a check. To graph the line whose equation is $y - 2x = 3$, using the three-point method, follow these steps:

Step	Example

1. Solve the original equation for y in terms of x.

$y - 2x = 3$
$y = 2x + 3$

2. Choose any three convenient values for x, and then calculate the corresponding values of y. The numbers -1, 0, and 1 are often good choices for x. It is helpful to organize your work in a table.

x	$y = 2x + 3$	(x, y)
-1	$y = 2(-1) + 3$ $= -2 + 3 + 1$	$(-1, 1)$
0	$y = 2(0) + 3$ $= 0 + 3 = 3$	$(0, 3)$
1	$y = 2(1) + 3$ $= 2 + 3 = 5$	$(1, 5)$

3. Graph the three points obtained in step 2. Use a straightedge to draw the line. Label the line with its equation, $y - 2x = 3$. If the three points do *not* lie on the same line, check your work since at least one of the points has been graphed incorrectly. Make certain that you have correctly numbered the axes and that each of the points has been accurately graphed. If the graphing of the points is correct, then return to step 2 and verify that the coordinates you are using are correct.

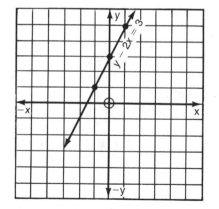

Intercept Method

An oblique line that does not pass through the origin intersects the x-axis at a value of x called the *x-intercept* and the y-axis at the value of y called the *y-intercept*. To graph the equation $2x - 3y = 12$, using the intercept method, follow these steps:

315

Step	Example
1. Find the x-intercept by setting $y = 0$ in the original equation and then solving for x. Note that $y = 0$ all along the x-axis.	$2x - 3y = 12$ $2x - 3(0) = 12$ $2x = 12$ $x = \dfrac{12}{2} = 6$ The x-intercept is 6.
2. Find the y-intercept by setting $x = 0$ in the original equation and then solving for y. Note that $x = 0$ all along the y-axis.	$2x - 3y = 12$ $2(0) - 3y = 12$ $-3y = 12$ $y = \dfrac{12}{-3} = -4$ The y-intercept is -4.
3. Use the intercepts to graph the points where the line crosses the axes, and then draw a line through these points. Label the line with its equation, $2x - 3y = 12$. ***Note 1:*** As a check, choose any convenient point that lies on the line and verify algebraically that its coordinates satisfy the equation. ***Note 2:*** Lines that have equations of the form $y = mx$ pass through the origin and, therefore, intersect the x- and y-axes at the same point, so this method cannot be used for such equations.	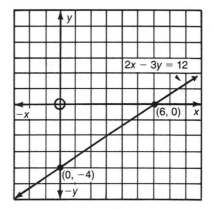

Slope-Intercept Method

To graph a linear equation such as $y = \frac{2}{5}x + 1$, using the slope-intercept method, follow these steps:

Step	Example

1. Use the *y*-intercept to graph the point where the line crosses the *y*-axis. The coordinate of the *y*-intercept of $y = \frac{2}{5}x + 1$ is 1, so the line crosses the *y*-axis at $A(0, 1)$.

2. Find another point of the given line, using the slope. The slope of $y = \frac{2}{5}x + 1$ is $\frac{2}{5}$. Start at $A(0, 1)$, and move up 2 units. Then move 5 units to the right. Label this point *B*. Note that the coordinates of *B* are $(5, 3)$.

3. Find another point that will serve as a check point by moving up 2 units from point *B*, and then moving 5 units to the right. Label this point *C*. The coordinates of *C* are $(10, 5)$.

4. Draw $\overline{AB}$ and label the line with its equation, $y = \frac{2}{5}x + 1$.

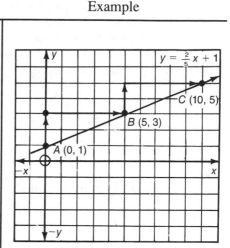

Example

Use the slope-intercept method to graph the line whose equation is $y + 3x = 5$.

Solution: Put the equation in slope-intercept form.
$$y + 3x = 5$$
$$y = -3x + 5$$

Obtain two points on the line by following these steps:

Step 1. Use the *y*-intercept, 5. The coordinates of the intersection of $y = -3x + 5$ with the *y*-axis are $A(0, 5)$.

Step 2. Graph another point of the given line, using the slope. The slope of $y = -3x + 5$ is -3 or $-3/1$. Since the numerator is *negative*, start at $A(0, 5)$ and move 2 units *down*. Then move 1 unit to the right. Label this point *B*. Note that the coordinates of *B* are $(1, 2)$.

Step 3. Starting at B, find another point that will serve as a check by moving 3 units down from point B, and then moving 1 unit to the right. Label this point C. The coordinates of C are $(2, -1)$.

Step 4. Draw $\overline{AB}$, which represents the graph of $y = -3x + 5$ (or $y + 3x = 5$).

Exercise Set 11.4

1–8. Using graph paper, draw the graph of each equation using the three-point method.

1. $y = 3x$

2. $y = -3x$

3. $y = x + 2$

4. $y = -2x + 1$

5. $y + 3 = x$

6. $y = 3x + 6$

7. $y = \frac{1}{2}x + 4$

8. $x - 2y = 8$

9–12. Using graph paper, draw the graph of each equation using the x- and y-intercepts method.

9. $x + y = 7$

10. $2y - x = 8$

11. $3y - 2x = 6$

12. $3x - 5y = 15$

13–16. Using graph paper, draw the graph of each equation using the slope-intercept method.

13. $y = 3x - 1$

14. $y = -2x + 4$

15. $2y - x = 10$

16. $x - 2y = 3$

17–20. Using graph paper, draw the graph of each equation using any method.

17. $y + 2x = 0$

18. $3y - 2x = 18$

19. $y = -\frac{1}{3}x + 2$

20. $8 - 2y = x$

21. **(a)** Graph $y - 2x = 7$.

(b) On the same set of axes, graph $x + y = -2$.

(c) Determine the coordinates of the point at which the two lines intersect.

11.5 WORKING WITH DIRECT VARIATION

===== △ KEY IDEAS △ =====

When two variables are related so that the ratio of their values always remains the same, the two variables are said to be in **direct variation**. If y varies directly as x, then the graph that describes this relationship is a line through the origin whose slope is called the *constant of variation*.

Expressing Direct Variation as an Equation

The equation $\frac{y}{x} = 2$ states that y "varies directly as" x since the ratio of y to x remains constant. The number 2 is called the *constant of variation*. The equation $\frac{y}{x} = 2$ can also be written in the equivalent form, $y = 2x$.

===== MATH FACTS =====

EQUATION OF DIRECT VARIATION

If y varies directly as x, then $y = kx$ where k is some fixed nonzero number called the **constant of variation**.

Algebraic Interpretation of Direct Variation

For an equation of the form $y = kx$, multiplying x by some fixed amount also multiplies y by the same fixed amount. For example, since the perimeter P of a square varies directly as the length of a side s of the square, $P = 4s$. If the length of a side of a square is doubled, then the perimeter of the square is also doubled. If the length of a side of a square is tripled, then the perimeter of the square is also tripled; and so forth.

Geometric Interpretation of Direct Variation

The equation $y = kx$ is a special case of the linear equation $y = mx + b$, with $b = 0$. Thus, a line through the origin always represents a direct variation between y and x. The slope of this line is the constant of variation.

Examples

1. Identify the table of values that shows that y varies directly as x and then write an equation that expresses the direct variation.

(1)

x	2	4	6
y	5	7	9

(3)

x	3	5	7
y	9	15	21

(2)

x	2	4	6
y	6	3	2

(4)

x	3	5	7
y	2	4	6

Solution: **Table (3)** is the only table in which for each ordered pair the ratio of y to x remains the same. That is,

$$\frac{y}{x} = \frac{9}{5} = \frac{15}{5} = \frac{21}{7} = 3 = k$$

Since the constant of variation is 3, an equation of this direct variation is **$y = 3x$.**

2. If y varies directly as x, and $y = 8$ when $x = 12$, find k and write an equation that expresses this variation.

Solution: Since y varies directly as x,

$$y = kx$$
$$8 = k(12)$$
$$\frac{8}{12} = \frac{12k}{12}$$
$$\frac{2}{3} = k$$

Since the constant of variation is $\frac{2}{3}$, an equation that expresses this variation is **$y = \frac{2}{3}x$.**

3. If y varies directly as x, and $y = 24$ when $x = 16$, find y when $x = 12$.

Solution:

Method 1: First find k and then an equation of variation.

Let $y = 24$ and $x = 16$:
$$y = kx$$
$$24 = k(16)$$
$$\frac{24}{16} = \frac{16k}{16}$$
$$\frac{3}{2} = k$$

Hence, an equation of direct variation is $y = \frac{3}{2}x$. If $x = 12$, then

$$y = \frac{3}{2}(12)$$
$$y = \mathbf{18}.$$

Method 2: Since the ratios of corresponding values of y to x are always equal, form and then solve the related proportion. Thus,

$$\frac{y}{x} = \frac{24}{16} = \frac{y}{12}$$

Equate the cross-products: $16y = (24)(12)$

$$y = \frac{288}{16}$$
$$y = \mathbf{18}.$$

4. If 20 identical coins weigh 42 grams, how many coins weigh 105 grams?

Solution: Since the coins are identical we can assume that their weight varies directly as their number. If x represents the number of coins that weigh 105 grams, then

$$\frac{\text{weight}}{\text{number}} = \frac{42}{20} = \frac{105}{x}$$

Equate the cross-product: $42x = 20(105)$

$$\frac{42x}{42} = \frac{2100}{42}$$
$$x = \mathbf{50}$$

Exercise Set 11.5

1. If y varies directly as x, and $y = 32$ when $x = 24$, find y when $x = 18$.

2. If f varies directly as g, and $f = 81$ when $g = 27$, find g when $f = 21$.

3–6. Determine whether the given table represents a direct variation between x *and* y. *If it does, write an equation that describes the direct variation.*

3.

x	2	4	6
y	3	6	9

5.

x	4	8	12
y	2	3	4

4.

x	1	3	5
y	2	8	14

6.

x	10	15	20
y	6	9	12

7. If 2.54 centimeters = 1 inch, how many centimeters are equivalent to 1 foot?

8. On a certain map 1.5 inches represents 100 miles. If two cities on this map are 12 inches apart, what is their distance in miles?

9. A recipe for four servings requires $\frac{2}{3}$ cup of sugar. How many cups of sugar are needed if the same recipe is used to prepare six servings?

10. If four pairs of socks cost $10.00, how many pairs of socks can be purchased for $15.00?

11.6 MOVING SETS OF POINTS IN THE PLANE

KEY IDEAS

The process of moving each point of a figure according to some given rule is called a **transformation**. Each point of the new figure corresponds to exactly one point of the original figure and is called the **image** of that point. *Reflections, translations, rotations,* and *dilations* are special types of transformations; each uses a different rule for locating the images of points of the figure that are undergoing the transformation.

Reflecting a Point in a Line

The image of point P under a transformation is usually denoted by P', read as "P prime." To reflect a point P in line l, follow these rules:

1. If point P does *not* lie on line l, then the *reflection* of P is the endpoint P' of $\overline{PP'}$ drawn so that line l is the perpendicular bisector of $\overline{PP'}$. P' is the *image* of point P. The original point P is called the preimage of P'.

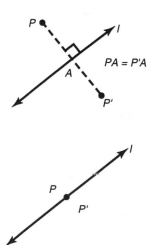

2. If point P lies on line l, then the reflection of point P is the point P', which coincides with P.

Example

1. In which figure is $\triangle A'B'C'$ a reflection of $\triangle ABC$ in line l?

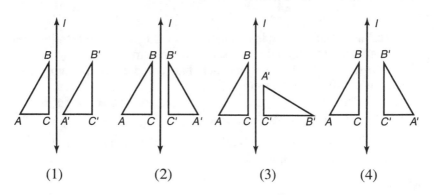

(1)　　　　　　(2)　　　　　　(3)　　　　　　(4)

Solution: In order for △ *A'B'C'* to be a reflection of △*ABC* in line *l,* the corresponding vertices of the two triangles must be on opposite sides of line *l* and the same distance from line *l.* Thus, in **Figure (2)** △ *A'B'C'* is a reflection of △ *ABC* in line *l.*

Transformations That Produce Congruent Figures

A reflection, translation, and rotation of △ *ABC* are shown in Figure 11.9.

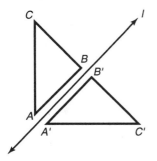

Reflection of △*ABC* in line *l*

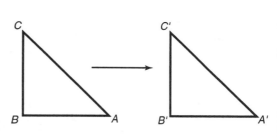

Translation of △*ABC*

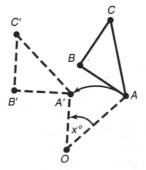

Counterclockwise rotation of △*ABC* x° about point O.
Figure 11.9   Transformations of △*ABC*

323

- A **reflection** of a figure in a line may be thought of as the "mirror image" of the figure in which the line serves as the "mirror."
- A **translation** "slides" a figure such that each point of the figure is moved the same distance in the same direction.
- A **rotation** "turns" a figure a fixed number of degrees about some fixed point called the *center of rotation.*

Since reflections, translations, and rotations do not change lengths and angle measures, they produce figures that are *congruent* to the original figures.

Example

2. In the accompanying diagram, sketch the counterclockwise rotation of rectangle *ABCD* **(a)** 90°, **(b)** 180°, and **(c)** 270° about point *A*.

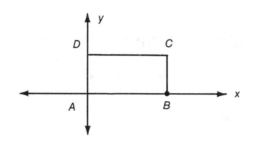

Solution:

(a) 90° Rotation

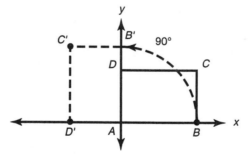

(b) 180° Rotation

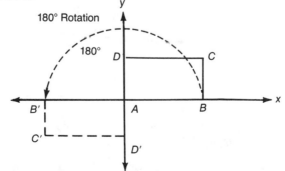

(c) 270° Rotation

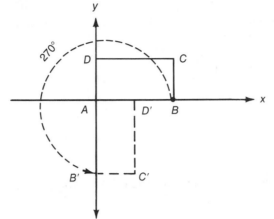

A Size Transformation

A **dilation** is a transformation that shrinks or magnifies the size of a figure by a fixed amount called the **scale factor**. Since a dilation preserves angle measure, it produces a figure that is *similar* to the original figure. For example, using a scale factor of 3, the dilation of a square is another square that is three times as large as the original square. Thus, if the length of each side of the original square is 2 inches, the length of each side of the new square is 3 × 2, or 6 inches.

Exercise Set 11.6

1. Which of the following types of transformations produces a figure similar to the original figure?
(1) Reflection   (2) Dilation   (3) Translation   (4) Rotation

2. Which property is *not* preserved under a line reflection?
(1) Collinearity of points   (3) Congruence of angles
(2) Congruence of line segments   (4) Orientation

3. Sketch the counterclockwise rotation of the figure in the accompanying diagram 180° about point *P*.

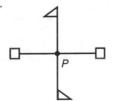

4. In the accompanying diagram, sketch the clockwise rotation of right triangle *ABC*:
(a) 90° about point *A*
(b) 180° about point *A*
(c) 270° about point *A*
(d) 360° about point *A*

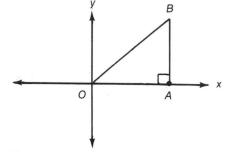

5. Draw equilateral triangle *ABC* and rotate it 120° clockwise about
(a) the center *O* of the triangle
(b) vertex *A*

6. Sketch the counterclockwise rotation of rectangle *ABCD* about point *E*.
(a) 90°   (b) 180°   (c) 270°

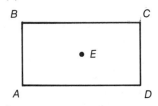

7. The perimeter of a square is 48. After a dilation, the perimeter of the new square is 72. What is the scale factor?

8. The three vertices of △*ABC* are located in quadrant II. The image of △*ABC* after a reflection in the *x*-axis is △*A'B'C'*. The image of △*A'B'C'* after a reflection in the *y*-axis is △*A"B"C"*. In which quadrant is △*A"B"C"* located?

9. The area of a circle is 16π. After the circle is dilated the area of the new circle is 64π. What is the scale factor?

10. Square *ABCD* is rotated in a clockwise direction about its center *O*. Name the vertex of the square that is the image of point *B* after *ABCD* is rotated:
(a) 90°   (b) 180°   (c) 270°

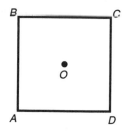

11. Regular hexagon *ABCDEF* is rotated in a counterclockwise direction about center *O*. Name the vertex of the hexagon that is the image of point *A* after *ABCDEF* is rotated
(a) 60°   **(b)** 180°   **(c)** 360°

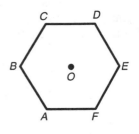

11.7   OBSERVING SYMMETRY

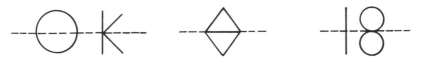

KEY IDEAS

There are many examples of symmetry in nature. People's faces, leaves, and butterflies have real or imaginary "lines of symmetry" that divide the figures into two parts that are mirror images. If a geometric shape has line symmetry, then it can be "folded" along the line of symmetry so that the two parts coincide.

Line Symmetry

A figure has line symmetry if a line can be drawn that divides the figure into two parts that are mirror images. The line of symmetry may be a horizontal line, a vertical line, or neither.

The shapes in Figure 11.10 have horizontal line symmetry.

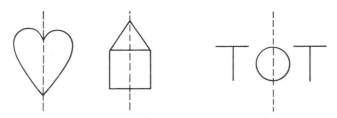

Figure 11.10   Horizontal Line Symmetry

The examples in Figure 11.11 have vertical line symmetry.

Figure 11.11   Vertical Line Symmetry

Figure 11.12 has a line of symmetry that is neither horizontal nor vertical.

Figure 11.13 illustrates that a figure may have both a horizontal and a vertical line of symmetry.

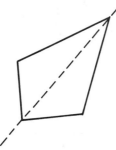

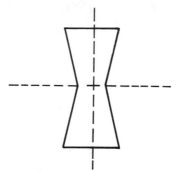

Figure 11.12 Neither Horizontal nor Vertical Line Symmetry

Figure 11.13 Both Horizontal and Vertical Line Symmetry

As shown in Figure 11.14, a figure may have more than one line of symmetry or may have no lines of symmetry.

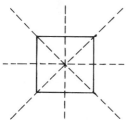

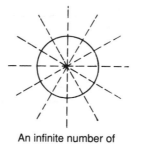

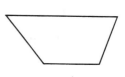

Four lines of symmetry

An infinite number of lines of symmetry

No lines of symmetry

Figure 11.14 Figures with More Than One or No Lines of Symmetry

Examples

1. Which letter has both a vertical and a horizontal line of symmetry?
(1) A (2) M (3) T (4) X

Solution: The correct choice is X since:

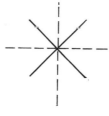

Horizontal line of symmetry

Vertical line of symmetry

The correct answer is **choice (4)**.

328

2. Which of the following triangles has line symmetry?
(1) Right   (2) Isosceles   (3) Obtuse   (4) Scalene

Solution: A line (angle bisector or altitude)
through the vertex angle of an isosceles triangle
divides the triangle into two mirror-image (that
is, congruent) triangles.
   The correct answer is **choice (2)**.

3. Which word has a horizontal line of symmetry?
(1) MOM   (2) EVE   (3) DAD   (4) BOB

Solution: The correct choice is BOB since ~~BOB~~.
   The correct answer is **choice (4)**.

Point Symmetry

A figure has **point symmetry** if, after undergoing a rotation of 180° in either
direction about a fixed point *P,* the figure coincides with itself. The figure in
Figure 11.15 has point symmetry.

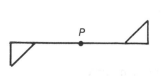

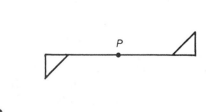

Original Figure

Clockwise
(or Counterclockwise)
Rotation of 90°
about Point *P*

Clockwise
(or Counterclockwise)
Rotation of 180°
about Point *P*

Figure 11.15   Point Symmetry

A right triangle (Figure 11.16) is an example of a figure that does *not* have point symmetry.

Original Triangle

Triangle after a 180° Clockwise
Rotation with Respect to Point *B*

Figure 11.16   Lack of Point Symmetry

Exercise Set 11.7

1. Determine the capital letters of the English alphabet that have
   (a) vertical but not horizontal line symmetry.
   (b) horizontal but not vertical line symmetry.
   (c) vertical and horizontal line symmetry.

2. Which letter has point symmetry?
   (1)   (2)   (3)   (4)

3. Which symbol has an image that coincides with itself after a rotation of 90°?
   (1)   (2)   (3)   (4)

4. What kind of symmetry does a square have?
   (1) Line symmetry only   (3) Both line and point symmetry
   (2) Point symmetry only   (4) Neither line nor point symmetry

5–12. Determine the number of lines of symmetry, if any, for each of the following figures or words:

5. a parallelogram   9. a right triangle

6. a trapezoid   10. OX

7. an equilateral triangle 11. a rhombus

8. WOW   12. a regular hexagon

13–18. Determine the kind of line symmetry, if any, for each numerical symbol.

13. 11   **14.** 38   **15.** 22   **16.** 101   **17.** 818   **18.** 100

19–24. Determine which of the following figures have point symmetry with respect to point P:

19.

22.

20.

23.

21.

24.

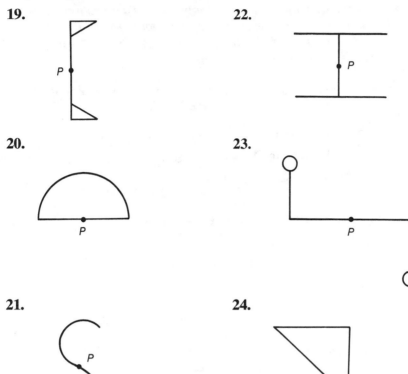

> ## REGENTS TUNE-UP: CHAPTER 11
>
> *Each of the questions in this section has appeared on a previous Course I Regents Examination. Here is an opportunity for you to review Chapter 11 and, at the same time, prepare for the Course I Regents Examination.*

1. The graph of which equation has a slope of 3 and a y-intercept of -2?
   (1) $y = 3x - 2$ (2) $y = 3x + 2$ (3) $y = 2x - 3$ (4) $y = 2x + 3$

2. If $(r, 3)$ is in the solution set of $2x + y = 7$, then the value of r must be:
   (1) 1   (2) 2   (3) 3   (4) 4

3. A point on the graph of $x + 3y = 13$ is:
   (1) $(4, 4)$ (2) $(-2, 3)$ (3) $(-5, 6)$ (4) $(4, -3)$

4. The graph of $y = 3x - 4$ is parallel to the graph of:
   (1) $y = 4x - 3$ (2) $y = 3x + 4$ (3) $y = -3x + 4$ (4) $y = 3$

5. What is the slope of the line whose equation is $2y = 3x - 7$?
   (1) $\dfrac{2}{3}$ (2) -2 (3) 3   (4) $\dfrac{3}{2}$

6. The coordinates of the vertices of a triangle are $(1, 1)$ and $(3, 1)$ and $(3, 5)$. The triangle formed is:
   (1) a right triangle   (3) an isosceles triangle
   (2) an obtuse triangle   (4) an equilateral triangle

7. The graph of $3x - y = 3$ intersects the x-axis at point:
   (1) $(1, 0)$ (2) $(0, 1)$ (3) $(0, 3)$ (4) $(3, 0)$

8. Which point does not lie on the graph of $x - 2y = 10$?
   (1) $(0, -5)$ (2) $(2, -4)$ (3) $(5, 0)$ (4) $(6, -2)$

9. Find the area of a triangle whose vertices are $(0, 0)$, $(0, 4)$, and $(5, 0)$.

10. If point $(2, 3)$ lies on the graph of the equation $2x + ky = -2$, find the value of k.

11. What is the point of intersection of the graphs of $x = 1$ and $y = 4$?

12. After each, write the number that makes the statement correct for each graph that is shown.
   (a) The equation of the graph shown is:
   (1) $y = x$
   (2) $y = -x$
   (3) $y = 2x$
   (4) $y = -2x$

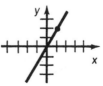

(b) The equation of the graph shown is:

(1) $x = 3$
(2) $x = -3$
(3) $y = 3$
(4) $y = -3$

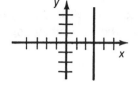

(c) The graph shown has a slope of:

(1) 1
(2) −1
(3) −4
(4) 4

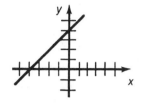

(d) The slope of the graph shown is:

(1) 1
(2) 2
(3) 0
(4) undefined

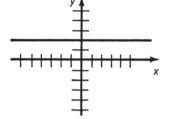

13. In the accompanying diagram, $\triangle A'B'C'$ is the image of $\triangle ABC$.

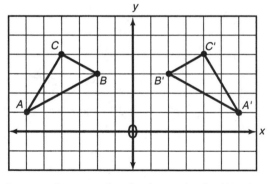

Which type of transaction is shown in the illustration?
(1) line reflection   (3) translation
(2) rotation   (4) dilation

14. Which letter has both point and line symmetry?
(1) **A**   (2) **M**   (3) **T**   (4) **X**

15. Which equation represents line l, shown in the accompanying diagram?

(1) $y = 2x + 3$

(2) $y = \dfrac{1}{2}x + 3$

(3) $y = 3x + \dfrac{1}{2}$

(4) $y = 3x + 2$

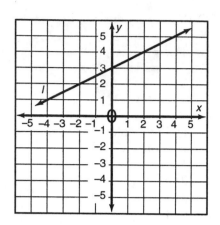

16. In the accompanying diagram, if $\triangle OAB$ is rotated counterclockwise 90° about point O, which figure represents the image of this rotation?

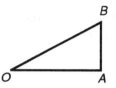

(1)   (2)   (3)   (4)

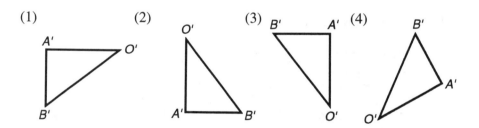

17. If x varies directly as y and $x = 2$ when $y = 12$, find y when $x = -1$.

18. Find the area of square $ABCD$ with vertices $A\,(-1, 2)$, $B\,(3, 2)$, $C\,(3, -2)$, and $D\,(-1, -2)$.

ANSWERS TO ODD-NUMBERED EXERCISES: CHAPTER 11

Section 11.1

1. $(-3, 2)$

3. 4

5. 6

7. 10

9. 15

11. 7.5

13. (a) $h = 8, k = 5$ **(b)** 40

15. 96

17. 54

Section 11.2

1. 0

3. $\dfrac{2}{7}$

5. 2

7. –2

9. $\overleftrightarrow{AB}$ is parallel to $\overleftrightarrow{CD}$

11. $\overleftrightarrow{AB}$ is parallel to $\overleftrightarrow{CD}$

13. 8

15. $x = 1, y = 4$

17. 3

19. (a) $\dfrac{2}{3}$ (b) 9

21. (a) $\dfrac{-2}{3}$ (b) –3

23. (a) 2 (b) 1

25. collinear

27. not collinear

29. 5

31. (a) –1 (b) 2

Section 11.3

1. (3)

3. (4)

5. (1)

7. 4

9. $m = 3, b = 1$

11. $m = \dfrac{2}{5}, b = -1$

13. $m = \dfrac{1}{2}, b = 10$

15. $y = -2x + 3$

17. $y = \dfrac{1}{3}x - 1$

19. $y = -2$

21. $y = x$

23. $y = -2x + 7$

25. $y = 1$

27. (a) $y = 4$
 (b) $y = -1$
 (c) $y = 3$
 (d) $y = -2$

29. (a) $\dfrac{1}{2}$

 (b) $y = \dfrac{1}{2}x - 3$

31. (a) parallel
 (b) parallel
 (c) not parallel
 (d) parallel

33. $y = -5x - 1$

35. $y = \dfrac{1}{2}x + \dfrac{9}{2}$

37. $y = -3x - \dfrac{10}{3}$

39. (a) $y = 2x + 2$
 (b) $y = -x + 2$

Section 11.4

1. Graph (–2, –6), (0, 0), and (2, 6)

3. Graph (–2, 0), (0, 2), and (2, 4)

5. Graph (–2, –5), (0, –3), and (2, –1)

7. Graph (–2, 3), (0, 4), and (2, 5)

9. Graph (0, 7) and (7, 0)

11. Graph (0, 2) and (–3, 0)

13. Graph (0, –1), (1, 2), and (2, 5)

15. Graph (0, 5), (2, 6), and (4, 7)

17. Graph (–2, 4), (0, 0), and (2, –4)

19. Graph (–3, 3), (0, 2), and (3, 1)

21. (a) Graph (–2, 3), (0, 7) and (4, 11)
 (b) Graph (0, –2,) and (–2, 0)
 (c) (–3, 1)

Section 11.5

1. 24

3. $y = \dfrac{3}{2}x$

5. y does not vary directly as x.

7. 30.48

9. 1

Section 11.6
1. (2)

3.

5. (a)

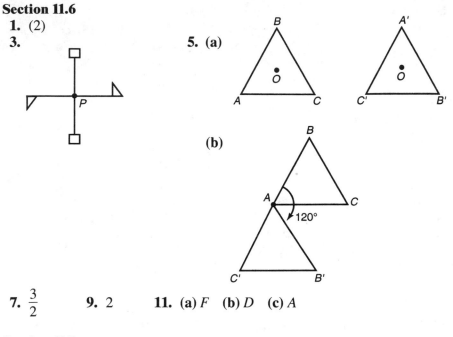

(b)

7. $\dfrac{3}{2}$ **9.** 2   **11.** (a) *F*   (b) *D*   (c) *A*

Section 11.7
1. (a) A,M,T,U,V,W,Y   (b) B,C,E,K   (c) H,I,O,X
3. (1)   **5.** 0   **7.** 3   **9.** 0   **11.** 2

13. vertical and horizontal line symmetry
15. no line symmetry
17. vertical and horizontal line symmetry
19. no point symmetry
21. point symmetry
23. point symmetry

Regents Tune-Up: Chapter 11
1. (1)   **5.** (4)   **9.** 10   **13.** (1)
3. (3)   **7.** (1)   **11.** (1, 4)   **15.** (2)
                                            17. −6

CHAPTER 12

SOLVING SYSTEMS OF EQUATIONS AND INEQUALITIES

12.1 GRAPHING SYSTEMS OF EQUATIONS

△
KEY IDEAS

Solving a group or *system* of linear equations means finding the set of all ordered pairs of numbers that make each of the equations true at the same time. One approach is to graph each equation on the same set of axes and then determine the coordinates of their points of intersection, if any.

Solving a System of Linear Equations Graphically

The graph of each equation of the system

$$y = 2x$$
$$x + y = 6$$

is shown in Figure 12.1. The lines intersect at point $(2, 4)$, so the solution set of this system of linear equations is $\{(2, 4)\}$, that is

$$(y = 2x) \wedge (x + y = 6) = (2, 4).$$

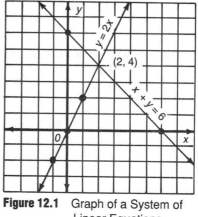

Figure 12.1 Graph of a System of Linear Equations

Examples

1. Solve the following system of equations graphically, and check by substitution:

$$2x + y = 5$$
$$y - x = -4$$

Solution: Graph each equation on the same set of axes. Using the slope-intercept method, you must first write each equation in the form $y = mx + b$:

$$2x + y = 5 \text{ becomes } y = -2x + 5$$
$$y - x = 4 \text{ becomes } y = x - 4$$

337

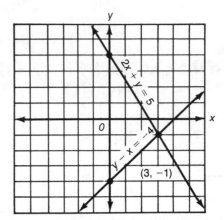

The solution is **(3, –1)**.

Check: Replace x by 3 and y by –1 in each of the *original* equations:

$$\begin{array}{c|c}
2x + y = 5 & y - x = 4 \\
\hline
2(3) + (-1) & -1 - 3 \\
6 - 1 & -4 = -4 \checkmark \\
5 = 5 \checkmark &
\end{array}$$

2. Solve by graphing:

$$6y - 3x = 18$$
$$2y = x + 1$$

Solution: Express each equation in the $y = mx + b$ form:

$$6y - 3x = 18$$
$$6y = 3x + 18$$
$$y = \frac{3x}{6} + \frac{18}{6}$$
$$= \frac{x}{2} + 3$$
$$2y = x + 1$$
$$y = \frac{x}{2} + \frac{1}{2}$$

The value of m for each line is $\frac{1}{2}$. Since the lines have the same slope but *different* y-intercepts, they are parallel and, therefore, do *not* intersect. There is *no* ordered pair of numbers that satisfies both equations simultaneously, so that the solution set is the empty set, which may be written as { } or ∅.

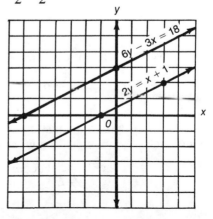

As Example 2 illustrates, systems of linear equations do not necessarily have one ordered pair in their solution sets. When the graphs of a pair of linear equations are drawn on the same set of axes, one of the following must be true:

1. The lines intersect in exactly one point, so that the system of equations has one ordered pair in its solution set.

2. The lines are parallel so that the system of equations has no ordered pairs in its solution set.

3. The lines coincide. This means that the equations graphed are equivalent. For example, consider the system of equations

$$y = 2x - 3$$
$$4x = 2y + 6.$$

Points $(-1, -5)$, $(0, -3)$ and $(1, -1)$ can be used to graph the first equation. These points also satisfy the second equation, so that the graphs of these equations are two lines that coincide. The solution set is the infinite set of ordered pairs that lie on the single line defined by these equations.

In this course we will restrict our attention to solving systems of equations consisting of lines that intersect at one point.

Exercise Set 12.1

1–12. Solve graphically, and check by substitution.

1. $2x + y = 8$
$y = x + 2$

2. $x + 2y = 7$
$y = 2x + 1$

3. $y = 3x + 1$
$x = y - 3$

4. $2x - y = 10$
$x + 2y = 10$

5. $2x + y = 1$
$5x + 3y = 4$

6. $x + y = 8$
$2x - y = 7$

7. $2x + y = 6$
$x - 2y = 8$

8. $2y = -5x$
$y - x = 7$

9. $2y = x + 6$
$y = 3x - 2$

10. $y - 2x = 5$
$x + 2y = 0$

11. $x - 3y = 9$
$2x + y = 4$
$3x - 2y = 4$
$3x + 2y = 8$

13. (a) On the same set of coordinate axes, graph the three lines whose equations are **(1)** $y = 2x + 1$, **(2)** $y = 1$, and **(3)** $x = 2$.
(b) Write the coordinates of the three vertices of the triangle formed by the lines graphed in part (a).
(c) Find the area of the triangle drawn in part (a).

14–17. Graph the triangle formed by the lines in each of the following systems of equations, and then determine its area:

14. $y = 2x$; $x = 5$; positive x-axis.

15. $y = x$; $y = -x$; $y = 8$

16. $y = -2x + 17$; $2y - 3x = 6$; $y = 3$

17. $x + y = 5$; $6y - 5x + 25 = 0$; $y = 5$

12.2 GRAPHING LINEAR INEQUALITIES

KEY IDEAS

In the accompanying graph, the line $y = 3$ serves as a *boundary* line dividing the coordinate plane into two *half-planes*. The region *above* the line $y = 3$ represents the solution set of the inequality $y > 3$ since every point in this region has a y-coordinate *greater than* 3. The region below the line $y = 3$ represents the solution set of the inequality $y < 3$ since the y-coordinate of every point in this region is *less than* 3. A half-plane that does *not* include the boundary line, such as $y < 3$ or $y > 3$, is referred to as an *open half-plane*. A *closed half-plane*, such as $y \le 3$ or $y \ge 3$, includes the boundary line.

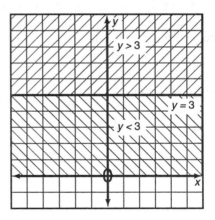

Graphing a Linear Inequality

To graph a linear inequality, proceed as follows:

1. Mentally replace the inequality with an equals symbol and graph the resulting equation.

2. If the inequality relation is < or >, draw the graph of the related equation as a broken or dashed line to indicate that points on the boundary line are *not* included in the solution set. If the inequality relation is ≤ or ≥, draw the boundary line as a continuous line which indicates that points on the line are included in the solution set.

3. Decide which half-plane represents the solution set. This can be done by choosing a test point on either side of the boundary line and determining which point makes the inequality a true statement.

Examples

1. Graph $y > 3x + 1$.

Solution: Step 1. Draw the graph of $y = 3x + 1$, using the slope-intercept method.

Step 2. Since the inequality is $>$, draw the boundary line as broken.

Step 3. Choose test points. Whenever the line does not contain the origin, it is usually convenient to use $(0, 0)$ as one of the test points.

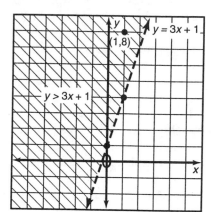

Test $(0, 0)$	Test $(2, 8)$
$y > 3x + 1$	$y > 3x + 1$
$0 > 3(0) + 1$	$8 > 3(2) + 1$
$0 > 1$ *False*	$8 > 6 + 1$ *True*

Step 4. Shade in the half-plane that contains $(2, 8)$, which is the region *above* the boundary line.

2. Which ordered pair is in the solution set of $x + 2y > 7$?
    (1) $(5, 1)$ (2) $(2, 6)$ (3) $(3, 1)$ (4) $(7, 0)$

Solution: The ordered pair $(2, 6)$ is correct since, if x is replaced by 2 and y is replaced by 6, then

$$x + 2y = 2 + 2(6) = 2 + 12 = 14,$$

which is greater than 7. Note that for choices (1) and (4) the value of $x + 2y = 7$, making the inequality $x + 2y > 7$ false.

The correct answer is **choice (2)**.

Graphing Systems of Inequalities

A system of inequalities can be solved graphically by graphing each of the inequalities on the same set of axes and then determining the region in which the solution set of the individual inequalities overlap.

Examples

3. Graph each of the following systems of inequalities, and identify the solution set:
    (a) $y > x$ **(b)** $y > 2$
         $x \le 5$ $y < -3$

Solution: **(a)** *Step 1.* Graph $y > x$. The open half-plane above the line $y = x$ represents the solution set of $y > x$.

Step 2. Graph $x \leq 5$ on the same set of axes. The closed half-plane to the left of $x = 5$ represents the solution set of $x \leq 5$.

The cross-hatched shading indicates the region where the solution sets of the two inequalities overlap.

Notice that an open circle appears at the point of intersection of the two boundary lines. Since this point is *not* a member of the solution set of $y > x$, it is not a member of the solution set of the *system* of inequalities.

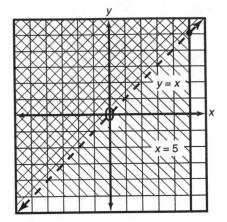

(b) As the accompanying graph illustrates, the solution sets of the two inequalities do *not* overlap, so the solution of this system of inequalities has no points.

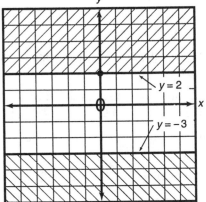

4. Graph the following systems of inequalities, and label the solution set S:

$$y + 3x > 6$$
$$y \leq 2x - 4$$

Solution: Step 1. Rewrite the first inequality as $y > -3x + 6$. Use the slope-intercept method to graph the line $y = -3x + 6$. Since the inequality relation is $<$, the graph of the related equation is drawn as a broken line.

Step 2. Determine the open-half plane that represents the solution set to $y + 3x > 6$. Shade in the region *above* and to the right of the boundary line.

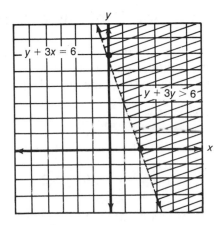

342

Step 3. On the same set of axes graph the line $y = 2x - 4$, using the slope-intercept method. Since the inequality relation is $\geq$, the graph of the related equation is drawn as a continuous line.

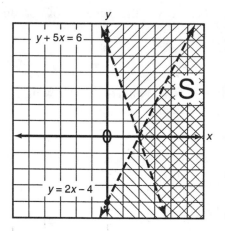

Step 4. Determine the closed half-plane that represents the solution set to $y \leq 2x - 4$. Shade in the region *below* and to the right of the boundary line.

Step 5. Label the region in which the shaded regions overlap as *S*. This cross-hatched area represents the solution set of the system of inequalities.

Keep in mind that any point in the region labeled as *S* is a member of the solution set of the conjunction $(y + 3x > 6) \wedge (y \leq 2x - 4)$.

Exercise Set 12.2

1–6. Graph each of the following inequalities:

1. $x < -1$ **4.** $y \geq 5x$

2. $y \geq 6$ **5.** $y - 3x < 1$

3. $y + x \leq 0$ **6.** $x - 3y \leq 9$

7. Which ordered pair is in the solution set of $x < 6 - y$?
 (1) (0, 6) (2) (6, 0) (3) (0, 5) (4) (7, 0)

8. Which is *not* a member of the solution set of $2x - 3y \geq 12$?
 (1) (0, −4) (2) (−3, −6) (3) (10, 3) (4) (6, 0)

9–16. **(a)** *On the same set of coordinate axes, graph the system of inequalities and label the solution set* S.
 (b) *Write the coordinates of a point in the solution set* S.
 (c) *Check algebraically that the coordinates of the point found in part (b) satisfy both inequalities.*

9. $2x + y < 6$ **13.** $2y \geq x - 6$
  $y \geq 2$ $x + y < 2$

10. $y < x - 5$ **14.** $y + 3x > 6$
  $y + 2x \geq 4$ $y \leq 2x - 4$

11. $y > 3x + 6$ **15.** $x + y < 8$
  $y \leq -2x + 4$ $y > x - 4$

12. $2y \geq x - 4$ **16.** $y + 2x > 7$
  $y < 3x$ $y + 3 \leq 3x$

17. **(a)** On the same set of coordinate axes, graph the following system of inequalities:

$$y < x + 4$$
$$x + y > 2$$

(b) Which of the following points is in the solution set of the graph drawn in answer to part (a)?
(1) (2, 3)   (2) (–5, 2)   (3) (0, 6)   (4) (–1, 0)

18. **(a)** On the same set of coordinate axes, graph the following system of inequalities:

$$y \le 3x + 4$$
$$y + x > 2$$

(b) Write the coordinates of one point that is *not* in the solution set of both inequalities.

12.3 SOLVING SYSTEMS OF EQUATIONS ALGEBRAICALLY: SUBSTITUTION METHOD

KEY IDEAS

The accompanying system of linear equations may be solved algebraically as well as graphically. Algebraic solutions of systems of linear equations are based on reducing the original system of equations to a single equation that contains one variable.

$$y = 2x \}$$
$$3x + y = 10$$

One way of solving the given system is to rewrite the second equation, replacing y with its equivalent, $2x$. The resulting equation has one variable, so it can be solved.

$$3x + 2x = 10$$
$$5x = 10$$
$$x = \frac{10}{5} = 2$$
$$y = 2x = 2(2) = 4$$

Using the Substitution Method

In some systems of linear equations one of the given equations expresses one variable in terms of the other variable. In this situation, the *substitution method* provides a convenient way of obtaining one equation with one variable so that the solution set can be easily found.

Example

1. Solve the following system of equations algebraically, and check:

$$x + 2y = 7$$
$$y = 2x + 1$$

Solution: Eliminate variable y in the first equation by substituting $2x + 1$ for y.

$$x + 2y = 7$$
$$x + 2(2x + 1) = 7$$
$$x + 4x + 2 = 7$$
$$5x + 2 = 7$$
$$5x = 7 - 2$$
$$5x = 5$$
$$x = \frac{5}{5} = 1$$

Find the corresponding value of y by substituting 1 for x in the second equation.

$$y = 2x + 1$$
$$= 2(1) + 1$$
$$= 2 + 1 = 3$$

The solution set is $\{(1, 3)\}$.
Check. Substitute 1 for x and 3 for y in each of the original equations.

$x + 2y = 7$
$1 + 2(3)$
$1 + 6$
$7 = 7\checkmark$

$y = 2x + 1$	
3	$2(1) + 1$
3	$2 + 1$
$3 = 3\checkmark$	

In some systems of linear equations one equation can easily be transformed into an equivalent equation in which one variable is expressed in terms of the other variable. After this has been accomplished, the substitution method can be used to find the solution set.

Examples

2. Solve the following system of equations algebraically, and check:

$$3x - 2y = 42$$
$$2x - y = 26$$

Solution: Solve the second equation for y in terms of x.

Subtract $2x$ from each side: $\quad -y = -2x + 26$
Divide each member of both
sides of the equation by -1: $\quad y = 2x - 26$

Substitute $2x - 26$ for y in the first equation.

$$
\begin{aligned}
3x - 2y &= 42 \\
3x - 2(2x - 26) &= 42 \\
3x - 4x + 52 &= 42 \\
-x + 52 &= 42 \\
-x &= 42 - 52 \\
&= -10 \\
x &= 10
\end{aligned}
$$

Find the corresponding value for y by substituting 10 for x in the equation $y = 2x - 6$.

$$
\begin{aligned}
y &= 2x - 26 \\
&= 2(10) - 26 \\
&= 20 - 26 = -6
\end{aligned}
$$

Since $x = 10$, and $y = -6$, the solution is **(10, –6)**. The check is left for you.

3. Solve the following system of equations algebraically, and check:

$$
\frac{x}{2} = y + 5
$$
$$
4y - x = 6
$$

Solution: Solve the first equation for x by multiplying each member of both sides of the equation by 2.

$$
2\left(\frac{x}{2}\right) = 2(y + 5)
$$
$$
x = 2y + 10
$$

In the second equation, replace x by $2y + 10$.

$$
\begin{aligned}
4y - x &= 6 \\
4y - (2y + 10) &= 6 \\
4y - 2y - 10 &= 6 \\
2y - 10 &= 6 \\
2y &= 6 + 10 \\
2y &= 16 \\
y &= \frac{16}{2} = 8
\end{aligned}
$$

Find the corresponding value for x by substituting 8 for y in the equation $x = 2y + 10$:

$$x = 2y + 10$$
$$= 2(8) + 10$$
$$= 16 + 10 = 26$$

Since $x = 26$ and $y = 8$, the solution is **(26, 8)**. The check is left for you.

4. Two angles are supplementary. If five times the measure of the smaller angle is subtracted from twice the measure of the larger angle, the result is 10. What is the measure of each angle?

Solution: Let x = measure of the smaller angle,
and y = measure of the larger angle.

Since the angles are supplementary, the sum of their measures is 180.

$$x + y = 180 \qquad \text{Equation (1)}$$

The second condition of the problem may be translated as follows:

$$2y - 5x = 10 \qquad \text{Equation (2)}.$$

From equation (1), $y = 180 - x$, so y may be eliminated in equation (2).

$$\begin{aligned}
2(180 - x) - 5x &= 10 \\
360 - 2x - 5x &= 10 \\
360 - 7x &= 10 \\
-7x &= 10 - 360 \\
x &= \frac{-350}{-7} = 50 \\
y = 180 - x &= 180 - 50 = 130
\end{aligned}$$

The measures of the two angles are **50** and **130**. Check that these numbers satisfy the conditions in the original statement of the problem.

Condition 1: Are the two angles supplementary? Yes, since $50 + 130 = 180$.

Condition 2: Is five times the measure of the smaller angle subtracted from twice the measure of the larger angle equal to 10? Yes, since $2(130) - 5(50) = 260 - 250 = 10$.

Note: This problem could have been solved with one variable by letting $180 - x$, rather than y, represent the measure of the larger angle. Then $2(180 - x) - 5x = 10$.

5. The denominator of a fraction is 7 more than its numerator. If the numerator is increased by 3 and the denominator is decreased by 2, the new fraction equals $\frac{4}{5}$. Find the original fraction.

Solution: Let x = numerator of the fraction,
and y = denominator of the fraction.

$$y = x + 7 \quad \text{Equation (1)}$$
$$\frac{x + 3}{y - 2} = \frac{4}{5} \quad \text{Equation (2)}$$

Cross-multiply:
$$5(x + 3) = 4(y - 2)$$
$$5x + 15 = 4y - 8$$
$$5x = 4y - 8 - 15$$
$$5x = 4y - 23$$
$$5x - 4y = -23$$

From equation (1), replace y with $x + 7$:

$$5x - 4(x + 7) = -23$$
$$5x - 4x - 28 = -23$$
$$x - 28 = -23$$
$$x = 28 - 23 = 5$$
$$y = x + 7 = 5 + 7 = 12$$

The original fraction is $\frac{5}{12}$. The check is left for you.

Note: This problem could have been solved using a single variable by letting $x + 7$, rather than y, represent the denominator of the given fraction.
Then

$$\frac{x + 3}{(x + 7) - 2} = \frac{4}{5}.$$

Digit Problems

Numbers such as 13, 48, and 69 are examples of two-digit numbers. The *tens* digit of 69 is 6, and the *units* digit of 69 is 9. An equivalent expression for 69 is $6 \cdot 10 + 9$. If the digits of the number 69 are interchanged, the resulting number is 96.

In general, if t represents the tens digit of a two-digit number and u represents the units digit, then the number may be represented by $10t + u$. The sum of the digits of a two-digit number may be represented by $t + u$. The number with its digits interchanged may be represented by $10u + t$.

Example

6. The units digit of a two-digit number exceeds the tens digit by 1. The original number is 2 less than five times the sum of the digits. Find the number.

Solution: Let t = tens digit of the two-digit number,
and u = units digit.
Then $10t + u$ = original number
Condition 1: Units digit exceeds the tens digit by 1.
Equation (1): $u = t + 1$
Condition 2: Number is 2 less than five times sum of the digits.

348

Equation (2): $10t + u = 5(t + u) - 2$

Simplify the second equation:
$$10t + u = 5t + 5u - 2$$
$$10t - 5t + u - 5u = -2$$
$$5t - 4u = -2$$

Equation (1) allows us to replace u by $t + 1$ in the preceding equation:

$$5t - 4(t + 1) = -2$$
$$5t - 4t - 4 = -2$$
$$t - 4 = -2$$
$$t = -2 + 4 = 2$$
$$u = t + 1 = 2 + 1 = 3$$

The number is $10t + u = 10(2) + 3 = \textbf{23}$. The check is left for you.

Exercise Set 12.3

1–14. Solve each of the following systems algebraically, and check:

1. $y = 2x$
$\quad 2x - 3y = 16$

2. $x = 3y$
$\quad 5x - 9y = 18$

3. $y + 5x = 0$
$\quad 3x - 2y = 26$

4. $\dfrac{y}{4} = x$
$\quad 2y - 3x = 7$

5. $5c - n = 0$
$\quad 3n - 4c = -19$

6. $a + b = 0$
$\quad 5a - 2b = 14$

7. $y - 1 = x$
$\quad -3x + 7y = -1$

8. $y - x = 1$
$\quad 7y - 11x = -5$

9. $y + 2x = 3$
$\quad 4x - 3y = 21$

10. $3y - 3x = 9$
$\quad 7y - 6x = 58$

11. $x + y = 28$
$\quad x - 3y = 0$

12. $0.3y - 0.2x = 1.5$
$\quad 0.75y = 2.25x$

13. $\dfrac{a}{2} - \dfrac{b}{5} = 1$
$\quad b - 2a = 3$

14. $\dfrac{x - 2}{3} = y$
$\quad 2x + 3y = -5$

15. Two angles are complementary. The difference between four times the measure of the smaller and the measure of the larger angle is 10. Find the degree measure of each angle.

16. The length of a side of one equilateral triangle exceeds twice the length of a side of another equilateral triangle by 1. If the difference in the perimeters of the triangles is 15, find the length of a side of each triangle.

17. The denominator of a fraction is 8 more than the numerator. If 5 is added to both the numerator and the denominator, the resulting fraction is equal to to $\frac{1}{2}$. Find the original fraction.

18. The denominator of a fraction is three times the numerator. If 8 is added to the numerator and 6 subtracted from the denominator, the value of the resulting fraction is $\frac{8}{9}$. Find the original fraction.

19. The tens digit of a two-digit number exceeds twice the units digit by 1. If 7 is added to the number, the result is equal to eight times the sum of the digits. Find the number.

20. The sum of the digits of a two-digit number is 7. If the digits are interchanged, the new number is 2 more than twice the original number. Find the original number.

12.4   SOLVING SYSTEMS OF EQUATIONS ALGEBRAICALLY: ADDITION METHOD

$\bigwedge$
KEY IDEAS
$\diagdown$

The accompanying system of linear equations is difficult to solve using the substitution method, but can be easily solved by adding corresponding sides of each equation.

The resulting equation does not contain variable y since the coefficients of y in the original equations are additive inverses (opposites), so their sum is 0.

$$2x - 9y = 17$$
$$\underline{5x + 9y = 11}$$
$$7x + 0 = 28$$
$$x = \frac{28}{7} = 4$$

Using the Addition Method

To use the *addition method* to solve a system of linear equations, write each equation in the form $Ax + By = C$ and then compare the coefficients of like variables. If the numerical coefficients of either the x terms or the y terms are additive inverses (opposites), then this variable can be eliminated by adding corresponding sides of both equations.

Examples

1. Solve the following system of equations algebraically, and check:

$$3y = 2x - 1$$
$$5y + 2x = 25$$

Solution: Write the first equation in the form $Ax + By = C$, and then compare the numerical coefficients of the like variables.

$$3y - 2x = -1$$
$$5y + 2x = 25$$

The coefficients of variable x are opposites, so this variable can be eliminated if the equations are added:

$$
\begin{array}{rl}
3y - 2x &= -1 \\
5y + 2x &= 25 \\
\hline
8y &= 24 \\
y &= \dfrac{24}{8} = 3
\end{array}
$$

Find the corresponding value of x by replacing y by 3 in one of the equations:

$$
\begin{array}{rl}
3y &= 2x - 1 \\
3(3) &= 2x - 1 \\
9 &= 2x - 1 \\
1 + 9 &= 2x \\
10 &= 2x \\
x &= \dfrac{10}{2} = 5
\end{array}
$$

Since $x = 5$ and $y = 3$ the solution is **(5, 3)**.
Check. Replace x by 5 and y by 3 in each of the original equations.

$$
\begin{array}{c|c}
\multicolumn{2}{c}{3y = 2x - 1} \\
\hline
3(3) & 2(5) - 1 \\
9 & 10 - 1 \\
9 & = 9\checkmark
\end{array}
\qquad
\begin{array}{c|c}
\multicolumn{2}{c}{5y + 2x = 25} \\
\hline
5(3) + 2(5) & \\
15 + 10 & \\
25 & = 25\checkmark
\end{array}
$$

2. The sum of two positive numbers is 27, and their difference is 13. What are the numbers?

Solution: Let x = larger of the two numbers,
and y = the other number.

351

The sum of the numbers is 27: $\qquad x + y = 27$
The difference of the numbers is 13: $\quad x - y = 13$
Add the equations: $\qquad\qquad\qquad\overline{2x \qquad = 40}$

$$x = \frac{40}{2} = 20$$

To find y, replace x by 20 in one of the equations:

$$x + y = 27$$
$$20 + y = 27$$
$$y = 27 - 20 = 7$$

The two numbers are **20** and **7**.

Check: Do these numbers satisfy the conditions in the original statement of the problem?

Condition 1: Is the sum of 20 and 7 equal to 27? Yes, since $20 + 7 = 27$.

Condition 2: Is the difference of 20 and 7 equal to 13? Yes, since $20 - 7 = 13$.

3. Solve the following system of equations algebraically, and check:

$$x = \frac{1}{2}y = 4$$
$$x + y = 7$$

Solution: Begin by clearing the first equation of fractions by multiplying each member by 2.

$$2(x) - 2\left(\frac{1}{2}y\right) = 2(4) \rightarrow 2x - y = 8$$

Rewrite the second equation: $\quad x + y = 7$
Add the equations: $\qquad\qquad\qquad\overline{3x = 15}$

$$x = \frac{15}{3} = 5$$

To find y, replace x by 5 in one of the equations.

$$x + y = 7$$
$$5 + y = 7$$
$$y = 7 - 5 = 2$$

Since $x = 5$ and $y = 2$ the solution is **(5, 2)**. The check is left for you.

Solving Systems of Equations Using Multipliers

Sometimes one or both of the equations in a system of linear equations must be multiplied by an appropriate number in order to obtain a pair of like variables that have opposite numerical coefficients.

Examples

4. Solve the following system of equations algebraically, and check:

$$3x + 2y = 4$$
$$9x + 2y = 16$$

Solution: The coefficients of y have the same value. Therefore mulitply each member of one of the equations by -1 so that the coefficient of y in the resulting equation will be -2.

Multiply the first equation by -1:	$-3 - 2y = -4$
Rewrite the second equation:	$9x + 2y = 16$
Add the equations:	$6x \quad\ = 12$

$$x = \frac{12}{6} = 2$$

Find the corresponding value for y by replacing x by 2 in one of the equations:

$$3x + 2y = 4$$
$$3(2) + 2y = 4$$
$$6 + 2y = 4$$
$$2y = 4 - 6$$
$$2y = -2$$
$$y = -\frac{2}{2} = -1$$

Since $x = 2$ and $y = -1$ the solution is **(2, –1)**. The check is left for you.

5. Solve the following system of equations algebraically, and check:

$$3x - 4y = -1$$
$$5x + 12y = 73$$

Solution: Compare the coefficients of like variables of the two equations. The coefficient of y in the second equation is an integer multiple of the coefficient of y in the first equation. Therefore variable y is the more likely choice for elimination. Multiply each member of the first equation by 3, since this will change the coefficient of y to -12, which is the opposite of the coefficient of y in the second equation.

Multiply the first equation by 3:	$(3)3x - (3)4y = (3)(-1)$
Simplify:	$9x - 12y = -3$
Rewrite the second equation:	$5x + 12y = 73$
Add the equations:	$14x \qquad\ = 70$

$$x = \frac{70}{14} = 5$$

Find the corresponding value of y by substituting 5 for x in one of the original equations:

$$3x - 4y = -1$$
$$3(5) - 4y = -1$$
$$15 - 4y = -1$$
$$-4y = -1 - 15$$
$$-4y = -16$$
$$y = \frac{-16}{-4} = 4$$

Since $x = 5$ and $y = 4$ the solution is **(5, 4)**. The check is left for you.

6. Solve the following system of equations algebraically, and check:

$$3x + 4y = 9$$
$$5x + 6y = 13$$

Solution: There is no clear advantage in choosing x or in choosing y as the variable to eliminate. To eliminate x, determine the lowest common multiple of the coefficients of x, which is 15. To obtain *opposite* numerical coefficients for x, multiply the first equation by 5 and the second equation by −3. This will produce equivalent equations having coefficients of x of 15 and −15. An equally acceptable approach is to multiply the first equation by −5 and the second equation by 3.

$$(5)3x + (5)4y = (5)9 \quad \rightarrow \quad 15x + 20y = 45$$
$$(-3)5x + (-3)6y = (-3)13 \quad \rightarrow \quad \underline{-15x - 18y = -39}$$

Add the equations: $\qquad\qquad\qquad\qquad\quad 2y = 6$

$$y = \frac{6}{2} = 3$$

Find the corresponding value of x by substituting 3 for y in one of the equations:

$$3x + 4y = 9$$
$$3x + 4(3) = 9$$
$$3x + 12 = 9$$
$$3x = 9 - 12$$
$$3x = -3$$
$$x = \frac{-3}{3} = -1$$

Since $x = -1$ and $y = 3$ the solution is **(−1, 3)**. The check is left for you.

Note: Another way of solving this system of equations is to eliminate y by multiplying the first equation by 3 and the second equation by −2. This produces equivalent equations having coefficients of y of 12 in the first equation and −12 in the second equation.

7. Six computer disks and two computer printer ribbons cost $19.00. Eight of the same disks and three of the same ribbons cost $27.00. What is the cost of one computer disk and the cost of one computer printer ribbon?

Solution: Let x = cost of one computer disk,
and y = cost of one computer printer ribbon.

$$6x + 2y = 19 \quad \text{Equation (1)}$$
$$8x + 3y = 27 \quad \text{Equation (2)}$$

Use the addition method. There is no clear advantage in choosing x or y as the variable to eliminate. To eliminate y, multiply equation (1) by 3 and equation (2) by –2. This will produce equivalent equations in which the coefficients of variable y are opposite numbers.

$$6x(3) + 2y(3) = 19(3) \quad \rightarrow \quad 18x + 6y = 57$$
$$8x(-2) + 3y(-2) = 27(-2) \quad \rightarrow \quad -16x - 6y = -54$$

Add the equations:
$$2x \quad = 3$$
$$x = \frac{3}{2} = 1.5$$

Find the value of y by replacing x with 1.5 in either of the equations.

$$6x + 2y = 19 \quad \text{Equation (1)}$$
$$6(1.5) + 2y = 19$$
$$9 + 2y = 19$$
$$2y = 19 - 9$$
$$y = \frac{10}{2} = 5$$

The cost of one computer disk is **$1.50**, and the cost of one computer printer ribbon is **$5.00**. The check is left for you.

Exercise Set 12.4

1–6. Solve for x.

1. $x + y = 4$
$x - y = 2$

2. $2x - y = 12$
$x + y = 3$

3. $2x + y = 7$
$3x = y + 8$

4. $3x + y = 13$
$x + y = 5$

5. $y - x = 6$
$x + y = 4$

6. $3x + 4y = -4$
$2x - y = -10$

7–12. Solve for y.

7. $3x + 2y = 7$
   $-3x + y = 8$

8. $2y + x = 8$
   $y + x = 5$

9. $3x - 2y = 12$
   $x + y = 4$

10. $4x - 3x = 15$
   $2x + 3y = 9$

11. $3x + 7y = 10$
   $3x - 2y = -8$

12. $7x - 3 = 3y$
   $7x - y = -1$

13. Which ordered pair is the solution of the following system of equations?

$$3x + y = 10$$
$$2x - y = 5$$

(1) $(1, 3)$ (2) $(5, -5)$ (3) $(3, 1)$ (4) $(-5, 5)$

14. The ordered pair $(2, -1)$ is the solution to which of the following systems of equations?

(1) $x = 2y$
   $3x + 4y = -10$
(2) $2x + 5y = -1$
   $x - y = 3$

(3) $5x + 4y = 6$
   $x + 2y = 0$
(4) $3x - 6y = 12$
   $5x - 10y = 4$

15–18. Solve algebraically for c *and* d *and check.*

15. $3c - d = 7$
   $c + 2d = 7$

16. $\dfrac{c}{2} + d = 5$
   $c - 2d = 8$

17. $0.4c + 1.5d = -1$
   $1.2c - d = 8$

18. $0.6c - 1.8d = 7.2$
   $0.4c - 0.9d = -3.9$

19–26. Use the addition method to solve the system of equations for x *and* y, *and check.*

19. $3x - 2y = 42$
   $2x - y = 26$

20. $2x + y = 6$
   $x - 3y = 10$

21. $3x - 2y = -1$
   $2x + 3y = 8$

22. $5x + 3y = 7$
   $6x + 5y = 17$

23. $3x = 7y - 48$
   $5x + 4y = 14$

24. $-3y + 7x = -15$
   $4x + 4y = -20$

25. $5 = 11x + 8y$
   $17 = 3x + 10y$

26. $5x + 3y = 3$
   $\dfrac{x}{3} + \dfrac{y}{2} = 2$

27. Solve algebraically for a *and* b, *and check.*

27. $\dfrac{a}{b+1} = \dfrac{2}{3}$

$a + b = 9$

28–33. Solve each of the following problems algebraically, using a system of equations:

28. The sum of two numbers is 21. The smaller number is one half of the larger number. Find the two numbers.

29. The difference between two positive numbers is 9. If four times the larger number is ten times the smaller, what are the two numbers?

30. Bob has 21 coins in dimes and quarters in his pocket. If the total value of these coins is $3.30, how many dimes and how many quarters does Bob have in his pocket?

31. Two angles are supplementary. The difference between the measure of the larger angle and twice the measure of the smaller angle is 15. Find the degree measure of each angle.

32. In a certain isosceles triangle three times the length of the base is equal to twice the length of a leg. If the perimeter of the triangle is 32, find the length of a leg and the length of the base.

33. Three shirts and two neckties cost $69. At the same prices, two shirts and three neckties cost $61. What is the cost of one shirt and one necktie?

REGENTS TUNE-UP: CHAPTER 12

Each of the questions in this section has appeared on a previous Course I Regents Examination. Here is an opportunity for you to review Chapter 12 and, at the same time, prepare for the Course I Regents Examination.

---

1. Solve the following system of equations for x:

$$3x + y = 5$$
$$2x - y = 0$$

2. Solve the following system of equations for y:

$$3x + 2y = 7$$
$$-3x + y = 8$$

3. Solve the following system of equations for x:

$$3x + y = 5$$
$$y = 5x - 3$$

4. Which ordered pair satisfies both of the following equations?
$$x + y = 5$$
$$y = 2$$
(1) (3, 2)   (2) (2, 3)   (3) (5, 0)   (4) (0, 5)

5. When drawn on the same set of axes, the graphs of the equations
$$y = x - 1 \quad \text{and} \quad x + y = 5$$
intersect at the point whose coordinates are
(1) (–5, 6)   (2) (2, 1)   (3) (3, 2)   (4) (4, 1)

6–7. Solve algebraically and check.

6. $2c - d = -1$
$c + 3d = 17$

7. $2x - 3y = 10$
$5x + 2y = 6$

8. Solve the following system of equations graphically, and check:
$$x + y = -3$$
$$2x - y = 6$$

9. **(a)** On the same set of coordinate axes, graph the following system of equations:
$$x + y = 10$$
$$y = 5$$

(b) Find the area of the trapezoid bounded by the x-axis, the y-axis, and the graphs drawn in part (a).

10. **(a)** On the same set of coordinate axes, graph the following system of inequalities:
$$y < 2x + 4$$
$$x + y \le 7$$

(b) On the basis of your answer to part (a), write the coordinates of a point that is *not* in the solution set of the system of inequalities.

11. **(a)** On a set of coordinate axes, graph the following system of equations:
$$y = 2x - 1$$
$$y - x = 1$$

(b) Solve algebraically the system of equations in part (a).

12. The accompanying diagram shows the graph of which inequality?

(1) $y > x - 1$

(2) $y \geq x - 1$

(3) $y < x - 1$

(4) $y \leq x - 1$

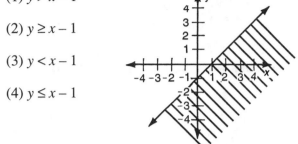

13. **(a)** On the same set of coordinate axes, graph the following system of inequalities.

$$y \leq \frac{1}{2}x - 3$$
$$y > -2x + 4$$

(b) Based on the graphs drawn in part (a), in which solution set(s) does the point whose coordinates are (0, 4) lie?

(1) $y \leq \frac{1}{2}x - 3$, only

(2) $y > -2x + 4$, only

(3) both $y \leq \frac{1}{2}x - 3$ and $y > -2x + 4$

(4) neither $y \leq \frac{1}{2}x - 3$ nor $y > -2x + 4$

14. **(a)** On the same set of coordinate axes, graph the following system of equations:

$$y = x + 4$$
$$x + y = 6$$
$$y = 2$$

(b) Find the area of the triangle whose vertices are the points of intersection of the lines graphed in part (a).

15. In $\triangle ABC$, the measure of angle B is twice the measure of angle A. If the measure of angle A is subtracted from the measure of angle C, the difference is 20. Find the measure of the *smallest* angle of the triangle.

16. On the same set of coordinate axes, graph the following system of inequalities:

$$y + x < 6$$
$$y \geq 2x + 3$$

(b) Using the graphs drawn in part (a), write the coordinates of a point in the solution set of the system of inequalities.

359

17. Solve the following system of equations and check:

$$3x + y = 4$$
$$2y = x - 6$$

ANSWERS TO ODD-NUMBERED EXERCISES: CHAPTER 12

Section 12.1

1. (2, 4)

3. (1, 4)

5. (−1, 3)

7. (4, −2)

9. (2, 4)

11. (3, −2)

13. **(b)** (0, 1), (2, 1), and (2, 5)

   (c) 4

15. 64

17. 27.5

Section 12.2

1.

3.

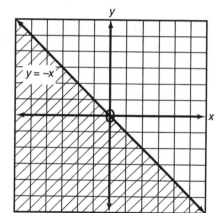

5.

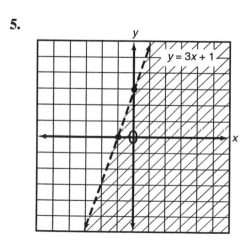

7. (3)

9. (a)   **(b)** (0, 3)

11. (a)   **(b)** (−3, 3)

13. (a)   **(b)** (−3, 3)

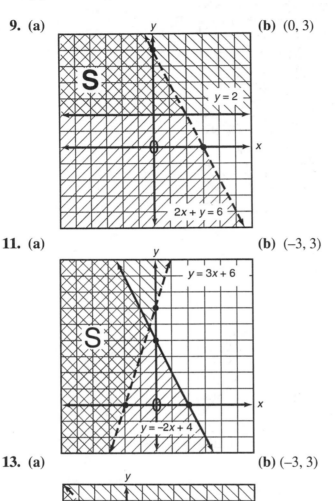

15. (a) 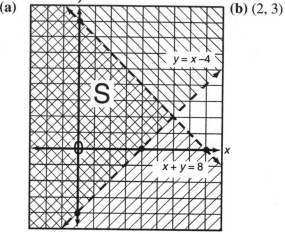 (b) (2, 3)

17. (a) (b) (1)

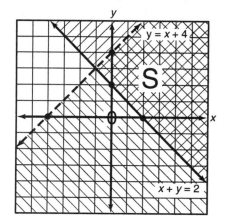

Section 12.3

1. $x = -4, y = -8$

3. $x = 2, y = -10$

5. $c = -1, n = -5$

7. $x = -2, y = -1$

9. $x = 3, y = -3$

11. $x = 21, y = 7$

13. $a = 16, b = 35$

15. 16, 74

17. $\dfrac{3}{11}$

19. 73

Section 12.4

1. $x = 3$

3. $x = 3$

5. $x = -1$

7. $y = 5$

9. $y = 0$

11. $y = 2$

13. (3)

15. $c = 3, d = 2$

17. $c = 5, d = -2$

19. $x = 10, y = -6$

21. $x = 1, y = 2$

23. $x = -2, y = 6$

25. $x = -1, y = 2$

27. $a = 4, b = 5$

29. $x = 15, y = 6$

31. $x = 125, y = 55$

33. necktie = \$9
shirt = \$17

Regents Tune-Up: Chapter 12

1. $x = 1$ **7.** $x = 2, y = -2$

3. $x = 1$ **9.** 37.5

5. (3)   **11.** **(b)** (2, 3)

13. (a)   **(b)** (4)

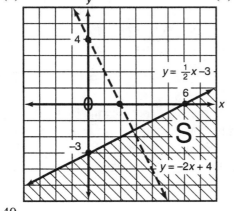

15. 40

17. $x = 2, y = -2$

PROBABILITY AND STATISTICS

CHAPTER 13

PROBABILITY

13.1 FINDING SIMPLE PROBABILITIES

=== $\bigwedge$ **KEY IDEAS** $\bigwedge$ ===

The mathematical *probability* of an event expresses the likelihood that the event will occur as a fraction whose value ranges from 0 to 1. The closer this fractional value is to 1, the greater the certainty that this event will occur. If an event can happen in r out of n equally likely ways, then the probability that this event will occur is $\frac{r}{n}$.

Probability Terms

The spinner shown in Figure 13.1 consists of five equal regions, numbered 1 through 5. An activity whose result is uncertain, like spinning the arrow shown, is called a **probability experiment**. Each of the possible results of a probability experiment is called an **outcome**. Although we do not know the region in which the spinner will stop, we know the set of all possible outcomes, called the **sample space**. For this experiment the sample space is the set of region numbers, $\{1, 2, 3, 4, 5\}$. Since each of the five regions are the same size, they are **equally likely** outcomes.

A possible *event* is spinning a 2. Another possible *event* is spinning an odd number. An **event** is any specific outcome or group of outcomes contained in the sample space. Suppose event E is the event of spinning an odd number. The probability that event E will occur is denoted by $P(E)$. Since three out of the five regions have odd numbers, $P(E) = \frac{3}{5}$.

Figure 13.1 Spinner with Five Equally Likely Outcomes

Definition of Probability

The outcomes that will make an event occur are the **favorable outcomes** or **successes** for that event. If *E* represents some event from a sample space that contains only equally likely outcomes, then

$$P(E) = \frac{\text{number of favorable outcomes}}{\text{total number of possible outcomes}}$$

Since the numerator of this probability fraction must always be less than or equal to the denominator, *P(E)* ranges in value from 0 to 1.

- *P(E)* = 0, if event *E* is impossible. In Figure 13.1, the probability of spinning a 6 is 0, since this is an impossible event.
- *P(E)* = 1, if event *E* is certain to occur. In Figure 13.1, the probability of spinning a number less than 6 is 1 since this event is certain to happen.

Examples

1. A **die** is a cube, each of whose six sides is marked with a different number of spots from one to six. Find the probability of rolling:
(**a**) 5 (**b**) an even number (**c**) a number greater than 4.

Solution: The sample space is {1, 2, 3, 4, 5, 6}, so the total number of possible outcomes is 6.

Let *X* = number on the die that shows up after the die is rolled.
(**a**) The sample space includes exactly *one* 5.

$$P(X = 5) = \frac{\text{Number of successes}}{\text{Number of total outcomes}} = \frac{1}{6}$$

(**b**) The sample space includes *three* even numbers (2, 4, and 6).

$$P(X = \text{even number}) = \frac{\text{Number of successes}}{\text{Number of total outcomes}} = \frac{3}{6} = \frac{1}{2}$$

(**c**) The sample space includes *two* numbers that are greater than 4 (5 and 6).

$$P(X > 4) = \frac{\text{Number of successes}}{\text{Number of total outcomes}} = \frac{2}{6} = \frac{1}{3}$$

2. An urn is a vase whose contents are not readily visible. The ratio of yellow to blue marbles in an urn is 2 : 3. If the urn contains only yellow and blue marbles that are identical except for their color, what is the probability that a marble drawn will be yellow?

Solution: Let 2*x* = the number of yellow marbles
and   3*x* = the number of blue marbles.
Then 2*x* + 3*x* = the total number of marbles.

Thus,

$$P(\text{yellow}) = \frac{2x}{2x + 3x} = \frac{2x}{5x} = \frac{2}{5}$$

3. What is the probability that a number picked at random (without looking) from the set $\{-2, -1, 0, 1, 2\}$ will satisfy the inequality $2x + 1 < 2$?

Solution: First solve the inequality:

$$2x + 1 < 2$$
$$\frac{2x}{2} < \frac{1}{2}$$
$$x < \frac{1}{2}$$

The sample space is $\{-2, -1, 0, 1, 2\}$. Since $-2, -1$, and 0 are each less than $\frac{1}{2}$, three out of the five numbers in the sample space are favorable outcomes. Thus, the required probability is $\frac{3}{5}$.

Problems with 52-Card Playing Decks

A standard deck of 52 playing cards is divided into four suits: hearts, diamonds, clubs, and spades. Hearts and diamonds are red cards, clubs and spades are black cards. Each suit contains 13 cards: 2, 3, 4, 5, 6, 7, 8, 9, 10, jack, queen, king, and ace. Jacks, queens, and kings are called *picture cards*. When a card is drawn from a playing deck, we always assume the deck is well-shuffled and each card is face down so that its identity is not known until it is selected and turned over.

Example

4. A single playing card is drawn from a standard deck of playing cards. Find the probability that the card is:
(**a**) an ace   (**b**) a club   (**c**) a red king

Solution: In each case, the sample space contains 52 possible outcomes since there are 52 different playing cards.
(**a**) A deck contains *four* aces, so

$$P(\text{ace}) = \frac{\text{Number of successes}}{\text{Number of total outcomes}} = \frac{4}{52} = \frac{1}{13}$$

(**b**) A deck contains *13* clubs, so

$$P(\text{club}) = \frac{\text{Number of successes}}{\text{Number of total outcomes}} = \frac{13}{52} = \frac{1}{4}$$

(c) There are four kings and *two* of these are red, so

$$P(\text{red king}) = \frac{\text{Number of successes}}{\text{Number of total outcomes}} = \frac{2}{52} = \frac{1}{26}$$

Probability That an Event Will Not Occur

Subtracting the probability that event E will happen from 1 gives the probability that event E will *not* happen. Thus,

$$P(\text{not E}) = 1 - P(E)$$

For example, since the probability of rolling a die and getting a two is $\frac{1}{6}$, the probability of *not* rolling a 2 is $1 - \frac{1}{6} = \frac{5}{6}$. If there is a 30% chance that it will rain tomorrow, then there is a 70% probability that it will *not* rain tomorrow since $1 - 30\% = 70\%$.

Theoretical vs. Empirical Probability

Dividing the number of favorable outcomes for an event by the total number of equally likely sample space outcomes gives the *theoretical probability* that the event will occur.

Not all probabilities can be determined theoretically. There are two possible outcomes that describe how a falling cone-shaped cup lands: point up or on its side. We cannot find the theoretical probability that the cup will land point up since the two possible outcomes are not equally likely to occur. An approximation of the theoretical probability that the cup will land point up can be obtained by dropping the cup many times and counting the number of times it lands point up. If the cup lands point up 27 times out of 100 tosses, then we can conclude that the *empirical probability* of point up is about $\frac{27}{100}$. An **empirical probability** value is a probability value that is calculated using experimental results obtained from direct observation. The more times a probability experiment is repeated, the closer we expect the empirical probability value will be to the theoretical probability value.

Example

5. In a production run of 10,000 light bulbs, a random sample of 100 bulbs is selected and then tested. If 2 bulbs from the sample are found to be defective, how many defective bulbs would we expect to find in the entire production run?

Solution: When selecting a bulb, there are two possible outcomes: defective and not defective. Since these outcomes are not equally likely, the probability of picking a defective bulb must be determined empirically. Since it is given that 2 out of a sample of 100 bulbs are defective, we can use the ratio $\frac{2}{100}$ = 0.02 as an approximation for the probability that a bulb selected from this production run will be defective. Using this probability value, the number of defective bulbs we would expect in a production run of 10,000 light bulbs is 0.02 × 10,000 = **200 bulbs**.

Exercise Set 13.1

1. A letter from the word POLYGON is selected at random. What is the probability that the letter is an O?

2. A bag contains three green marbles and five white marbles. One marble is drawn at random. Find the probability of drawing:
   (a) a white marble   (b) *not* a white marble   (c) a blue marble

3. Find the probability of rolling a single die and obtaining:
   (a) a 3   (d) a prime number
   (b) an odd number   (e) an even number that is > 4
   (c) a number greater than 2   (f) a number less than 1

4. A letter is selected at random from the alphabet. What is the probability that it is *not* a vowel?

5. One of the angles of a right triangle is selected at random. What is the probability of each of the following?
   (a) The angle is obtuse.   (b) The angle is acute.

6. If all the letters of the word GEOMETRY are placed in a hat, what is the probability of drawing at random a letter that is a vowel?

7. There are 13 boys and 17 girls in a class. If a teacher calls on a student at random, what is the probability that the student is a girl?

8. The numbers from 1 to 20, inclusive, are written on individual slips of paper and placed in a hat. What is the probability of selecting each of the following?
   (a) An even number   (d) A number divisible by 5 and 10
   (b) A number divisible by 5   (e) A number that is at least 16
   (c) A number divisible by 10   (f) A prime number

9. A single playing card is drawn from a standard deck of cards. Find the probability that the card is:
   (a) a diamond   (c) the 2 of hearts
   (b) a 5   (d) a picture card

10. The probability that the Cougars will win when playing the Bengals at basketball is 60%. What is the probability that the Bengals will win when they play the Cougars on Saturday?

11. Given four geometric figures: an equiangular triangle, a square, a trapezoid, and a rhombus. If one of the figures is selected at random, what is the probability that the figure will be equilateral?

12. On a test the probability of getting the correct answer to a certain question is represented by $\frac{x}{10}$. Which *cannot* be a value of x?
(1) 11   (2) 0   (3) 1   (4) 10

13. An urn contains only red, white, and blue marbles. If the ratio of red to white to blue marbles is $1 : 3 : 8$, what is the probability that a marble drawn at random is *not* red?

14. What is the probability that a number picked at random from the set $\{-3, -2, -1, 0, 1, 2, 3\}$ will satisfy the inequality $1 - 3x < 5$?

15. A jar contains x red marbles, $2x - 1$ blue marbles, and $2x + 1$ white marbles. One marble is drawn at random.
(**a**) Express in terms of x the total number of marbles in the jar.
(**b**) Express in terms of x the probability of drawing a blue marble.
(**c**) If the probability of drawing a blue marble is $\frac{1}{3}$, find the value of x.
(**d**) What is the probability of *not* drawing a red marble?

16. If the probability that an event will occur is $\frac{x}{4}$ and $x \neq 0$, what is the probability that this event will *not* occur?
(1) $\dfrac{1-x}{4}$ (2) $\dfrac{4-x}{4}$ (3) $\dfrac{4-x}{x}$ (4) $\dfrac{4}{x}$

13.2   FINDING PROBABILITIES BY COUNTING SAMPLE SPACE OUTCOMES

KEY IDEAS

A probability experiment may consist of two or more activities performed in succession or at the same time. We can describe the sample space by listing all possible ordered pairs of the form

(outcome of first activity, outcome of second activity).

Another way of specifying the sample space is by drawing a tree diagram in which the branches show the different possible sequences of outcomes for the different activities.

Listing Outcomes

Suppose that a fair coin is tossed, and then a die is rolled. A *fair* coin is a coin that when tossed has equally likely chances of coming up heads (H) and tails (T). The sample space of this experiment may be described as a set of 12 pairs of outcomes:

$$\{(H, 1), (H, 2), (H, 3), (H, 4), (H, 5), (H, 6),$$
$$(T, 1), (T, 2), (T, 3), (T, 4), (T, 5), (T, 6)\}.$$

The first member of each pair represents the results of the coin toss, and the second member of the pair is a possible outcome of rolling the die.

Example: What is the probability of getting a head and rolling a 4? Since there are 12 possible outcomes in the sample space and only one of these outcomes is (H, 4), $P(H \text{ and } 4) = \frac{1}{12}$.

Example: What is the probability of getting a tail and rolling an odd number? The outcomes (T, 1), (T, 3) and (T, 5) satisfy the conditions of the problem, so the probability of getting a tail and rolling an odd number is $\frac{3}{12}$ or $\frac{1}{4}$.

Example: What is the probability of getting a head *or* rolling a 6? The successful outcomes are (H, 1), (H, 2), (H, 3), (H, 4), (H, 5), (H, 6), and (T, 6), so $P(H \text{ or } 6) = \frac{7}{12}$.

Tree Diagrams

The sample space for the experiment described above can also be represented by using the tree diagram shown in Figure 13.1. The two primary branches correspond to the two possible outcomes of flipping a coin. Each primary branch has six secondary branches, which reflect the six different possible outcomes of rolling a die. Each of the 12 primary-secondary branches represents a possible outcome of the two events.

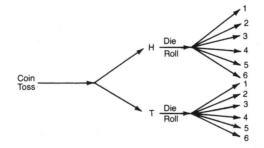

Example

1. A bag contains one red marble, one blue marble, and one green marble. Henry selects a marble at random from the bag, notes its color, and then places it back in the bag. Henry then selects another marble and notes its color.

(**a**) List the sample space as a set of ordered pairs that show all possible outcomes.

(**b**) Draw a tree diagram that shows all possible outcomes.

(**c**) What is the probability that Henry will select the red marble both times?

(**d**) What is the probability that two red marbles will not be selected?

(**e**) What is the probability that Henry will pick the blue marble *at least* once?

(**f**) What is the probability that Henry will select a marble of the same color both times?

Solution: Let R = red marble, B = blue marble, and G = green marble.

(**a**) The first member of the ordered pair represents the color of the first marble selected, and the second member represents the color of the second marble selected.

(R, R), (R, B), (R, G)
(B, R), (B, B), (B, G)
(G, R), (G, B), (G, G)

(**b**)

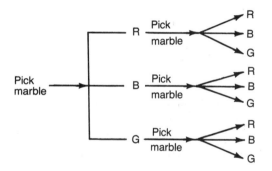

(**c**) The outcome (R, R) appears exactly once in a total of nine possible outcomes, so $P(\text{both red}) = \frac{1}{9}$.

(**d**) $P(\text{neither red}) = 1 - P(2 \text{ red})$

$$= 1 - \frac{1}{9}$$

$$= \frac{8}{9}$$

(**e**) There are five different outcomes in which the blue marble appears: (R, B), (B, R), (B, B), (B, G), and (G, B). Therefore the probability that Henry will select at least one blue marble is $\frac{5}{9}$.

(**f**) There are three different outcomes in which the color of the marble is the same: (R, R), (B, B) and (G, G). Therefore the probability that Henry will select a marble of the same color both times is equal to $\frac{3}{9}$ or $\frac{1}{3}$.

The Counting Principle

Notice that in the coin and die experiment the first activity has two possible outcomes, the second activity has six possible outcomes, and the total number of different ways in which both activities can occur is 2×6 or 12 ways. The generalization of this result, called the *counting principle*, may be stated as follows:

================================ **MATH FACTS** ================================

COUNTING PRINCIPLE

If one activity can be performed in p ways and another activity in q ways, then there are $p \times q$ possible ways in which both activities may be performed.

The counting principle may also be extended for more than two events; see part (b) of Example 2.

Examples

2. A man has five different shirts and four different neckties.
(**a**) In how many ways can he choose a shirt and a necktie?
(**b**) If the man has three different sport jackets, how many different outfits consisting of a sport jacket, shirt, and a tie are possible?

Solution:
(**a**) Use the counting principle: $5 \times 4 = 20$. There are **20** different ways in which he can choose a shirt and a necktie.
(**b**) Use the counting principle for three activities: $5 \times 4 \times 3 = 60$. There are **60** different outfits.

3. Find the probability that when two dice are rolled the sum of the numbers showing is:
(**a**) 7	(**c**) less than 13	(**e**) at least 10
(**b**) 1	(**d**) not greater than 5	(**f**) at most 3

Solution: The counting principle tells us that there is a total of 6×6 or 36 possible outcomes. These outcomes may be described as a set of ordered pairs in which the first member represents the number showing on one die and the second member represents the number showing on the other die.

$$(1, 1), (1, 2), (1, 3), (1, 4), (1, 5), (1, 6)$$
$$(2, 1), (2, 2), (2, 3), (2, 4), (2, 5), (2, 6)$$
$$(3, 1), (3, 2), (3, 3), (3, 4), (3, 5), (3, 6)$$
$$(4, 1), (4, 2), (4, 3), (4, 4), (4, 5), (4, 6)$$
$$(5, 1), (5, 2), (5, 3), (5, 4), (5, 5), (5, 6)$$
$$(6, 1), (6, 2), (6, 3), (6, 4), (6, 5), (6, 6)$$

(a) There are six successes: $(1, 6), (2, 5), (3, 4), (4, 3), (5, 2),$ and $(6, 1)$.

$$P(\text{sum} = 7) = \frac{6}{36} = \frac{1}{6}$$

(b) There are no outcomes in which the sum of the numbers is 1.

$$P(\text{sum} = 1) \frac{0}{36} = 0$$

(c) The outcome $(6, 6)$ has the largest sum, 12. Therefore each of the 36 outcomes has a sum that is less than 13.

$$P(\text{sum} < 13) = \frac{36}{36} = 1$$

(d) Finding outcomes whose sum is not greater than 5 is equivalent to finding outcomes whose sum is less than or equal to 5. There are ten such outcomes: $(1, 1), (1, 2), (1, 3), (1, 4), (2, 1), (2, 2), (2, 3), (3, 1), (3, 2),$ and $(4, 1)$.

$$P(\text{sum not} > 5) = P(\text{sum} \leq 5) = \frac{10}{36} = \frac{5}{18}$$

(e) Finding outcomes whose sum is at least 10 is equivalent to finding outcomes whose sum is greater than or equal to 10. There are six such outcomes: $(4, 6), (5, 5), (5, 6), (6, 4), (6, 5),$ and $(6, 6)$.

$$P(\text{sum at least } 10) = P(\text{sum} \geq 10) = \frac{6}{36} = \frac{1}{6}$$

(f) Finding outcomes whose sum is at most 3 is equivalent to finding outcomes whose sum is less than or equal to 3. There are three such outcomes: $(1, 1), (1, 2),$ and $(2, 1)$.

$$P(\text{sum is at most } 3) = P(\text{sum} \leq 3) = \frac{3}{36} = \frac{1}{12}$$

Exercise Set 13.2

1. A large parking lot has five entrances and seven exits. In how many different ways can a driver enter the parking lot and exit from it?

2. An ice cream parlor makes a sundae using one of six different flavors of ice cream, one of three different flavors of syrup, and one of four different toppings. What is the total number of different sundaes that this ice cream parlor sells?

3. Marcy has five skirts, six blouses, and three scarves. How many outfits can Marcy create consisting of one skirt, one blouse, and one scarf?

4. An experiment consists of tossing a fair coin, and then picking a card at random from a standard deck of playing cards. How many outcomes are contained in the sample space?

5. Two dice are rolled. Find the probability that:
   (**a**) their sum is 8.
   (**b**) the dice show the same number.
   (**c**) their sum is at least 9.
   (**d**) their sum is at most 6.
   (**e**) their sum is a prime number.

6. A fair coin and a fair die are tossed simultaneously.
   (**a**) Draw a tree diagram, or list the sample space showing all possible outcomes.
   (**b**) What is the probability of getting a head and an even number?
   (**c**) What is the probability of getting a 7 and a head?
   (**d**) What is the probability of getting a 5 or a tail?

7. A penny, a nickel, and a dime are in a box. Bob randomly selects a coin, notes its value, and returns it to the box. He then randomly selects another coin from the box.
   (**a**) Show all possible outcomes by drawing a tree diagram or by representing the possible outcomes as a set of ordered pairs.
   (**b**) What is the probability that the same coin will be drawn both times?
   (**c**) What is the probability that a nickel will be drawn *at least* once?
   (**d**) What is the probability that the total value of both coins that are selected will exceed 11¢?

8. The assembly committee of the River High School student council consists of four students whose ages are 14, 15, 16, and 17, respectively. One student will be chosen at random to be chairperson, and then, from the remaining three, one will be chosen at random to be the recording secretary.
   (**a**) Draw a tree diagram or list the sample space showing all possible outcomes after both drawings.
   (**b**) Find the probability that:
      (**1**) the chairperson is older than the recording secretary.
      (**2**) both students chosen are under the age of 16.
      (**3**) both students chosen are the same age.

9. A certain game requires spinning a marker and then tossing a single die. The marker must stop over one of three equally likely colors (red, green, blue), and the die is a standard fair six-sided type.

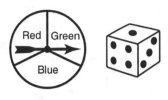

(**a**) Draw a tree diagram or list the sample space of all possible pairs of outcomes for spinning the marker and tossing the die.

(**b**) What is the probability of obtaining the result (red, 4)?

(**c**) What is the probability of obtaining the result (green, 8)?

(**d**) What is the probability of obtaining the result (blue, even number)?

10. A silver dollar, a half-dollar, a quarter, a dime, and a nickel are in a box. One coin is drawn at random. Without replacing the first coin, a second coin is drawn.

(**a**) Draw a tree diagram or list the sample space showing all possible outcomes for this experiment.

(**b**) Find the probability that:

    (1) the value of the first coin drawn will be greater than the value of the second coin drawn.

    (2) the sum of the values of the two coins drawn will be greater than $1.00.

    (3) the sum of the values of the two coins drawn will be exactly $1.00.

11. The first step of an experiment is to pick one number from the set $\{1, 2, 3\}$. The second step of the experiment is to pick one number from the set $\{1, 4, 9\}$.

(**a**) Draw a tree diagram or list the sample space of all possible pairs of outcomes.

(**b**) Determine the probability that:

    (**1**) both numbers will be the same.

    (**2**) the second number will be the square of the first.

    (**3**) both numbers will be odd.

13.3 ADDING PROBABILITIES

△ KEY IDEAS △

When two events have no outcomes in common, then adding their probabilities gives the probability that one event *or* the other event will occur. If the two events have outcomes in common, then we must subtract from this sum the probability that the common outcomes will occur.

Probability of Mutually Exclusive Events

Suppose a purse contains 8 pennies, 3 dimes, and 2 quarters. Picking a penny *or* picking a quarter are *mutually exclusive* events since they cannot happen at the same time. The probability of picking a penny or picking a quarter is the sum of their individual probabilities. Since there is a total of 13 coins in the purse,

$$P(\text{penny } or \text{ quarter}) = \frac{8}{13} + \frac{2}{13} = \frac{10}{13}$$

MATH FACTS

PROBABILITY OF MUTUALLY EXCLUSIVE EVENTS

Two events are **mutually exclusive** if they have no outcomes in common. If *A* and *B* are mutually exclusive events, then

$$P(A \text{ or } B) = P(A) + P(B)$$

Example

1. A single die is tossed. What is the probability of rolling a 3 or an even number?

Solution: Since 3 and an even number cannot be rolled at the same time, these events are mutually exclusive. Thus,

$$P(3 \text{ } or \text{ even number}) = P(3) + P(\text{even number})$$
$$= \frac{1}{6} + \frac{3}{6}$$
$$= \frac{4}{6}$$

Probability of Inclusive Events

Suppose a single die is tossed. The probability of rolling a 3 is $\frac{1}{6}$. The probability of rolling an odd number is $\frac{3}{6}$. The probability of rolling a 3 *or* an odd number is *not* $\frac{1}{6} + \frac{3}{6}$. Since 3 is an odd number, the sum $\frac{1}{6} + \frac{3}{6}$ counts the same outcome, rolling a 3, twice. To avoid counting the same outcome twice, we need to subtract the probability that this common outcome occurs. Thus,

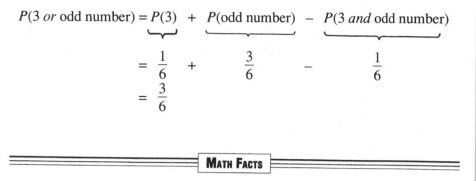

$$P(3 \text{ or odd number}) = P(3) + P(\text{odd number}) - P(3 \text{ and odd number})$$

$$= \frac{1}{6} + \frac{3}{6} - \frac{1}{6}$$

$$= \frac{3}{6}$$

MATH FACTS

PROBABILITY OF INCLUSIVE EVENTS

Two events are **inclusive** if they have one or more outcomes in common. If A and B are inclusive events, then

$$P(A \text{ or } B) = P(A) + P(B) - P(A \text{ and } B)$$

where $P(A \text{ and } B)$ represents the probability that any outcomes common to events A and B will occur. If A and B are mutually exclusive events, then events A and B have no outcomes in common so $P(A \text{ and } B) = 0$.

Example

2. If a letter is picked at random from the English alphabet, what is the probability that it is contained in the word "FACTOR" or in the word "PRODUCT"?

Solution: If event A is picking a letter from the English alphabet that is contained in the word "FACTOR" and event B is picking a letter from the English alphabet that is contained in the word "PRODUCT," then

$$P(A) = \frac{6}{26} \text{ and } P(B) = \frac{7}{26}$$

Since "FACTOR" and "PRODUCT" have the four letters C, T, O, and R in common, $P(A \text{ and } B) = \frac{4}{26}$. Thus,

$$P(A \text{ or } B) = P(A) + P(B) - P(A \text{ and } B)$$

$$= \frac{6}{26} + \frac{7}{26} - \frac{4}{26}$$

$$= \frac{9}{26}$$

Exercise Set 13.3

1. A letter is selected at random from the word TRAPEZOID. Find the probability that the letter:
   (a) is a vowel or a D
   (b) is the first or the last letter of the alphabet
   (c) is a vowel or the letter E
   (d) is a vowel and the letter A
   (e) has vertical and horizontal line symmetry
   (f) has vertical or horizontal line symmetry

2. If $P(A) = 0.7$, $P(B) = 0.5$, and $P(A$ and $B) = 0.35$, then what is $P(A$ or $B)$?

3. Find the probability that, when a single die is rolled, the number rolled is:
   (a) odd and greater than 3
   (b) odd or greater than 3
   (c) prime and greater than 3
   (d) prime or greater than 3
   (e) prime and less than 3
   (f) prime or even

4. One number is selected from the set of integers from 1 to 10, inclusive. Find the probability that the number selected is:
   (a) even and prime
   (b) even or prime
   (c) prime and greater than 5
   (d) divisible by 3 or by 5
   (e) divisible by 3 and by 5
   (f) even or greater than 5

5. A card is selected at random from a standard playing deck. Find the probability that the card selected is a:
   (a) red ace
   (b) red card or an ace
   (c) club or a diamond
   (d) black card or a picture card

6. A letter is picked at random from the English alphabet. Find the probability that the letter is contained in the words:
   (a) "ALGORITHM" and "FRACTION"
   (b) "ALGORITHM" or "FRACTION"
   (c) "STATISTICS" and "PROBABILITY"
   (d) "STATISTICS" or "PROBABILITY"

13.4 MULTIPLYING PROBABILITIES

Multiplying the probability that event *A* will occur by the probability that event *B* will occur gives the probability that events *A* and *B* will both occur. Sometimes the notation *P(A, B)* is used to indicate the probability that events *A* and *B* occur with event *B* occurring after event *A*.

Probability of Independent Events

Suppose a coin is tossed twice. Tossing a head on the first toss and tossing a head on the second toss are *independent* events since the result of the first toss does not affect the result of the second toss. The probability of tossing two heads is the product of the probabilities of each event. Thus,

$$P(\text{Head, Head}) = \frac{1}{2} \cdot \frac{1}{2} = \frac{1}{4}$$

MATH FACTS

PROBABILITY OF INDEPENDENT EVENTS

Two events are **independent** if the first event does not affect the probability of the second event. If *A* and *B* are independent events, then

$$P(A \text{ and } B) = P(A) \cdot P(B)$$

Examples

1. If a coin is tossed and then a die is rolled, what is the probability of getting a *head* and a number *less than* 3?

Solution: $P(\text{H}) = \frac{1}{2}$. There are two numbers less than 3 (1 and 2), so $P(N < 3) = \frac{2}{6} = \frac{1}{3}$.

$$P(\text{H and } N < 3) = P(\text{H}) \times P(N < 3)$$
$$= \frac{1}{2} \times \frac{1}{3}$$
$$= \frac{1}{6}$$

379

2. (a) What is the probability of rolling a number greater than 4 on two successive rolls of the die?

(b) What is the probability of rolling a 6 and then a different number on the second roll?

Solution: The sample space for each roll is $\{1, 2, 3, 4, 5, 6\}$ so the total number of possible outcomes is six. Let $X =$ the number that is rolled.

(a) Since the favorable outcomes are 5 and 6, $P(X > 4) = \frac{2}{6} = \frac{1}{3}$. To find the probability of rolling a number greater than 4 on the first roll *and* on the second roll, we need to multiply probabilities. Thus,

$$P(X > 4 \text{ and } X > 4) = P(X > 4 \cdot P(X > 4) = \frac{1}{3} \cdot \frac{1}{3} = \frac{1}{9}$$

(b) Since $P(X = 1) = \frac{1}{6}$ and $P(X \neq 1) = \frac{5}{6}$

$$P(X = 1 \text{ and } X \neq 1) = \frac{1}{6} \cdot \frac{5}{6} = \frac{5}{36}$$

Probability of Dependent Events

If event A is drawing a king from a deck of cards and event B is drawing another card from the same deck that is also king, then A and B may or may not be independent events. If the first card is put back in the deck before the second card is drawn, then events A and B are independent. If the first card is *not* replaced, then the sample space changes for event B. In this case, events A and B are *dependent*.

Example

3. Find the probability of picking two kings from a deck of cards when the first card

(a) is not replaced   **(b)** is replaced.

Solution: **(a)** When the first card is not replaced in the deck, we assume a favorable outcome is removed from the original sample space of 52 cards. For the second draw there are 51 cards left in the deck, three of which are kings. Thus,

$$P(\text{king, king}) = P(\text{first king}) \times P(\text{second king})$$
$$= \frac{4}{52} \times \frac{3}{51}$$
$$= \frac{1}{13} \times \frac{1}{17}$$
$$= \frac{1}{221}$$

(b) Since the first card is replaced, the two selections are independent events and have the same sample space. Thus,

$$P(\text{king, king}) = P(\text{first king}) \times P(\text{second king})$$

$$= \frac{4}{52} \times \frac{4}{52}$$

$$= \frac{1}{13} \times \frac{1}{13}$$

$$= \mathbf{\frac{1}{169}}$$

MATH FACTS

PROBABILITY OF DEPENDENT EVENTS

Two events are dependent if one event affects the probability that the other event will occur. If *A* and *B* are dependent events, then the probability that *B* will occur after *A* occurs is

$$P(A,B) = P(A) \cdot P(B \text{ given } A \text{ occurs})$$

Examples

4. A pants pocket contains 3 dimes and 5 quarters. A second pants pocket contains 1 nickel, 2 dimes, and 3 quarters. Two coins are randomly drawn in succession and without replacing the first coin.

(a) If one coin is drawn from each pocket, what is the probability that two dimes will be selected?

(b) If both coins are drawn from the pocket that contains 3 dimes and 5 quarters, what is the probability that two dimes will be selected?

Solution: **(a)** If a pocket contains 3 dimes and 5 quarters, then

$$P(\text{dime}) = \frac{3}{3+5} = \frac{3}{8}$$

If another pocket contains 1 nickel, 2 dimes, and 3 quarters, then

$$P(\text{dime}) = \frac{2}{1+2+3} = \frac{2}{6} = \frac{1}{3}$$

The probability of drawing two dimes in succession is the product of the two individual probabilities. Thus,

$$P(\text{dime, dime}) = \frac{3}{8} \cdot \frac{1}{3} = \mathbf{\frac{1}{8}}$$

(b) The probability of drawing a dime from the first pocket is $\frac{3}{8}$. After this coin is drawn, 7 coins, of which 2 are dimes, are left in the pocket. Therefore, the probability of drawing the second dime *after* the first dime is drawn is $\frac{2}{7}$. The probability of drawing two dimes in succession from the first pants pocket is the product of the two individual probabilities. Thus,

$$P(\text{dime, dime}) = \frac{3}{8} \cdot \frac{2}{7} = \frac{6}{56} \text{ or } \frac{3}{28}$$

5. Each of the letters of the word TRIANGLE is written on a slip of paper and then placed in a hat. Two slips of paper are randomly selected in succession and without replacement. Find the probability that:
   (a) both letters have vertical line symmetry
   (b) both letters have the same type of line symmetry
   (c) both letters have vertical *and* horizontal line symmetry.

Solution: Let $P(V, V)$ represent the probability of picking two letters with vertical line symmetry and $P(H, H)$ represent the probability of picking two letters with horizontal line symmetry.

(a) Since the letters, T, I, and A have vertical line symmetry, three out of the eight letters of the word TRIANGLE have vertical line symmetry. On the first draw, the probability of picking one of these three letters is $\frac{3}{8}$. Since the first letter picked is not replaced, on the second draw the probability of picking one of the remaining two letters with vertical line symmetry is $\frac{2}{7}$. Hence,

$$P(V, V) = \frac{3}{8} \cdot \frac{2}{7} = \frac{6}{56}$$

(b) Since the letters I and E have horizontal line symmetry, two out of the eight letters of the word TRIANGLE have horizontal line symmetry. On the first draw, the probability of picking one of these two letters is $\frac{2}{8}$ and, without replacement, the probability of picking the remaining letter with horizontal line symmetry is $\frac{1}{7}$. Hence,

$$P(H, H) = \frac{2}{8} \cdot \frac{1}{7} = \frac{2}{56}$$

From part (a), $P(V, V) = \frac{6}{56}$. Thus,

$$P(\text{two letters with same symmetry}) = P(V, V) + P(H, H).$$
$$= \frac{6}{56} + \frac{2}{56}$$
$$= \frac{8}{56} \text{ or } \frac{1}{7}.$$

(c) The letter I is the only letter of the word TRIANGLE that has both vertical and horizontal line symmetry. Since picking two letters with both types of symmetry is an impossibility, the probability of this event happening is **0**.

6. A jar contains white marbles and blue marbles. The number of white marbles is 3 more than twice the number of blue marbles. The probability of selecting a blue marble from the jar is $\frac{2}{7}$.

(**a**) How many marbles of each color are in the jar?

(**b**) If two marbles are selected from the jar in succession and without replacement, what is the probability that *at least one* of the two marbles will be blue?

Solution: Let x = the number of blue marbles.

Then $2x + 3$ = the number of white marbles.

Write an equation that expresses in terms of x the probability of selecting a blue marble.

$$P(\text{Blue}) = \frac{\text{number of blue marbles}}{\text{total number of marbles}} = \frac{x}{x + (2x + 3)} = \frac{x}{3x + 3}$$

Since $P(\text{Blue}) = \frac{2}{7}$,

$$\frac{2}{7} = \frac{x}{3x + 3}$$

Since in a proportion the product of the means equals the product of the extremes, make the cross-products equal:

$$7x = 2(3x + 3)$$
$$7x = 6x + 6$$
$$x = 6.$$

Thus, the number of blue marbles is **6** and the number of white marbles is $2x + 3 = 2(6) + 3 = \mathbf{15}$.

(**b**) At least one of the two marbles will be blue when: a blue and then a white marble is selected, or if a white and then a blue marble is selected, or if two blue marbles are selected. On the first draw there is a total of 21 marbles in the jar. With no replacement, the jar contains 20 marbles for the second draw. Hence,

$$P(\text{Blue, White}) = \frac{6}{21} \cdot \frac{15}{20}$$

$+$

$$P(\text{White, Blue}) = \frac{15}{21} \cdot \frac{6}{20}$$

$+$

$$P(\text{Blue, Blue}) = \frac{6}{21} \cdot \frac{5}{20}$$

$$P(at \ least \ \text{one blue}) = \frac{6}{21} \cdot \frac{15}{20} + \frac{15}{21} \cdot \frac{6}{20} + \frac{6}{21} \cdot \frac{5}{20}$$
$$= \frac{90}{420} + \frac{90}{420} + \frac{30}{420}$$
$$= \frac{\mathbf{210}}{\mathbf{420}} \ \text{or} \ \frac{1}{2}.$$

Exercise Set 13.4

1. The integers from 1 to 10, inclusive, are written on slips of paper and placed in an urn. Two slips of paper are drawn without replacement. Find the probability that:
   (a) both numbers are even
   (b) the first number is even and the second number is odd
   (c) both numbers are prime
   (d) both numbers are divisible by 3
   (e) both numbers are at least 5
   (f) both numbers have horizontal line symmetry

2. Answer each part of Problem 1 assuming replacement.

3. The letters of the word PARALLELOGRAM are written on individual slips of paper and placed in an urn. Find the probability of drawing two letters at random and obtaining:
   (a) an L on both selections, assuming replacement after the first pick
   (b) an L on both selections without replacement
   (c) two letters that are *not* vowels, assuming replacement after the first pick
   (d) two letters that are *not* vowels without replacement
   (e) two letters that are the same, assuming replacement after the first pick
   (f) two letters that are the same without replacement
   (g) two letters with vertical line symmetry without replacement

4. Two cards are drawn at random without replacement from a standard deck of playing cards. Find the probability that the two cards:
   (a) are both spades   (c) are picture cards
   (b) are in the same suit   (d) have the same face value

5. Answer each part of Problem 4 assuming replacement.

6. John has 10 navy blue socks and 14 black socks in a drawer. If John selects two socks at random, what is the probability they will be the same color?

7. Mary chose at random one of the four numbers 1, 2, 3, and 6. She then chose at random one of the two numbers 1 and 5. Find the probability that Mary chose:
   (a) an even number first, followed by an odd number
   (b) *at least* one even number
   (c) the same two numbers
   (d) two even numbers

8. A coach has to purchase uniforms for a team. A uniform consists of one pair of pants and one shirt. The colors available for the pants are black and white. The colors available for the shirt are green, orange, and yellow. Find the probability that in the uniform the coach chooses:
(**a**) the pants are black and the shirt is orange
(**b**) the shirt is green
(**c**) the pants and the shirt are of different colors

9. Three coins are tossed simultaneously.
(**a**) In how many different, equally likely ways can these coins fall?
(**b**) What is the probability that all three coins will come up the same?
(**c**) In how many ways can a tail and two heads appear?
(**d**) What is the probability of two tails and one head?

10. Jill picks a letter from the word "WIN" at random and then picks a letter from the word "GAME" at random. Find the probability that in the two selections
(**a**) one of the letters will be an "N" and the other letter will be an "M"
(**b**) one of the letters will be a "W" and the other letter will be a vowel
(**c**) *at least* one letter chosen will be a vowel
(**d**) neither letter will be a vowel
(**e**) both letters will have vertical line symmetry
(**f**) both letters will have horizontal line symmetry
(**g**) the same letter will be chosen

13.5 COUNTING ARRANGEMENTS OF OBJECTS: PERMUTATIONS

$$\bigwedge\text{ Key Ideas }\bigwedge$$

A **permutation** is an arrangement of objects in which order matters. A special notation is useful when discussing permutations. The product of the integers from n to 1, inclusive, is called n **factorial** and is written as $n!$ For example,

$$5! = 5 \cdot 4 \cdot 3 \cdot 2 \cdot 1 = 120.$$

Note that n is defined only if n is a positive integer. 0! is defined to be equal to 1. Alternatively, $n!$ may be written as $_nP_n$. For example,

$$_4P_4 = 4! = 4 \cdot 3 \cdot 2 \cdot 1 = 24.$$

Arranging *N* Objects in *N* Available Positions

Consider the number of different ways in which the letters A, H, and W can be arranged in a row. There are *three* available letters that can be placed in position (1).

$$\frac{3}{(1)} \quad \frac{}{(2)} \quad \frac{}{(3)}$$

Once the first position is filled, then either of the *two* letters remaining can be inserted in position (2).

$$\frac{3}{(1)} \quad \frac{2}{(2)} \quad \frac{}{(3)}$$

There is *one* letter left, so it must be placed in position (3).

$$\frac{3}{(1)} \quad \frac{2}{(2)} \quad \frac{1}{(3)}$$

The counting principle may be applied in this instance, giving $3 \cdot 2 \cdot 1$ or 6 ways in which the three letters can be arranged. Also note that $3! = 3 \cdot 2 \cdot 1 = 6$. In general, we may state that:

MATH FACTS

n-FACTORIAL ARRANGEMENTS

n! represents the number of different ways *n* objects can be arranged in *n* available positions.

Since the number of available positions is equal to the number of objects being permuted, each object is used in every arrangement. This process is symbolized by the notation $_nP_n$, which is read as "the permutation of *n* objects taken *n* at a time." The notations $_nP_n$ and $n!$ are mathematically equivalent.

Examples

1. In how many different ways can the letters of the word SQUARE be arranged?

Solution: There are six letters to be arranged.

$$_6P_6 = 6! = 6 \cdot 5 \cdot 4 \cdot 3 \cdot 2 \cdot 1 = 720.$$

The letters of the word SQUARE can be arranged in **720** different ways.

2. In how many ways can six students be arranged in a line if one particular student must be placed first?

Solution: There is only one choice for the first position. Each of the remaining five positions may be filled by any of the remaining five students. Therefore the students can be arranged in

$$1 \cdot {}_5P_5 = 1 \cdot 5 \cdot 4 \cdot 3 \cdot 2 \cdot 1 = \mathbf{120} \text{ ways.}$$

3. In how many different ways can the digits 1, 3, 5, and 7 be arranged to form a four-digit number if repetition of digits
(**a**) is allowed   (**b**) is *not* allowed

Solution: (**a**) There are $4 \cdot 4 \cdot 4 \cdot 4 = \mathbf{256}$ ways in which the digits may be arranged, allowing for repetition of digits.
   (**b**) Since each digit can be used only once, there are ${}_4P_4$ or $4 \cdot 3 \cdot 2 \cdot 1 = \mathbf{24}$ ways in which the digits may be arranged without repeating a digit.

4. How many even four-digit numbers can be formed using the digits 1, 2, 3, and 9 if repetition is not allowed?

Solution: An integer is even if it ends in an even number. Therefore the last digit of the number must be 2. The first three positions of the number may be filled in ${}_3P_3$ ways, so the number of different even numbers that can be formed using these digits is ${}_3P_3 \cdot 1 = 3 \cdot 2 \cdot 1 \cdot 1 = \mathbf{6}$.

5. The letters L, O, G, I, and C are randomly arranged to form a five-letter word. What is the probability that the word "LOGIC" will be formed?

Solution: The five letters L, O, G, I, and C can be arranged in $5! = 120$ different ways. Since 120 different five-letter words can be formed and exactly one of these words is LOGIC, the probability that the word "LOGIC" will be formed is $\frac{1}{120}$.

Arranging *N* Objects in Fewer Than *N* Positions

Sometimes the number of objects to be arranged is greater than the number of positions that are allocated, so that not all of the objects are used in each possible arrangement. For example, in how many different ways can five students be seated in three chairs arranged in a row? Here the number of students ("objects" that are being permuted) exceeds the number of available positions. The first seat may be filled by any one of the five students. Once this seat is filled, there are four students who can be assigned the second seat, and then any of the three remaining students can take the third seat:

$$\frac{5}{\text{Seat 1}} \times \frac{4}{\text{Seat 2}} \times \frac{3}{\text{Seat 3}} = 60 \text{ ways.}$$

Therefore, five "objects" can be arranged in three positions by using the three greatest factors of 5! This is indicated by using the notation ${}_5P_3$:

$$_5P_3 = 5 \cdot 4 \cdot 3 = \mathbf{60}.$$

In general, $_nP_r$ is read as "the permutation of n objects taken r at a time" and is equal to the r greatest factors of $n!$:

$$_nP_r = n(n-1)(n-2)(n-3)\dots(n-r+1).$$

Example

6. How many three-digit numbers greater than 500 can be formed from the digits 1, 2, 3, 4, 5, and 6:
(**a**) without repetition of digits?
(**b**) with repetition of digits allowed?

Solution: (**a**) Since the number being formed must be greater than 500, the first digit may be either a 5 or a 6. Any one of the remaining five digits can then be used for the second position, leaving four available digits for the last position:

$$\underset{\text{first digit}}{2} \times \underset{\text{second digit}}{5} \times \underset{\text{third digit}}{4} = \mathbf{40} \text{ different numbers.}$$

(**b**) Again, the first digit may be either a 5 or a 6. Since repetition of digits is allowed, six digits are available for the second and the third position:

$$\underset{\text{first digit}}{2} \times \underset{\text{second digit}}{6} \times \underset{\text{third digit}}{6} = \mathbf{72} \text{ different numbers.}$$

Exercise Set 13.5

1. Evaluate each of the following:
(**a**) 6!   (**c**) $\dfrac{9!}{4!}$ (**e**) $\dfrac{7!}{6!}$ (**g**) $_5P_2$ (**i**) $_5P_4$

(**b**) $(9-1)!$ (**d**) $_4P_4$ (**f**) $\dfrac{(4+1)!}{(5-2)!}$ (**h**) $_7P_3$ (**j**) $2(_3P_3)$

2. Show that $_8P_3$ and $\dfrac{8!}{(8-3)!}$ are equivalent.

3–10. Find the number of ways in which each of the following activities can be performed:

3. Arranging a chemistry book, a calculus book, a history book, and a poetry book on a shelf.

4. Forming a three-digit number using the digits 1, 3, and 5.

5. Seating seven students in a row of seven chairs.

6. Forming six-letter arrangements using letters of the word SQUARE.

7. Forming four-letter arrangements using letters of the word SQUARE without using the same letter twice.

8. Forming a three-digit number using the digits 2, 4, 6, 8 in which repetition of digits is not allowed.

9. Forming a three-digit number using the digits 2, 4, 6, 8 in which repetition of digits is allowed.

10. Arranging the letters of the word TRIANGLE so that a vowel comes first.

11. How many three-digit numbers less than 400 can be formed from the digits 1, 2, 3, 4, and 5:
    (a) without repetition of digits?
    (b) with repetition of digits allowed?

12. How many four-digit numbers greater than 1000 can be formed from the digits 0, 1, 2, 3, 4, and 5:
    (a) without repetition of digits?
    (b) with repetition of digits allowed?

13. In how many ways can an odd number be formed from the digits 3, 4, 5, 6, and 8 if:
    (a) all the digits are used without repetition.
    (b) all the digits are used with repetition.
    (c) three of the digits are used without repetition.
    (d) three of the digits are used with repetition.

14. What is the probability that, when Allan, Barbara, John, Steve, and George line up, Barbara is first?

15. What is the probability that, when one red, one white, one blue, one green, and one orange marble are placed in a line, the red marble is first and the blue marble is last?

16. How many integers greater than 300 but less than 9999 can be formed from the digits 0, 1, 2, 3, 4, and 5?

REGENTS TUNE-UP: CHAPTER 13

Each of the questions in this section has appeared on a previous Course I Regents Examination. Here is an opportunity for you to review Chapter 13 and, at the same time, prepare for the Course I Regents Examination.

1. In the accompanying figure, the spinner has five equal sections numbered 1 through 5. If the arrow is equally likely to land on any of the sections, what is the probability that it will land on an even number on the next spin?

2. If the probability that Jones will win the election is 0.6, what is the probability that Jones will not win the election?

3. A purse contains three pennies, two nickels, four dimes, and five quarters. If one coin is drawn at random from the purse, what is the probability of drawing a dime?

4. The probability that an event will *not* occur is 7/12. What is the probability that the event will occur?

5. How many different arrangements of four digits can be formed from the digits 2, 5, 6, and 7 if each digit is used only once in each arrangement?

6. There are three ways of going from town A to town B and six ways of going from town B to town C. Find the total number of ways in which a person can go from town A to town B to town C.

7. A fair coin and a fair die are tossed simultaneously. What is the total number of possible outcomes in the sample space?

8. If one card is drawn from a standard deck of 52 playing cards, what is the probability that the card is a red 7?

9. What integer does $\dfrac{4!}{3!}$ equal?

10. From a standard deck of 52 cards, one card is drawn at random. What is the probability that the card is *not a heart*?

11. A six-sided fair die is rolled. What is the probability of rolling a 3 or a 6?

12. In how many different ways can the subjects math, English, social studies, and science be scheduled during the first four periods of the school day?

13. Mary has 2 blouses (1 red and 1 blue) and 3 pairs of slacks (1 yellow, 1 white, and 1 green). The tree diagram represents the outfits she can wear. If Mary chooses 1 blouse and 1 pair of slacks at random, what is the probability that the outfit she chooses will include a pair of green slacks?

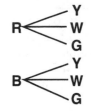

14. An urn contains three red marbles and two green marbles. One marble is randomly selected, its color is noted, and it is *not* replaced. A second marble is then selected and its color is noted.
(**a**) Draw a tree diagram or list the sample space showing all possible outcomes.
(**b**) Find the probability that:
(1) both marbles selected are green.
(2) neither marble selected is green.
(3) at least one marble selected is green.

15. A quarter, a dime, a nickel, and a penny are in a box. Ann draws one coin, replaces it, and draws a coin once again.
(**a**) Draw a tree diagram or list the sample of all possible pairs of outcomes for this experiment.
(**b**) What is the probability that the dime will be drawn at least once?
(**c**) What is the probability that the total value of the coins drawn will be exactly 30¢?

16. In a certain class, there are four students in the first row: three girls, Ann, Barbara, and Cathy, and one boy, David. The teacher called one of these students to the board to solve a problem. When the problem was done, the teacher called one of the remaining students in the first row to do a second problem at the board.
(**a**) Draw a tree diagram or list the sample space of all possible pairs of names for calling two students to the board.
(**b**) Find the probability that the teacher called Ann first and Barbara second.
(**c**) Find the probability that the teacher called two girls to the board.
(**d**) Find the probability that David was one of the two students called.

17. The diagram shown represents an arrow attached to a cardboard disk. The arrow is free to spin, but cannot land on a line. The disk is divided into three regions of equal area, one of which is red and the other is blue.

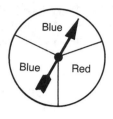

(**a**) For any one spin, what is the probability of the arrow:

    (**1**) landing on red   (**2**) landing on blue

(**b**) The arrow is spun twice and each outcome is recorded. What is the probability of the arrow:

    (**1**) landing on red on the first spin and blue on the second spin

    (**2**) landing on blue on both spins

    (**3**) *not* landing on blue on either spin

    (**4**) landing on the same color on both spins

18. Each card below is printed with either a vowel (A, E, I, O, or U) or a consonant (a letter other than a vowel). The cards are either plain or striped.

(**a**) The cards are shuffled and a card is drawn at random. Find the probability that the card drawn will be:

    (**1**) a vowel   (**2**) striped and a vowel   (**3**) striped or a vowel

(**b**) One card is drawn and the letter is noted. The card is then replaced and a second card is drawn. Find the probability that both cards drawn will be vowels.

(**c**) Two cards are drawn without replacement. What is the probability that the two cards drawn will both be vowels?

19. (**a**) How many different ways can the letters of the word "CHORD" be arranged?

(**b**) How many of the arrangements in part (a) begin with either an "H" or an "O"?

(**c**) If one of the arrangements in part (a) is selected at random, what is the probability it will begin with "C"?

(**d**) If one letter is selected at random from the letters of the word "CHORD," find the probability that it will have:
(1) horizontal line symmetry, only
(2) vertical line symmetry, only
(3) both horizontal line and vertical line symmetry
(4) neither horizontal line nor vertical line symmetry

ANSWERS TO ODD-NUMBERED EXERCISES: CHAPTER 13

Section 13.1

1. $\dfrac{2}{7}$

3. (a) $\dfrac{1}{6}$ (d) $\dfrac{1}{2}$

 (b) $\dfrac{1}{2}$ (e) $\dfrac{1}{6}$

 (c) $\dfrac{4}{6}$ (f) 0

5. (a) 0 (b) $\dfrac{2}{3}$

7. $\dfrac{17}{30}$

9. (a) $\dfrac{13}{52}$ (c) $\dfrac{1}{52}$

 (b) $\dfrac{4}{52}$ (d) $\dfrac{12}{52}$

11. $\dfrac{3}{4}$

13. $\dfrac{11}{12}$

15. (a) $5x$ (c) 3

 (b) $\dfrac{2x-1}{5x}$ (d) $\dfrac{12}{15}$

Section 13.2

1. 35 **3.** 90 **5.** (a) $\dfrac{5}{36}$ (b) $\dfrac{6}{36}$ (c) $\dfrac{10}{36}$ (d) $\dfrac{15}{36}$ (e) $\dfrac{15}{36}$

7. (a) (P, N), (P, D), (P, P), (N, P), (N, D), (N, N), (D, N), (D, P), (D, D)
 (b) $\dfrac{3}{9}$ (c) $\dfrac{5}{9}$ (d) $\dfrac{3}{9}$

9. (a) (R, 1), (R, 2), (R, 3), (R, 4), (R, 5), (R, 6), (G, 1), (G, 2), (G, 3), (G, 4), (G, 5), (G, 6), (B, 1), (B, 2), (B, 3), (B, 4), (B, 5), (B, 6).
 (b) $\dfrac{1}{18}$ (c) 0 (d) $\dfrac{3}{18}$

11. (a) (1, 1), (1, 4), (1, 9), (2, 1), (2, 4), (2, 9), (3, 1), (3, 4), (3, 9)
 (b) (1) $\dfrac{1}{9}$ (2) $\dfrac{3}{9}$ (3) $\dfrac{4}{9}$

Section 13.3

1. (a) $\dfrac{5}{9}$ (b) $\dfrac{2}{9}$ (c) $\dfrac{3}{9}$ (d) $\dfrac{1}{9}$ (e) $\dfrac{2}{9}$ (f) $\dfrac{6}{9}$

3. (a) $\dfrac{1}{6}$ (b) $\dfrac{5}{6}$ (c) $\dfrac{1}{6}$ (d) $\dfrac{5}{6}$ (e) $\dfrac{1}{6}$ (f) $\dfrac{5}{6}$

5. (a) $\dfrac{2}{52}$ (b) $\dfrac{28}{52}$ (c) $\dfrac{26}{52}$ (d) $\dfrac{20}{52}$

Section 13.4

1. (a) $\dfrac{20}{90}$ (b) $\dfrac{25}{90}$ (c) $\dfrac{12}{90}$ (d) $\dfrac{6}{90}$ (e) $\dfrac{30}{90}$ (f) $\dfrac{12}{90}$

3. (a) $\dfrac{9}{169}$ (b) $\dfrac{6}{156}$ (c) $\dfrac{64}{169}$ (d) $\dfrac{56}{156}$ (e) $\dfrac{22}{169}$ (f) $\dfrac{14}{156}$ (g) $\dfrac{20}{156}$

5. (a) $\dfrac{1}{16}$ (b) $\dfrac{1}{4}$ (c) $\dfrac{9}{169}$ (d) $\dfrac{1}{13}$

7. (a) $\dfrac{1}{2}$ (b) $\dfrac{1}{2}$ (c) $\dfrac{1}{8}$ (d) 0

9. (a) 8   (b) $\dfrac{1}{4}$ (c) 3   (d) $\dfrac{3}{8}$

Section 13.5

1. (a) 720
   (b) 40,320
   (c) 15,120
   (d) 24
   (e) 7
   (f) 20
   (g) 20
   (h) 210
   (i) 120   (j) 12

3. 24
5. 5040
7. 360
9. 64
11. (a) 36   (b) 75
13. (a) 24   (c) 12
    (b) 625   (d) 25
15. $\dfrac{6}{120}$

Regents Tune-Up: Chapter 13

1. $\frac{2}{5}$ **3.** $\frac{4}{14}$ **5.** 24   **7.** 12   **9.** 4   **11.** $\frac{2}{6}$ **13.** $\frac{2}{6}$

15. **(a)** (Q, Q), (Q, D), (Q, N), (Q, P), (D, D), (D, Q), (D, N), (D, P), (N, N), (N, Q), (N, D), (N, P), (P, P), (P, Q), (N, P, D), (P, N)

     (b) $\frac{7}{16}$ **(c)** $\frac{2}{16}$

17. **(a) (1)** $\frac{1}{3}$ **(2)** $\frac{2}{3}$

     (b) (1) $\frac{2}{9}$ **(2)** $\frac{4}{9}$ **(3)** $\frac{1}{9}$ **(4)** $\frac{5}{9}$

19. **(a)** 120   **(b)** 48   **(c)** $\frac{24}{120}$ **(d) (1)** $\frac{2}{5}$ **(3)** $\frac{2}{5}$

                                                 (2) 0   **(4)** $\frac{1}{5}$

CHAPTER 14

STATISTICS

14.1 FINDING THE MEAN, MEDIAN, AND MODE

△
KEY IDEAS
△ △

Statistics is the branch of mathematics that is involved with methods of collecting, organizing, displaying, and interpreting *data*. **Data** are simply facts and figures. Numerical data may be referred to as *data values, scores,* or *measures*. The *mean, mode,* and *median* are single numbers that help describe how the individual scores in a set are distributed in value.

The Mean

The **arithmetic mean** is another name for the average of a set of scores. The mean of a set of scores is found by dividing the sum of the scores by the number of scores. For example, the mean of 76, 82, and 85 is 81 since

$$\text{Mean} = \frac{76 + 82 + 85}{3} = \frac{243}{3} = \textbf{81}.$$

Notice that, since the individual scores are fairly close in value, the mean of 81 represents a central value about which the scores in the original data set are clustered. The mean is an example of a **measure of central tendency**.

If a set of scores includes a number that differs by a large amount from the other numbers in the group, then the mean is *not* a good measure of central tendency. For example, the mean of the set of numbers 1, 2, 5, and 200 is 52 since

$$\text{Mean} = \frac{1 + 2 + 5 + 200}{4} = \frac{208}{4} = \textbf{52}.$$

In this case, the mean does not provide useful information on how the individual data values in the set are distributed.

The Mode

The **mode** of a set of data values is the number in the set that appears most frequently. For example, in the set of numbers

$$19, 23, 19, 18, 27, 19, 15, 23, 16$$

the number 19 occurs three times, 23 occurs two times, and each of the remaining numbers appears only once. The mode of this set of numbers is 19.

A set of numbers may have *more than one* mode. The set 15, 9, 8, 9, 11, 15 has two modes, 15 and 9.

If every number in a set appears the same number of times, then the set of numbers has *no* mode. The set of numbers 2, 4, 6, 8, 10 has no mode.

The Median

The mean, mode, and median are measures of central tendency. The **median** of a set of data values is the "middle" value after the data values are arranged in size order. To find the median of the set of numbers 18, 11, 50, 23, 37, arrange the numbers in increasing (or decreasing) order:

$$11, 18, \mathbf{23}, 37, 50$$

Median is the middle score.

The "middle" value is 23, so 23 is the median of this group of numbers. Notice that two numbers in the set are below the median and two numbers in the set are above the median. The median always divides a set of numbers into two groups that have the same number of values.

The following set of numbers contains an even number of values:

$$6, 17, 18, 22, 23, 31$$

Median is the average of the two middle values.

The median of this set is the *average* of the two middle values:

$$\text{Median} = \frac{18 + 22}{2} = \frac{40}{2} = \mathbf{20}$$

Examples

1. The average of a set of four numbers is 78. If three of the numbers in the set are 71, 74, and 83, what is the fourth number?

Solution: Let x = fourth number of the set.

$$\frac{71 + 74 + 83 + x}{4} = 78$$
$$71 + 74 + 83 + x = 4(78)$$
$$228 + x = 312$$
$$x = 312 - 228 = \mathbf{84}$$

2. If a group of data consists of the numbers 2, 2, 5, 6, 15, which of the following is true?

(1) Median > mean   (3) Mode < median

(2) Mean = mode   (4) Median = mode

Solution: Mean = $\dfrac{2 + 2 + 5 + 6 + 15}{5} = \dfrac{30}{5} = 6$

Mode = 2

Median = 5

The mode is less than the median. The correct answer is **choice (3)**.

Quartiles

Quartiles are numbers that separate a set of scores arranged in size order into *four* groups that contain the same number of scores.

If a set of scores is arranged in increasing size order,

- the first or **lower quartile**, denoted by Q_1, is the median of those values that fall below the median for the whole set of scores;
- the second or **middle quartile**, denoted by Q_2, is the median for the whole set of scores;
- the third or **upper quartile**, denoted by Q_3, is the median of those values that fall above the median for the whole set of scores.

For the ordered set of scores

$$\underbrace{4, 13, 15,}\ 19, \underbrace{23, 27, 30}$$

below median   above median

the median (Q_2) for the whole set of scores is 19. The lower quartile (Q_1) is the median of the three scores to the left of 19 and the upper quartile (Q_3) is the median of the three scores to the right of 19. Hence, $Q_1 = 13$, $Q_2 = 19$, and $Q_3 = 27$.

Examples

3. Find the lower, second, and upper quartiles for the following set of scores: 33, 18, 65, 25, 21, 78, 84, 65, 52, 45, 60, and 72.

Solution: After arranging the scores in increasing order, find the median for the whole set. Then find the median of each subgroup.

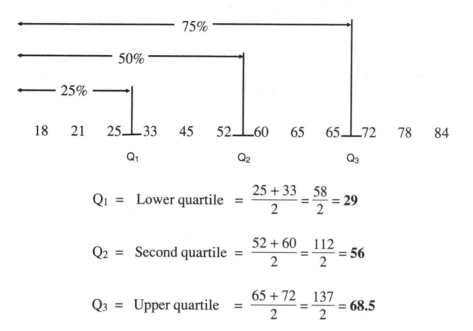

$$Q_1 \; = \; \text{Lower quartile} \; = \; \frac{25+33}{2} = \frac{58}{2} = \mathbf{29}$$

$$Q_2 \; = \; \text{Second quartile} \; = \; \frac{52+60}{2} = \frac{112}{2} = \mathbf{56}$$

$$Q_3 \; = \; \text{Upper quartile} \; = \; \frac{65+72}{2} = \frac{137}{2} = \mathbf{68.5}$$

Since the total number of scores is a multiple of 4, exactly 25% of the original scores are at or below the lower quartile mark of 29; 50% of the original scores are at or below the second quartile (median) mark of 56; 75% of the original scores are at or below the upper quartile mark of 68.5.

4. If a number is randomly selected from the set 8, 5, 13, 5, 7, and 10, find the probability that the number will be:
(**a**) greater than the mean
(**b**) greater than or equal to the mode
(**c**) exactly equal to the median

Solution: (**a**) Mean $= \dfrac{8+5+13+5+7+10}{6} = 8.$

Since 2 out of the 6 numbers in the set are greater than 8, the probability that the number selected will be greater than the mean is $\frac{2}{6}$ or $\frac{1}{3}$.

(**b**) The mode is 5. Since 6 out of the 6 numbers in the set are greater than or equal to 5, the probability that the number selected will be greater than or equal to the mode is $\frac{6}{6}$ or **1**.

(**c**) To find the median, first arrange the numbers in size order.

$$5, 5, 7, 8, 10, 13$$

The median is the middle score. Since there is an even number of scores,

$$\text{Median} = \frac{7+8}{2} = 7.5$$

Since no number in the set is equal to 7.5, the probability that the number selected will be equal to the median is **0**.

Percentiles

Suppose 20 out of 25 students had test scores less than or equal to 92. Since $\frac{20}{25} = 80\%$, 80% of the students had a test score of 92 or less. In this example, 92 represents the 80th *percentile*. In general, the **pth percentile** is that value for which $p\%$ of the data values fall at or below. Thus,

- the lower quartile corresponds to the 25th percentile;
- the median corresponds to the 50th percentile; and
- the upper quartile corresponds to the 75th percentile.

Exercise Set 14.1

1–4. Find the mean, the median, and the mode (if any) of each set of numbers.

1. 6, 9, 4, 2, 9

2. 34, 45, 68, 72, 51, 49

3. 13, 14, 15, 16, 87, 86, 85, 84

4. $\frac{1}{2}$, 0.6, $\frac{3}{4}$, 0.2, $\frac{1}{5}$, 0.3

5–6. Express, in terms of x, the mean of

5. $(2x - 1)$ and $(6x + 1)$

6. $x, (7x - 1),$ and $(4x + 7)$

7. The average of five numbers is 84. If four of the numbers are 71, 81, 94, and 77, what is the fifth number?

8. The average of a set of four numbers is 79. If the sum of three of the numbers is 231, what is the other number of the set?

9. Susan received 78, 89, and 82 on her first three exams in mathematics. What is the lowest score she can receive on her next mathematics exam and have an exam average of at least 85?

10. What is the mean number of sides in a group of polygons that consists of two triangles, one parallelogram, and one hexagon?

11. If a group of data consists of the numbers 8, 13, 8, 7, 4, 8 which is true?
(1) Median > mean   (3) Mode < median
(2) Mean = mode   (4) Median = mode

12. For the group of data 3, 3, 5, 8, 18, which is true?
(1) Median > mean   (3) Mean > mode
(2) Mode > mean   (4) Median = mode

13. A number is selected at random from the set, 2, 2, 2, 2, 3, 3, 3, 7. Find the probability that the number selected is:
(**a**) the mode   (**c**) the median
(**b**) the mean   (**d**) the upper quartile

14. Eight of Mr. Smith's students weigh 79, 60, 80, 50, 55, 100, 80, and 72 pounds. If one of the students is picked at random, find the probability that the student's weight will be:
(**a**) greater than the mean
(**b**) exactly equal to the median
(**c**) less than the mode

15. In a group of test scores, 9 scores are *above* the 40th percentile. How many scores are in the original group of test scores?

16. The midterm exam scores for a class were 59, 65, 65, 67, 71, 70, 73, 75, 76, 76, 79, 81, 83, 84, 85, 85, 88, 89, 90, 92, 95, 96, 99, 100, and 100. Find the:
(**a**) median   (**c**) upper quartile
(**b**) lower quartile   (**d**) 68th percentile

17. In the accompanying table, the frequency column gives the number of times the score on the same line occurs. Find the:

(**a**) total number of data values in the table
(**b**) median
(**c**) lower quartile
(**d**) upper quartile
(**e**) score that is closest to the 48th percentile
(**f**) approximate percentile that corresponds
      to a score of 19

Score	Frequency
8	5
13	1
19	1
22	6
28	1
30	5
35	1
39	1
40	4
48	2

14.2 CONSTRUCTING FREQUENCY TABLES AND HISTOGRAMS

KEY IDEAS

The number of times a particular data value occurs in a data list is called the **frequency** of the data value. Large numbers of data may be more efficiently handled if the data are organized into a *frequency table* that lists each data value and its frequency. A *histogram* is a type of bar graph that provides a more visually appealing representation of the data contained in the frequency table.

Making Frequency Tables

Data can be grouped by frequency for further analysis by inspecting the original list of data and *tallying* each occurrence of a data value. For example, suppose that a class of 25 students received the following scores on a mathematics exam: 58, 70, 60, 65, 68, 70, 90, 70, 72, 74, 70, 70, 75, 78, 80, 96, 75, 80, 83, 80, 83, 88, 90, 65, 75. Here is a more convenient way of representing the same set of data:

Exam Score	Tally	Frequency
58	/	1
60	/	1
65	//	2
68	/	1
70	//////	5
72	/	1
74	/	1
75	///	3
78	/	1
80	///	3
83	//	2
88	/	1
90	//	2
96	/	1
		Sum = 25

The mode is 70 since it has a frequency of 5, which is the largest number in the frequency column. There are 25 scores; the median is the thirteenth score since 12 scores are below this score and 12 scores are above it. By adding the entries in the frequency column, you see that the twelfth score is 74. Since the next three scores are each 75, the *thirteenth* score is 75. The median, therefore, is **75**.

Making Frequency Tables Using Intervals

Sometimes data are more easily managed if they are first divided into intervals having a convenient width. In Figure 14.1 the data in the preceding example is grouped by frequency within intervals having a uniform width of 10 points. The intervals must be selected so that they accommodate both the lowest and the highest score. Since the lowest score is 58, the first interval is 50–59. The highest score is 96, so the last interval is 90–99.

Interval	Tally	Frequency
50–59	/	1
60–69	////	4
70–79	ʜʜʜ ʜʜʜ /	11
80–89	ʜʜʜ /	6
90–99	///	3
		Sum = 25

Figure 14.1 Frequency Table with Intervals

A quick glance at the table gives a good idea of how the test grades are distributed, with most of the scores falling between 70 and 79.

Drawing Frequency Histograms

Here is how to draw a frequency histogram based on the frequency table in Figure 14.1.

Step 1. Using graph paper, draw the coordinate axes in the first quadrant. Call the vertical axis "Frequency," and the horizontal axis "Test scores."

Step 2. Label the vertical axis in units of 1.

Step 3. Label the horizontal axis so that each interval has the *same* width. The width may be of any convenient size.

Step 4. For each interval, draw vertical bars next to one another. The frequency of each interval determines the bar height.

Figure 14.2 shows the completed frequency histogram.

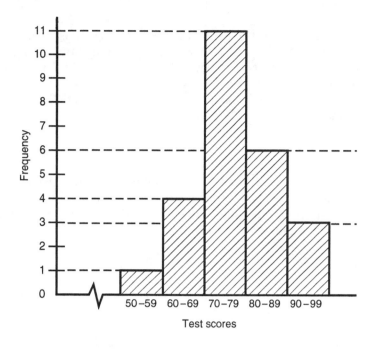

Figure 14.2 Drawing a Frequency Histogram

When drawing a histogram based on a frequency table in which the data are organized into intervals, remember that:

- the intervals are located on the horizontal axis and, for the same histogram, always have the same width.
- the *frequency* of an interval is the number of data values that fall into that interval. Frequencies are always located on the vertical axis.
- the height of each rectangle represents the number of data values that are contained in the interval that is the base of that rectangle. For example, in Figure 14.2 the interval 70–79 is the base of the rectangle with the greatest height so that this interval contains more data values than any other interval. Since the height of this rectangle corresponds to a frequency of 11, the interval 70–79 contains 11 of the original data values.

Examples

1. For the accompanying frequency table, answer the following questions:

(**a**) Which interval contains the median?

(**b**) Which interval contains the lower quartile?

(**c**) Which interval contains the upper quartile?

Interval	Frequency
1–15	1
16–30	2
31–45	6
46–60	4
61–75	2
76–90	5

405

Solution: (**a**) There is a total of 20 scores. The median is the middle value, so it lies between the 10th and 11th scores. By accumulating frequencies, we know that the fourth interval contains the 10th, 11th, 12th, and 13th scores. Therefore, the interval **46–60** contains the median.

(**b**) The lower quartile is the number at or below which 25% of the scores fall. Since there is a total of 20 scores,

$$25\% \text{ of } 20 = \frac{1}{4} \times 20 = 5.$$

Adding frequencies shows that the 5th score is found in the third interval, so the lower quartile is contained in the interval **31–45**.

(**c**) The upper quartile is the number at or below which 75% of the scores fall. Since there is a total of 20 scores,

$$75\% \text{ of } 20 = \frac{3}{4} \times 20 = 15.$$

Adding frequencies reveals that the 15th score is found in the fifth interval, so the upper quartile is contained in the interval **61–75**.

2. The table at the right represents the distribution of ages of principals in a school district.

(**a**) Using this table, draw a frequency histogram.

(**b**) Find the interval that contains the: (**1**) lower quartile.

(**2**) median.

(**3**) upper quartile.

(**c**) What percent of the principals are older than 43?

(**d**) What is the probability that if a principal is chosen at random the age of the principal will be less than 28?

Interval	Frequency
68–75	2
60–67	6
52–59	5
44–51	8
36–43	5
28–35	2

Solution:

(a)

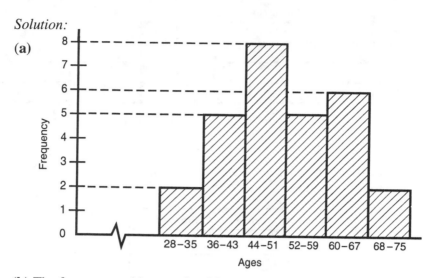

(b) The frequency table contains 28 data values.

(1) In an ordered list of data, the lower quartile is that value at or below which $\frac{1}{4}$ of the data values fall. Since $\frac{1}{4} \times 28 = 7$, the lower quartile is the 7th lowest data value in the frequency table. Adding the frequencies in the bottom two intervals gives 7. Thus, the lower quartile is contained in the second interval, **36–43**.

(2) In an ordered list of data, the median is that value which has the same number of data values above it as below it. Hence, the median is located midway between the 14th and 15th data values. Summing the frequencies in the bottom three intervals gives 15. Thus, the median is contained in the third interval, **44–51**.

(3) Since the lower quartile is 7 data values up from the bottom of the frequency table, the upper quartile is 7 data values down from the top of the same table, which places it in the interval **60–67**.

(c) Summing the frequencies in the intervals above 36–43 gives 8 + 5 + 6 + 2 or 21. Thus, the ages of 21 out of the 28 principals fall in the intervals above 36–43. Since $\frac{21}{28} = \frac{3}{4} = 75\%$, **75%** of the principals are older than 43.

(d) Since no principal is younger than 28, the probability that the age of a principal chosen at random will be less than 28 is **0**.

Exercise Set 14.2

1. Based on the accompanying histogram, find
  (a) the total number of test scores
  (b) the percent of students who had test scores less than 76
  (c) the interval that contains the 80th percentile
  (d) the probability that a score picked at random is between 65 and 70

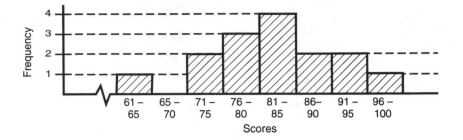

2. The distribution of grades in a college mathematics class is given in the accompanying table.

Grade	Frequency
A	2
B	3
C	10
D	4
F	1

(a) Using graph paper, draw a frequency histogram.
(b) How many students are in the class?
(c) Which grade is the mode?
(d) Which grade represents the lower quartile?
(e) What percent of students received a grade of B or higher?
(f) If a student from the class is selected at random, what is the probability that the student received a grade of at least C?
(g) If a student from the class is selected at random, what is the probability that the student did *not* receive a grade of F?

3. The points scored by Rosa in 20 basketball games are 35, 33, 27, 35, 29, 37, 32, 35, 35, 32, 23, 37, 32, 29, 26, 30, 28, 31, 29, 35.
(a) Find the mode.
(b) Copy and complete the table below.

Interval	Tally	Frequency
35–37		
32–34		
29–31		
26–28		
23–25		

(c) Construct a frequency histogram based on the table completed in part (b).

(d) In which interval does the median lie?

4. The following data represent the heights of 14 students in a certain class: 65, 63, 68, 59, 74, 59, 68, 61, 64, 60, 69, 72, 55, 64.
   (a) Copy and complete the table below:

Interval	Number (Frequency)
55–58	
59–62	
63–66	
67–70	
71–71	

(b) On graph paper, construct a frequency histogram based on the data.
(c) The median is contained in which interval?

5. A class record showed the following number of misspelled words in each of 25 essays.

Misspelled Words	Frequency (Number of Essays)
0	1
1	0
2	3
3	5
4	4
5	9
6	3

(a) On graph paper, construct a frequency histogram based on the data.
(b) Find the mode number of misspelled words.
(c) Find the mean number of misspelled words.
(d) Find the median number of misspelled words.

6. The following table represents the ages of the teachers at a school.

Interval	Number (f)
53–57	4
48–52	8
43–47	6
38–42	4
33–37	2
28–32	4
23–27	2

(a) In which interval is the median?
(b) A teacher is chosen at random from this school. What is the probability that the teacher's age is in the interval 33–37?
(c) What is the probability that the age of a teacher from this school is less than 38?
(d) What is the probability that a teacher from this school is older than 57?
(e) What percent of the teachers are in the interval 43–47?

7. The table below represents the distribution of grades in a college mathematics class. A, B, C, and D are passing grades, and x represents a positive integer.
(a) If 29 students are in this class, find x.
(b) What is the median grade?
(c) If a student is selected at random from the class, find the probability that the student's grade is
  (1) *not* a passing grade
  (2) equal to the mode
  (3) at least a C

Grade	Frequency
A	x
B	$2x - 3$
C	$x + 1$
D	5
F	$x - 4$

8. The table below gives the distribution of test scores for a class of 20 students.

Test Score Interval	Number of Students (frequency)
91–100	1
81–90	3
71–80	3
61–70	7
51–60	6

(**a**) Draw a frequency histogram for the given data.
(**b**) Which interval contains the median?
(**c**) Which interval contains the lower quartile?
(**d**) What is the probability that a student selected at random scored above 90?

14.3 CONSTRUCTING CUMULATIVE FREQUENCY TABLES AND HISTOGRAMS

KEY IDEAS

Cumulative frequency tables and histograms display, for each given interval, the sum of the number of scores in all preceding intervals up to and including those contained in the interval being studied.

Drawing Cumulative Frequency Histograms

The first two columns of Table 14.1 represent a frequency table that shows the distribution of 25 student test scores. The last two columns of this table contain the cumulative frequency table for this set of scores.

TABLE 14.1 COMPLETING A CUMULATIVE FREQUENCY TABLE

Interval	Frequency	Cumulative Frequency		Interval
50–59	1	1		50–59
60–69	4	5	(4 + 1 = 5)	50–69
70–79	11	16	(11 + 5 = 16)	50–79
80–89	6	22	(6 + 16 = 22)	50–89
90–99	3	25	(3 + 22 = 25)	50–99

Observe that for each interval after the first, the entries in the cumulative frequency column are obtained by adding the frequency entry that appears on the same line to the cumulative frequency of the preceding line. Since the first frequency has no entry before it, nothing is added to 1 to obtain the cumulative frequency for the first interval.

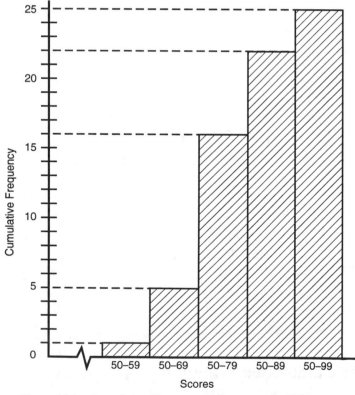

Figure 14.3 Cumulative Frequency Histogram for Table 14.1

Figure 14.3 shows the cumulative histogram for the data contained in the cumulative frequency table (see Table 14.1). The difference in the heights of consecutive rectangles gives the number of scores in the corresponding frequency interval. For example, since the difference in the heights of the rectangles whose bases are the cumulative frequency intervals 50–79 and 50–89 is 22 – 16 = 6, there are 6 scores in the frequency interval 80–89.

Examples

1. Table 1 shown below represents the distribution of the SAT math scores for 60 students at State High School. Table 2 is the cumulative frequency table for the same set of scores.

TABLE 1

Scores	Frequency
710–800	4
610–700	10
510–600	15
410–500	18
310–400	11
210–300	2

TABLE 2

Scores	Cumulative Frequency
210–800	
210–700	
210–600	
210–500	
210–400	
210–300	2

(**a**) Copy and then complete the cumulative frequency table (Table 2).

(**b**) Using the table completed in part (a), draw a cumulative frequency histogram.

(**c**) Using frequency table (Table 1), determine which interval contains the:
(**1**) median   (**2**) upper quartile.

(**d**) What percent of students scored above 700 or below 310?

(**e**) If a student is selected at random, what is the probability of choosing a student who scored higher than 500?

Solution: (**a**) On each line of the cumulative frequency table after the first, add the frequency from the corresponding line of the frequency table to the cumulative frequency on the preceding line.

TABLE 2

Scores	Cumulative Frequency
210–800	56 + 4 = 60
210–700	46 + 10 = 56
210–600	31 + 15 = 46
210–500	13 + 18 = 31
210–400	2 + 11 = 13
210–300	2

413

(b)

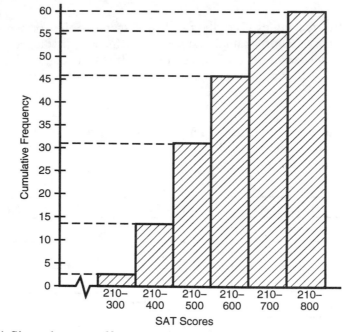

SAT Scores

(c) **(1)** Since there are 60 scores, the median is located between the 30th and 31st scores when the scores are arranged in size order. Counting up from the first line of Table 1 shows that the 30th and 31st lowest scores are contained in the third frequency interval, 410–500.

The median is contained in the interval **410–500**.

(2) The upper quartile is that value at or below which 75% of the data values fall. Since

$$75\% \text{ of } 60 = 0.75 \times 60 = 45,$$

the upper quartile is the 45th data value. Counting up from the first line of Table 1 shows that that 45th score from the lowest is contained in the fourth frequency interval, 510–600.

The upper quartile is contained in the interval **510–600**.

(**d**) Using Table 1, 4 students scored above 700 and 2 students scored below 310. Thus, 6 out of 60 students scored above 700 or below 310. Since $\frac{6}{60} = \frac{1}{10} = 10\%$, **10%** of the students scored above 700 or below 310.

(**e**) Using Table 1, 15 + 10 + 4, or a total of 29 out of 60 students scored higher than 500. Hence, the probability of choosing a student who scored higher than 500 is $\frac{29}{60}$.

2. Table 1 below shows the cumulative frequency of the ages of 35 people standing on a movie theater line.

TABLE 1

Ages	Cumulative Frequency
10–19	2
10–29	17
10–39	27
10–49	32
10–59	32
10–69	35

TABLE 2

Interval	Frequency
10–19	2
20–29	
30–39	
40–49	
50–59	
60–69	

(**a**) Based on the data given in the cumulative frequency table (Table 1), copy and then complete the frequency table (Table 2).

(**b**) Using the frequency table obtained in part (a), determine the interval in which the median occurs.

Solution: (**a**) The frequency on each line after the first is obtained by taking the difference of the cumulative frequency that appears on the corresponding line of the cumulative frequency table and the entry that appears below it.

Interval	Frequency
10–19	2
20–29	17 – 2 = 15
30–39	27 – 17 = 10
40–49	32 – 27 = 5
50–59	32 – 32 = 0
60–69	35 – 32 = 3

(b) Since there are 35 people, the median age is the age of the 18th person when the people are arranged in order of their ages. Counting down from the top line of the frequency table shows that the first two intervals contain 2 + 15 or 17 of the lowest ages. Hence, the 18th age is contained in the next interval, 30–39.

The median occurs in the interval **30–39**.

Exercise Set 14.3

1. The frequency table below shows the distribution of scores of 30 students on a test.

(a)

Scores	Frequency
91–100	3
81–90	11
71–80	8
61–70	6
51–60	1
41–50	1

(b)

Scores	Cumulative Frequency
41–100	
41–90	
41–80	
41–70	
41–60	
41–50	1

(a) Using the data in the frequency table, draw a frequency histogram.
(b) Copy the table and fill in the column headed "Cumulative Frequency."
(c) Using the data in the "Cumulative Frequency" column of the table, draw a cumulative frequency histogram.

2. Table 1 represents the cumulative distribution of a set of student test scores. Table 2 is a frequency table for the same set of scores.

TABLE 1

Scores	Cumulative Frequency
45–100	30
45–90	27
45–80	16
45–70	8
45–60	2
45–50	1

TABLE 2

Scores	Frequency
91–100	
81–90	
71–80	
61–70	
51–60	
45–50	1

(**a**) Draw a cumulative frequency histogram.
(**b**) Copy and then complete Table 2.
(**c**) Draw a frequency histogram.
(**d**) Using Table 2, determine which interval contains the:
    (**1**) lower quartile   (**2**) median   (**3**) upper quartile
(**e**) What percent of the students scored between 61 and 70?

3. The table below shows the distribution of scores that 100 students received on a standardized test.

(**a**) Copy and complete the cumulative frequency table below.

Scores	Frequency
91–100	15
81–90	26
71–80	23
61–70	15
51–60	11
41–50	5
31–40	3
21–30	2

Scores	Cumulative Frequency
21–100	
21–90	
21–80	
21–70	
21–60	
21–50	
21–40	
21–30	

(**b**) Using the table completed in part (a), draw a cumulative frequency histogram.

(**c**) Based on the frequency table, which interval contains the upper quartile?
    (1) 31–40   (3) 71–80
    (2) 61–70   (4) 81–90

(**d**) What percent of the students scored less than 51?

4. The graph below is a frequency histogram that shows the distribution of student test scores.

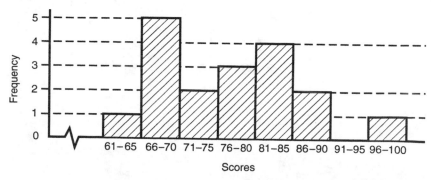

(a) Based on the graph, copy and then complete the cumulative frequency table at the right.

(b) Using the table completed in part (a), construct a cumulative frequency histogram.

(c) If a student score is selected at random, what is the probability that the score is *at least* 86?

Scores	Number of Students
61–100	
61–95	
61–90	
61–85	
61–80	
61–75	
61–70	
61–65	

5. The graph below is a cumulative frequency histogram that shows the distribution of the heights in inches of 24 high school students.

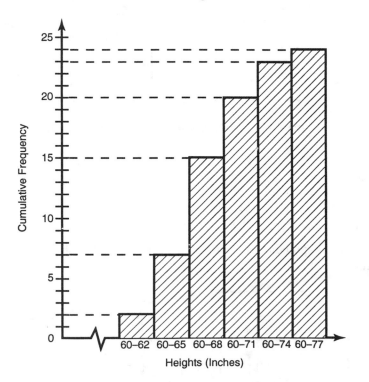

(a) Based on the graph, copy and then complete the frequency table at the right.

(b) Using the table completed in part (a),

(1) Draw a frequency histogram.

(2) Determine the interval that contains the lower quartile.

(3) Determine the interval that contains the upper quartile.

(c) If a student is selected at random, what is the probability that the student is *at most* 68 inches tall?

Heights	Number of Students
60–62	
63–65	
66–68	
69–71	
72–74	
75–77	

REGENTS TUNE-UP: CHAPTER 14

Each of the questions in this section has appeared on a previous Course I Regents Examination. Here is an opportunity for you to review Chapter 14 and, at the same time, prepare for the Course I Regents Examination.

---

1. The test results from a certain examination were 60, 65, 65, 70, 75, 90, and 95. What is the median test score?

2. If student heights are 176 cm, 172 cm, 160 cm, and 160 cm, what is the mean height of these students?

3. Express, in terms of x, the mean of $(2x - 8)$ and $(6x + 4)$.

4. A student received test scores of 82, 94, and 96. What must she receive as a fourth test score so that the mean of her four scores will be exactly 90?

5. Find the mode of the following data: 10, 12, 14, 20, 14, 15, 12, 17, 12.

6. Express the mean of $(2x + 1)$, $(x + 1)$, and $(3x - 8)$ in terms of x.

7. On a math test, a score of 60 was the lower quartile (25th percentile). If 20 students took the test, how many students received scores of 60 or below?

8. Five girls in a club reported the number of boxes of cookies that they sold: 20, 20, 40, 50, and 70. Which is true?

(1) The median is 20.　　(3) The median is equal to the mean.
(2) The mean is 20.　　(4) The median is equal to the mode.

9. For which set of data do the mean, median, and mode all have the same value?
(1) 1, 3, 3, 3, 5   (3) 1, 1, 1, 2, 5
(2) 1, 1, 2, 5, 6   (4) 1, 1, 3, 5, 10

10. For the data 2, 2, 4, 5, 12, which statement is true?
(1) mean = median   (3) mean < mode
(2) mean > mode   (4) mode = median

11. On a mathematics test, Bob scored at the 80th percentile. Which statement is true?
(1) Bob scored 80% on his test.
(2) Bob answered 80 questions correctly.
(3) Eighty percent of the students who took the test had the same score as Bob.
(4) Eighty percent of the students who took the test had a score equal to or less than Bob's score.

12. On a quiz taken by 24 students, the 75th percentile was 84. How many students scored higher than 84?
(1) 6   (2) 12   (3) 18   (4) 24

13. In the set of scores below, how many scores are less than the mean?
32, 40, 42, 52, 59

14. Given the set of numbers $\{30, 42, x, 50, 54, 80\}$, if the median is 49, what is the value of x?

15. The following data are test scores for a class of 16 students: 96, 83, 91, 77, 58, 88, 80, 62, 89, 100, 87, 93, 64, 98, 88, 86.
(**a**) Copy and complete the following table.

Interval	Number (frequency)
91–100	
81–90	
71–80	
61–70	
51–60	

(**b**) On graph paper, construct a frequency histogram based on the data.
(**c**) Which interval contains the median?
(**d**) Which interval contains the lower quartile?

16. The frequency histogram below shows the distribution of scores on a math test.

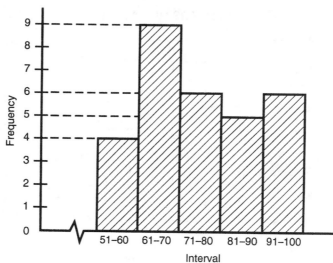

(**a**) Copy and complete the tables below:

Scores	Frequency
51–60	
61–70	
71–80	
81–90	
91–100	

Scores	Cumulative Frequency
51–60	
51–70	
51–80	
51–90	
51–100	

(**b**) How many students took the math test?
(**c**) Which interval of the frequency table contains the upper quartile?
(**d**) How many students scored above 80?
(**e**) Using the table completed in part (a), draw a cumulative frequency histogram.

17. The frequency histogram shows the weights of the members of a Junior Varsity football team at Union High School.

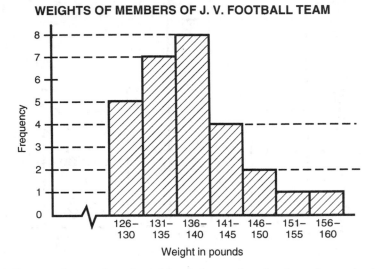

WEIGHTS OF MEMBERS OF J. V. FOOTBALL TEAM

(a) Copy and complete the tables below using the data shown in the frequency histogram.

Interval	Frequency
156–160	
151–155	
146–150	
141–145	
136–140	
131–135	
126–130	5

Interval	Cumulative Frequency
126–160	
126–155	
126–150	
126–145	
126–140	
126–135	
126–130	5

(b) Which interval of the frequency table contains the upper quartile?
(c) If one member of the team is selected at random, what is the probability the member will weigh less than 146 pounds?
(d) What percent of the team weighs at least 141 pounds but less than 156 pounds?

Section 14.1

1. mean = 6
   median = 6
   mode = 9

3. mean = 50
   median = 50
   no mode

5. $4x$

7. 97

9. 91

11. (2)

13. (a) $\frac{1}{2}$ (c) 0

   (b) $\frac{3}{8}$ (d) $\frac{3}{8}$

15. 15

17. (a) 27
   (b) 28
   (c) 19
   (d) 39
   (e) 22
   (f) 26

Section 14.2

1. (a) 15
   (b) 20%
   (c) 86–90
   (d) 0

3. (a) 35
   (b)

Interval	Tally	Frequency
35–37	/////	6
32–34	////	4
29–31	////	5
26–28	///	3
23–25	//	2

(c)

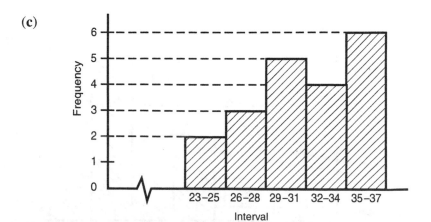

(d) 29–31

5. (a)

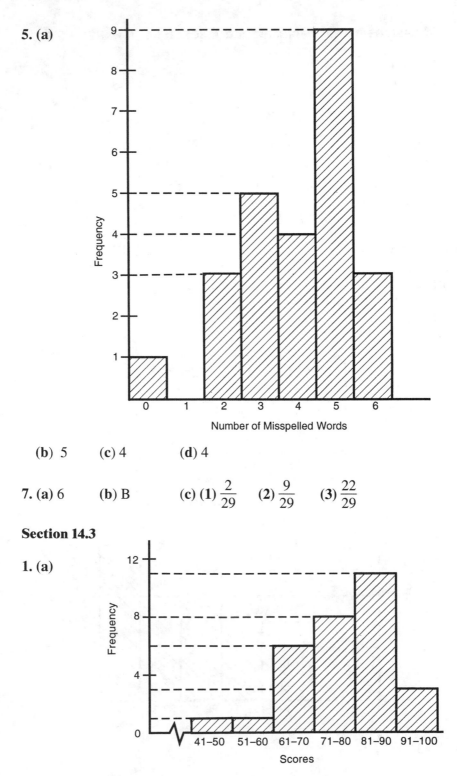

(b) 5   **(c)** 4   **(d)** 4

7. (a) 6   **(b)** B   **(c) (1)** $\dfrac{2}{29}$ **(2)** $\dfrac{9}{29}$ **(3)** $\dfrac{22}{29}$

Section 14.3

1. (a)

424

(b)

Scores	Cumulative Frequency
41–100	30
41–90	27
41–80	16
41–70	8
41–60	2
41–50	1

(c)

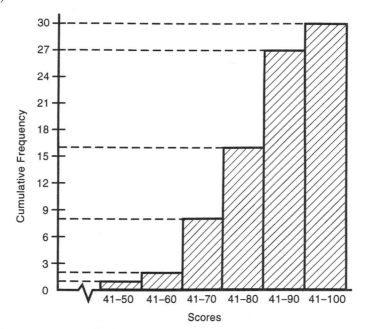

3. (a)

Scores	Cumulative Frequency
21–100	100
21–90	85
21–80	59
21–70	36
21–60	21
21–50	10
21–40	5
21–30	2

(b)

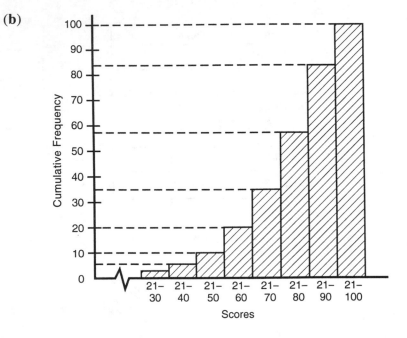

(c) (4)   **(d)** 10%

5. (a)

Height (inches)	Number of Students
60–62	2
63–65	5
66–68	8
69–71	5
72–74	3
75–77	1

(b) (1)

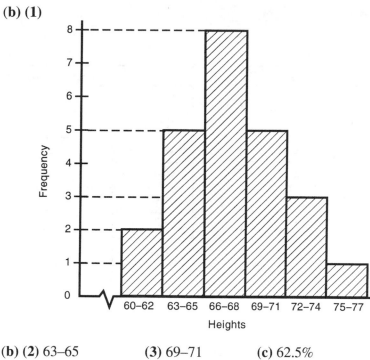

Heights

(b) (2) 63–65   **(3)** 69–71   **(c)** 62.5%

Regents Tune-Up: Chapter 14

1. 70
3. $4x - 2$
5. 12
7. 5
9. (1)
11. (4)
13. 3

15. (a)

Interval	Number
91–100	5
81–90	6
71–80	2
61–70	2
51–60	1

(b)

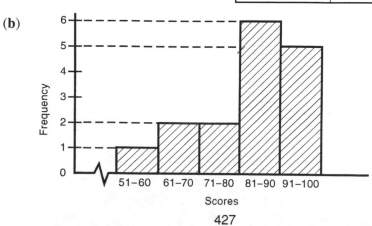

Scores

(c) 81–90   (d) 71–80

17. (a)

Interval	Frequency
156–160	1
151–155	1
146–150	2
141–145	4
136–140	8
131–135	7
126–130	5

Interval	Cumulative Frequency
126–160	28
126–155	27
126–150	26
126–145	24
126–140	20
126–135	12
126–130	5

(b) 141–145   (c) $\frac{6}{7}$ (d) 25%

GLOSSARY

Abscissa The x-coordinate of a point in the coordinate plane.

Absolute value The absolute value of a number x, denoted by $|x|$, is its distance from zero on the number line. Thus, the absolute value of a nonzero number is the number or its opposite—whichever is nonnegative.

Acute angle An angle whose degree measure is less than 90.

Acute triangle A triangle that contains three acute angles.

Additive inverse The opposite of a number. The additive inverse of a number x is $-x$ since $x + (-x) = 0$.

Adjacent angles Two angles that have the same vertex, share a common side, but do not have any interior points in common.

Alternate interior angles Two interior angles that lie on opposite sides of a transversal.

Altitude A segment that is perpendicular to the side to which it is drawn.

Angle The union of two rays that have the same endpoint.

Antecedent The part of a conditional statement that follows the word "if." Sometimes the term "hypothesis" is used in place of "antecedent."

Associative property The mathematical law that states that the order in which three numbers are added or multiplied does not matter.

Average See *Mean*.

Axes The two number lines that intersect at right angles to form a coordinate plane.

Base angles of an isosceles triangle The two congruent angles that include the base of an isosceles triangle.

Base of a power The number that is repeated as a factor in a product. In $3^4 = 3 \cdot 3 \cdot 3 \cdot 3 = 81$, 3 is the base and 4 is the exponent since 3 appears as a factor 4 times.

Biconditional A statement of the form p if and only if q, denoted by $p \leftrightarrow q$, which is true only when p and q have the same truth value.

Binomial A polynomial with two unlike terms.

Bisector of an angle A ray that divides the angle into two angles that have the same degree measure.

Bisector of a segment A line, ray, or segment that contains the midpoint of the given segment.

Coefficient The number that multiplies the literal factors of a monomial. The coefficient of $-5x^2y$ is -5.

Collinear points Points that lie on the same line.

Commutative property The mathematical law that states that the order in which two numbers are added or multiplied does not matter.

Complementary angles Two angles whose degree measures add up to 90.

Compound statement A statement in logic that is formed by combining two or more simple statements using logical connectives.

Conditional statement A statement that has the form "If p then q," denoted by $p \rightarrow q$. It is always true except in the case in which p is true and q is false.

Congruent angles Angles that have the same degree measure.

Congruent figures Figures that have the same size and the same shape. The symbol for congruence is $\cong$.

Congruent polygons Two polygons having the same number of sides are congruent if their vertices can be paired so that all corresponding sides have the same length and all corresponding angles have the same degree measure.

Congruent triangles Two triangles are congruent if any one of the following conditions is true: (1) the sides of one triangle are congruent to the corresponding sides of the other triangle ($SSS \cong SSS$); (2) two sides and the included angle of one triangle are congruent to the corresponding parts of the other triangle ($SAS \cong SAS$); (3) two angles and the included side of one triangle are congruent to the corresponding parts of the other triangle ($ASA \cong ASA$); (4) two angles and the side opposite one of these angles of one triangle are congruent to the corresponding parts of the other triangle ($AAS \cong AAS$).

Congruent segments Line segments that have the same length.

Conjugate pair The sum and difference of the same two terms, as in $a + b$ and $a - b$.

Conjunct Each of the individual statements that comprise a conjunction.

Conjunction A statement of the form p and q, denoted by $p \wedge q$, which is true only when p and q are true at the same time.

Consequent The part of a conditional statement that follows the word "then." Sometimes the word "conclusion" is used in place of "consequent."

Constant A quantity that is fixed in value. In the equation $y = x + 3$, x and y are variables and 3 is a constant.

Contradiction A compound statement that is always false.

Contrapositive The conditional statement formed by negating and then interchanging both parts of a conditional statement. The contrapositive of $p \rightarrow q$ is $\sim q \rightarrow \sim p$.

Converse The conditional statement formed by interchanging both parts of a conditional statement. The converse of $p \rightarrow q$ is $q \rightarrow p$.

Coordinate The real number that corresponds to the position of a point on a number line.

Coordinate plane The region formed by a horizontal number line and vertical number line intersecting at their zero points.

Corresponding angles Pairs of angles that lie on the same side of a transveral, one of which is an interior angle and the other is an exterior angle. In congruent or similar polygons, corresponding angles are pairs of angles formed at matching vertices.

Corresponding sides In congruent or similar polygons, corresponding sides are pairs of sides whose endpoints are matching vertices.

Cumulative frequency The sum of all frequencies from a given data point up to and including another data point.

Cumulative frequency histogram A histogram whose bar heights represent the cumulative frequency at stated intervals.

Data Items of information.

Degree A unit of angle measure that is defined as 1/360th of one complete rotation of a ray about its vertex.

Degree of a monomial The sum of the exponents of its variable factors.

Degree of a polynomial The greatest degree of its monomial terms.

Dilation A transformation in which a figure is enlarged or reduced in size based on a center and a scale factor.

Direct variation A set of ordered pairs in which the ratio of the second member to the first member of each ordered pair is the same nonzero number. Thus, if y varies directly as x then $\frac{y}{x} = k$ or, equivalently, $y = kx$, where k is a nonzero number called the *constant of variation*.

Disjunct Each of the statements that comprise a disjunction.

Disjunction A statement of the form p or q, denoted by $p \vee q$, which is true when p is true, q is true, or both p and q are true.

Distributive property of multiplication over addition For any real numbers a, b, and c, $a(b + c) = ab + ac$ and $(b + c)a = ba + ca$.

Domain The set of all possible replacements for a variable.

Equation A statement that two quantities have the same value.

Equilateral triangle A triangle whose three sides have the same length.

Equivalent equations Two equations that have the same solution set. Thus, $2x = 6$ and $x = 3$ are equivalent equations.

Event A particular subset of outcomes from the set of all possible outcomes of a probability experiment. In flipping a coin, one event is getting a head; another event is getting a tail.

Exponent A number that indicates how many times another number, called the base, is used as a factor. In 3^4, the number 4 is the exponent and tells the number of times the base 3 is used as a factor in a product. Thus, $3^4 = 3 \cdot 3 \cdot 3 \cdot 3 = 81$.

Extremes In the proportion $\frac{a}{b} = \frac{c}{d}$, the terms a and d are the *extremes*.

Factor A number or variable that is being multiplied in a product.

Factoring The process by which a number or polynomial is written as the product of two or more terms.

Factoring completely Factoring a number or polynomial into its prime factors.

Factorial n Denoted by $n!$ and defined for any positive integer n as the product of consecutive integers from n to 1. Thus, $5! = 5 \cdot 4 \cdot 3 \cdot 2 \cdot 1 = 120$.

FOIL The rule for multiplying two binomials horizontally by forming the sum of the products of the first terms (F), the outer terms (O), the inner terms (I), and the last terms (L) of each binomial.

Formula An equation that shows how one quantity depends on one or more other quantities.

Fraction A number that has the form $\frac{a}{b}$, where a is a real number called the numerator and b is a real number called the *denominator*, provided $b \neq 0$.

Frequency The number of times a data value appears in a list.

Fundamental Counting Principle If event A can occur in m ways and event B can occur in n ways, then both events can occur in m times n ways.

Graph of an equation The set of all points on a number line or in the coordinate plane that are solutions of the equation.

Greatest Common Factor (GCF) The GCF of two or more monomials is the monomial with the greatest coefficient and the variable factors of the greatest degree that are common to all the given monomials. The GCF of $8a^2b$ and $20ab^2$ is $4ab$.

Histogram A vertical bar graph whose bars are adjacent to each other.

Hypotenuse The side of a right triangle that is opposite the right angle.

Identity An equation that is true for all possible replacements of the variable. The equation $2(x + 1) = 2x + 2$ is an identity.

Image In a geometric transformation, the point or figure that corresponds to the original point or figure.

Inequality A sentence that compares two quantities using a comparison symbol such as < (is less than), ≤ (is less than or equal to), > (is greater than), ≥ (is greater than or equal to), or ≠ (is unequal to).

Integer A number from the set $\{\ldots-3, -2, -1, 0, 1, 2, 3, \ldots\}$.

Inverse The statement formed by negating both the antecedent and consequent of a conditional statement. Thus, the inverse of $p \to q$ is $\sim p \to \sim q$.

Irrational number A number that cannot be expressed as the quotient of two integers.

Isosceles triangle A triangle in which at least two sides have the same length.

Leg of a right triangle A side of a right triangle that is not opposite the right angle.

Like terms Terms that differ only in their numerical coefficient.

Line Although an undefined term in geometry, it can be described as a continuous set of points that describes a straight path that extends indefinitely in two opposite directions.

Linear equation   An equation in which the greatest exponent of a variable is 1. A linear equation in two variables can be put into the form $Ax + By = C$, where A, B, and C are constants and A and B are not both zero.

Line reflection   A transformation in which each point P that is not on line l is paired with a point P' on the opposite side of line l so that line l is the perpendicular bisector of $\overline{PP'}$. If P is on line l, then P is paired with itself.

Line segment   Part of a line that consists of two different points on the line, called *endpoints,* and all points on the line that are between them.

Line symmetry   A figure has line symmetry when a line l divides the figure into two parts such that each part is the reflection of the other part in line l.

Logical connectives   The conjunction ($\wedge$), disjunction ($\vee$), conditional ($\rightarrow$), and biconditional ($\leftrightarrow$) of two statements.

Mean   The mean or average of a set of n data values is the sum of the data values divided by n.

Means   In the proportion $\frac{a}{b} = \frac{c}{d}$, the terms b and c are the *means.*

Median   The middle value when a set of data values are arranged in size order.

Mode   The data value that occurs most frequently in a given set of data.

Monomial   A number, variable, or their product.

Multiplicative inverse   The reciprocal of a nonzero number.

Negation   The negation of statement p is the statement, denoted by $\sim p$, that has the opposite truth value of p.

Obtuse angle   An angle whose degree measure is greater than 90 and less than 180.

Obtuse triangle   A triangle that contains an obtuse angle.

Open sentence   A sentence whose truth value cannot be determined until its placeholders are replaced with values from the replacement set.

Opposite rays   Two rays that have the same endpoint and form a line.

Ordered pair   Two numbers that are written in a definite order.

Ordinate   The y-coordinate of a point in the coordinate plane.

Origin   The zero point on a number line.

Outcome   A possible result in a probability experiment.

Parallel lines   Lines in the same plane that do not intersect.

Parallelogram   A quadrilateral that has two pairs of parallel sides.

Perfect square   A rational number whose square root is rational.

Permutation   An ordered arrangement of objects.

Perpendicular lines   Lines that intersect at right angles.

Plane   Although undefined in geometry, it can be described as a flat surface that extends indefinitely in all directions.

Point   Although undefined in geometry, it can be described as indicating location with no size.

Polygon   A simple closed curve whose sides are line segments.

Polynomial   A monomial or the sum or difference of two or more monomials.

Postulate   A statement whose truth is accepted without proof.

Power   A product of identical factors written in the form x^n where the number n, called the *exponent,* gives the number of times the common factor x, called the *base,* is used in the product.

Prime factorization   The factorization of a polynomial into factors each of which are divisible only by itself and 1 (or -1).

Probability of an event   The number of ways in which the event can occur divided by the total number of possible outcomes.

Proportion   An equation that states that two ratios are equal. In the proportion

$\frac{a}{b} = \frac{c}{d}$, the product of the means equals the product of the extremes. Thus, $b \cdot c = a \cdot d$.

Pythagorean theorem The square of the length of the hypotenuse of a right triangle is equal to the sum of the squares of the lengths of the legs of the right triangle.

Quadrant One of four rectangular regions into which the coordinate plane is divided.

Quadratic equation An equation that can be put into the form $ax^2 + bx + c = 0$, provided $a \neq 0$.

Quadratic polynomial A polynomial whose degree is 2.

Quadrilateral A polygon with four sides.

Radical (square root) sign The symbol $\sqrt{}$ that denotes the positive square root of a nonnegative number.

Radicand The expression that appears underneath a radical sign.

Ratio A comparison of two numbers by division. The ratio of a to b is the fraction $\frac{a}{b}$, provided $b \neq 0$.

Rational number A number that can be written in the form $\frac{a}{b}$ where a and b are integers and $b \neq 0$. Decimals in which a set of digits endlessly repeat, like .25000... $\left(= \frac{1}{4}\right)$ and .33333... $\left(= \frac{1}{3}\right)$ represent rational numbers.

Ray The part of a line that consists of a point, called an *endpoint*, and the set of points on one side of the endpoint.

Real number A number that is a member of a set that consists of all rational and irrational numbers.

Reciprocal The reciprocal of a nonzero number x is $\frac{1}{x}$.

Rectangle A parallelogram with four right angles.

Replacement set The set of values that a variable may have.

Rhombus A parallelogram with four sides that have the same length.

Right angle An angle whose degree measure is 90.

Right triangle A triangle that contains a right angle.

Root A number that makes an equation a true statement.

Rotation A transformation in which a point or figure is moved about a fixed point a given number of degrees.

Rotational symmetry A figure has rotational symmetry if it can be rotated so that its image coincides with the original figure.

Scalene triangle A triangle in which no two sides have the same length.

Scientific notation A number is in scientific notation when it is expressed as the product of a number between 1 and 10 and a power of 10. The number 81,000 written in scientific notation is 8.1×10^4.

Similar polygons Two polygons with the same number of sides are similar if their vertices can be paired so that corresponding angles have the same measure and the lengths or corresponding sides are in proportion.

Similar triangles If two triangles have two pairs of corresponding angles that have the same degree measure, then the triangles are similar.

Slope A measure of the steepness of a nonvertical line. The slope of a horizontal line is 0, and the slope of a vertical line is undefined.

Slope formula The slope of a nonvertical line that contains the points (x_1, y_1) and (x_2, y_2) is given by the formula $\frac{y_2 - y_1}{x_2 - x_1}$.

Slope-intercept form An equation of a line that has the form $y = mx + b$ where m is the slope of the line and b is the y-intercept.

Solution Any value from the replacement set of a variable that makes an open sentence true.

Solution set The collection of all values from the replacement set of a variable that makes an open sentence true.

Square A rectangle whose four sides have the same length.

Square root The square root of a non-negative number n is one of two identical numbers whose product is n. Thus, $\sqrt{9} = 3$ because $3 \times 3 = 9$.

Statement Any sentence whose truth value can be assessed as true or false, but not both.

Success Any favorable outcome of a probability experiment.

Supplementary angles Two angles are supplementary if the sum of their degree measures is 180.

System of equations A set of equations whose solution is the set of values that make each of the equations true at the same time.

Tautology A compound statement that is true regardless of the truth values of its component statements.

Theorem A generalization in mathematics that can be proved.

Transformation The process of "moving" each point of a figure according to some given rule.

Translation A transformation in which each point of a figure is moved the same distance and in the same direction.

Transversal A line that intersects two other lines in two different points.

Trapezoid A quadrilateral with exactly one pair of parallel sides.

Tree diagram A diagram whose branches describe the different possible outcomes in a probability experiment.

Triangle A polygon with three sides.

Trinomial A polynomial with three unlike terms.

Truth value Either true or false, but not both.

Undefined term A term that can be described but not defined. The terms *point, line,* and *plane* are undefined in geometry.

Unlike terms Terms that do not have the same variable factors.

Variable The symbol, usually a letter, that represents an unspecified member of a given set called the *replacement set.*

Vertical angles A pair of nonadjacent angles formed by two intersecting lines.

x-axis The horizontal axis in the coordinate plane.

x-coordinate The first number in an ordered pair.

y-axis The vertical axis in the coordinate plane.

y-coordinate The second number in an ordered pair.

y-intercept The y-coordinate of the point at which the graph of an equation intersects the y-axis.

REGENTS EXAMINATION

Three-Year Sequence for High School Mathematics-Course I

Part I

Answer 30 questions from this part. Each correct answer will receive 2 credits. No partial credit will be allowed. Write your answers in the spaces provided on the separate answer sheet. Where applicable, answers may be left in terms of π or in radical form.   [60]

1 Let p represent "Mr. Ladd teaches mathematics" and let q represent "Mr. Ladd is the football coach." Write in symbolic form: "Mr. Ladd teaches mathematics and Mr. Ladd is not the football coach."

2 In the accompanying diagram, $\overleftrightarrow{AB}$ intersects $\overleftrightarrow{CD}$ at E. If m$\angle AEC$ = 40 and m$\angle DEA$ = 4x, what is the value of x?

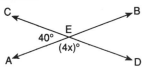

3 Solve for x: $8x - 7 = 5(x - 2)$

4 In the accompanying diagram of $\triangle ABC$, the measure of exterior angle BCD is 120° and m$\angle BAC$ = 50. Find m$\angle ABC$.

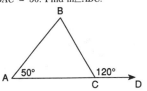

5 In the accompanying diagram, which triangle is the image of $\triangle 2$ after a reflection in the x-axis?

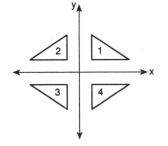

6 If $x = -2$ and $y = 4$, find the numerical value of the expression $7y - 3x$.

7 Solve for x: $\frac{2}{3}x - 6 = 14$

8 From $y^2 + 5y - 7$, subtract $y^2 - 3y - 4$.

9 In the accompanying diagram, $\overleftrightarrow{AB}$ is parallel to $\overleftrightarrow{CD}$, m$\angle BAC$ = 46, and m$\angle BCD$ = 65. Find the measure of $\angle ACB$.

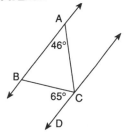

10 Solve for x: $\frac{6}{x + 2} = \frac{2}{x}$, $x \neq 0, -2$

11 What is the slope of the line whose equation is $y = \frac{3}{4}x + 4$?

12 Express $\frac{4x}{2} - \frac{5x}{4}$ as a single fraction in simplest form.

13 For which value of x is the expression $\frac{2}{x + 3}$ undefined?

14 Solve for x: $\begin{array}{l} 2x - y = 3 \\ x + y = 3 \end{array}$

15 A circle has a radius of 5. Express, in terms of π, the circumference of the circle.

16 In symbolic form, write the contrapositive of $p \rightarrow \sim q$.

17 A basketball team consists of 15 girls. The accompanying table shows the number of points each player scored in one season. Which interval contains the median for these data?

Interval	Frequency
101–120	1
81–100	4
61–80	2
41–60	3
21–40	3
0–20	2

18 The lengths of the sides of a triangle are 6, 8, and 12. If the length of the shortest side of a similar triangle is 10, what is the length of its longest side?

Directions (19–35): For *each* question chosen, write on the separate answer sheet the *numeral* preceding the word or expression that best completes the statement or answers the question.

19 What is the product of $4xy^3$ and $3xy^2$?
(1) $12xy^5$ (3) $12x^2y^5$
(2) $12xy^6$ (4) $12x^2y^6$

20 During a half hour of television programming, eight minutes is used for commercials. If a television set is turned on at a random time during the half hour, what is the probability that a commercial is *not* being shown?
(1) 1   (3) $\frac{8}{30}$
(2) $\frac{22}{30}$ (4) 0

21 Which inequality is represented by the graph below?

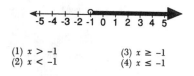

(1) $x > -1$ (3) $x \geq -1$
(2) $x < -1$ (4) $x \leq -1$

22 Which equation is represented by this graph of line ℓ?

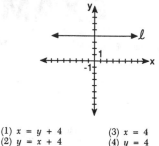

(1) $x = y + 4$ (3) $x = 4$
(2) $y = x + 4$ (4) $y = 4$

23 The expression $\sqrt{75}$ is equal to
(1) $2\sqrt{5}$ (3) $5\sqrt{2}$
(2) $3\sqrt{5}$ (4) $5\sqrt{3}$

24 In the accompanying diagram of $\triangle ABC$, $\overline{CD} \perp \overline{AB}$, $AB = 12y$, and $CD = 8y$.

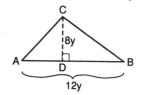

The area of $\triangle ABC$ can be expressed as
(1) $96y^2$ (3) $24y^2$
(2) $48y^2$ (4) $20y^2$

25 In the truth table below, which statement should be the heading for column 3?

Column 1	Column 2	Column 3
p	q	?
T	T	T
T	F	T
F	T	F
F	F	T

(1) $p \vee \sim q$ (3) $\sim p \rightarrow q$
(2) $q \vee \sim p$ (4) $\sim q \rightarrow p$

26 The graph of which inequality is shown in the accompanying diagram?

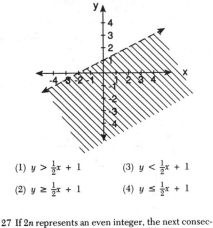

(1) $y > \frac{1}{2}x + 1$ (3) $y < \frac{1}{2}x + 1$

(2) $y \geq \frac{1}{2}x + 1$ (4) $y \leq \frac{1}{2}x + 1$

27 If $2n$ represents an even integer, the next consecutive even integer is represented by

(1) $4n$ (3) $2n + 1$
(2) $n + 2$ (4) $2n + 2$

28 Which is a rational number?

(1) $\sqrt{13}$ (3) $\sqrt{3}$
(2) π (4) $\sqrt{0.16}$

29 The ages of the members of a family are 46, 48, 21, 12, 15, and 8. What is the mean age?

(1) $16\frac{1}{2}$ (3) 25

(2) 18   (4) 36

30 What is the supplement of an angle that measures $3x°$?

(1) $90° - 3x°$ (3) $180° - 3x°$
(2) $3x° - 90°$ (4) $3x° - 180°$

31 The product of $2x - 3$ and $x + 4$ can be expressed as

(1) $2x^2 + 5x - 12$ (3) $2x^2 + x - 12$
(2) $3x + 1$ (4) $2x^2 - 12$

32 When the expressions $x^2 - 9$ and $x^2 - 5x + 6$ are factored, a common factor is

(1) $x + 3$ (3) $x - 2$
(2) $x - 3$ (4) x^2

33 How many different ways can six different plants be arranged side by side on a shelf?

(1) 6   (3) 720
(2) 36   (4) 5,040

34 A wire reaches from the top of a 13-meter telephone pole to a point on the ground 9 meters from the base of the pole. What is the length of the wire to the *nearest tenth* of a meter?

(1) 15.6   (3) 16.0
(2) 15.8   (4) 16.2

35 If the angles of a triangle are represented by $x, 3x + 20$, and $6x$, the triangle must be

(1) obtuse   (3) acute
(2) right   (4) isosceles

Part II

Answer four questions from this part. Clearly indicate the necessary steps, including appropriate formula substitutions, diagrams, graphs, charts, etc. Calculations that may be obtained by mental arithmetic or the calculator do not need to be shown.   [40]

36 Solve the following system of equations graphically and check:

$$3y = -2x + 12$$
$$y - x = 4$$ [8,2]

37 Without looking, Liz chooses one chip from a box containing four chips numbered 1 through 4. Next she chooses one chip from a second box containing four chips lettered a, b, c, and d.

a Draw a tree diagram or list the sample space of all possible outcomes.   [3]

b Find the probability that Liz chose
   (1) an odd number   [1]
   (2) an even number and the letter c   [2]
   (3) a number less than 4 or the letter a   [2]
   (4) a number greater than 3 and a letter from the word "cobra"   [2]

38 In the accompanying diagram, ABCD is a trapezoid with bases $\overline{AED}$ and $\overline{BC}$, $\overline{AB}$ is perpendicular to $\overline{AD}$ and is the diameter of semicircle O, AB = 16, BC = 17, ED = 14, and BE = 20.

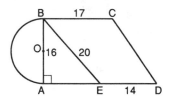

a Find the length of $\overline{AE}$.   [3]

b Find, to the nearest tenth, the area of the entire figure. [Use π = 3.14.]   [7]

39 For a class project, 20 students recorded the number of hours of television that they each watched in one week: 5, 12, 29, 23, 35, 8, 41, 40, 13, 16, 31, 29, 18, 28, 15, 32, 38, 26, 20, 22.

a On your answer sheet, copy and complete the tables below to find the frequency and cumulative frequency in each interval.   [4]

Interval	Tally	Frequency
0–9		
10–19		
20–29		
30–39		
40–49		

Interval	Cumulative Frequency
0–9	
0–19	
0–29	
0–39	
0–49	

b Using the cumulative frequency table completed in part a, construct a cumulative frequency histogram.   [4]

c In one week, what percent of the 20 students watched television more than 9 hours but less than 20 hours?   [2]

40 Solve the following system of equations algebraically and check:

$$3x - 5y = -6$$
$$2x - 3y = -5$$ [8,2]

41 Find two consecutive negative integers such that the product is 42. [Only an algebraic solution will be accepted.]   [4,6]

42 A landscaper has two gardens: one is a square and the other is a rectangle. The width of the rectangular garden is 5 yards less than a side of the square one, and the length of the rectangular garden is 3 yards more than a side of the square garden. If the sum of the areas of both gardens is 165 square yards, find the measure of a side of the square garden. [Show or explain the procedure used to obtain your answer.]   [10]

REGENTS EXAMINATION

Three-Year Sequence for High School Mathematics-Course I

Part I

Answer 30 questions from this part. Each correct answer will receive 2 credits. No partial credit will be allowed. Write your answers in the spaces provided on the separate answer sheet. Where applicable, answers may be left in terms of π or in radical form.

1 What is the mode of the following set of numbers?

$$6, 7, 8, 3, 5, 8, 1$$

2 The probability that an event will occur is $\frac{4}{9}$. What is the probability that the event will *not* occur?

3 In the accompanying diagram, $\overleftrightarrow{AB}$ and $\overleftrightarrow{CD}$ intersect at E, $m\angle AEC = 5x - 20$, and $m\angle DEB = 2x + 10$. Find the value of x.

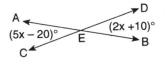

4 In the accompanying diagram, $\overline{AC} \cong \overline{BC}$ and $m\angle A = 80$. Find $m\angle C$.

5 Express $(a + 3)(a - 4)$ as a trinomial.

6 Let p represent "x is an even number" and let q represent "x is greater than 15." Using p and q, write in symbolic form: "x is an even number or x is not greater than 15."

7 What is the converse of $p \rightarrow q$?

8 Solve for the positive value of x:
$$3x^2 - 27 = 0$$

9 In the accompanying diagram of rhombus $ABCD$, the lengths of sides $\overline{AB}$ and $\overline{BC}$ are represented by $3x - 8$ and $2x + 1$, respectively. Find the value of x.

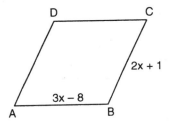

10 Solve for x: $2(x + 1) = 3(4 - x)$

11 In the accompanying diagram, $\triangle ABC$ is similar to $\triangle RST$, $BC = 12$, $AC = 15$, and $ST = 8$. Find RT.

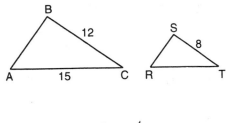

12 Solve for x: $\dfrac{x}{x + 2} = \dfrac{4}{5}$, $x \neq -2$

13 Factor: $2x^2 - 5x + 2$

14 If $(k,4)$ is a point on the graph of the equation $4x + 2y = 4$, what is the value of k?

15 Find the number of square units in the area of parallelogram *ABCD*.

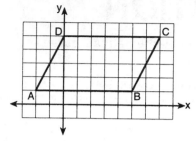

16 Solve for x: $0.5x + 1.8 = -0.7$

17 Two angles are complementary. If the measure of one angle is 20° more than the measure of the second angle, what is the number of degrees in the measure of the *smaller* angle?

18 The cumulative frequency histogram below shows the scores that 24 students received on an English test. How many students had scores between 71 and 80?

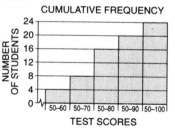

CUMULATIVE FREQUENCY

TEST SCORES

Directions (19–35): For *each* question chosen, write on the separate answer sheet the *numeral* preceding the word or expression that best completes the statement or answers the question.

19 Which letter has both point and line symmetry?
(1) **A**   (3) **M**
(2) **H**   (4) **S**

20 In the accompanying diagram of $\triangle ABC$, m$\angle B$ = 80.

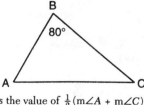

What is the value of $\frac{1}{2}(m\angle A + m\angle C)$?
(1) 40   (3) 100
(2) 50   (4) 140

21 If $4x + y = H$, then x is equal to
(1) $\frac{H}{4} - y$ (3) $\frac{H + y}{4}$
(2) $\frac{H}{4} + y$ (4) $\frac{H - y}{4}$

22 When a number is chosen at random from the set {1,2,3,4,5,6}, which event has the greatest probability of occurring?
(1) choosing an even number
(2) choosing a prime number
(3) choosing a number greater than 3
(4) *not* choosing either 1 or 6

23 Which equation could be used to solve the problem below?

If three times a number is increased by 24, the result is 4 less than seven times the number.

(1) $3(x + 24) = 7x - 4$ (3) $3x + 24 = 7x - 4$
(2) $3x + 24 = 4 - 7x$ (4) $27x = 7x - 4$

24 A member of the solution set of $-1 \le x < 4$ is
(1) -1 (3) 5
(2) -2 (4) 4

25 The reciprocal of $-\frac{1}{x}$, $x \ne 0$, is
(1) $\frac{1}{x}$ (3) $-x$
(2) x (4) $1 + \frac{1}{x}$

26 In the accompanying diagram, $\angle ACD$ is an exterior angle of $\triangle ABC$, m$\angle A$ = 3x, m$\angle ACD$ = 5x, and m$\angle B$ = 50.

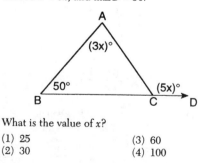

What is the value of x?
(1) 25   (3) 60
(2) 30   (4) 100

27 Expressed in scientific notation, 0.0000047 is equivalent to
(1) 4.7×10^{-6} (3) 4.7×10^{6}
(2) 0.47×10^{-5} (4) 47×10^{6}

28 Which graph represents the inequality $x > 3$?

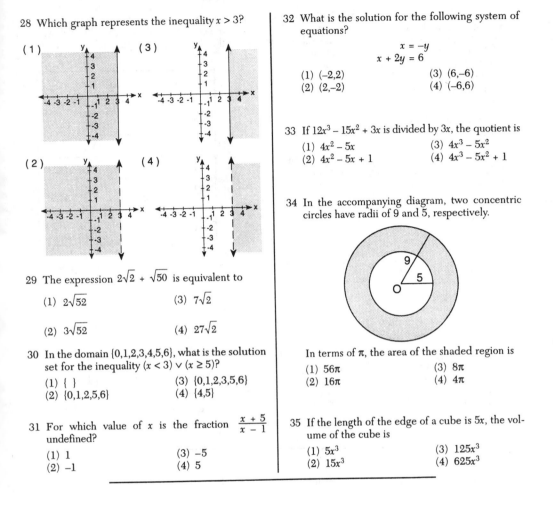

(1)

(3)

(2)

(4)

29 The expression $2\sqrt{2} + \sqrt{50}$ is equivalent to

(1) $2\sqrt{52}$

(3) $7\sqrt{2}$

(2) $3\sqrt{52}$

(4) $27\sqrt{2}$

30 In the domain {0,1,2,3,4,5,6}, what is the solution set for the inequality $(x < 3) \lor (x \geq 5)$?

(1) { }
(2) {0,1,2,5,6}
(3) {0,1,2,3,5,6}
(4) {4,5}

31 For which value of x is the fraction $\dfrac{x + 5}{x - 1}$ undefined?

(1) 1
(2) −1
(3) −5
(4) 5

32 What is the solution for the following system of equations?

$$x = -y$$
$$x + 2y = 6$$

(1) (−2,2)
(2) (2,−2)
(3) (6,−6)
(4) (−6,6)

33 If $12x^3 - 15x^2 + 3x$ is divided by $3x$, the quotient is

(1) $4x^2 - 5x$
(2) $4x^2 - 5x + 1$
(3) $4x^3 - 5x^2$
(4) $4x^3 - 5x^2 + 1$

34 In the accompanying diagram, two concentric circles have radii of 9 and 5, respectively.

In terms of π, the area of the shaded region is

(1) 56π
(2) 16π
(3) 8π
(4) 4π

35 If the length of the edge of a cube is $5x$, the volume of the cube is

(1) $5x^3$
(2) $15x^3$
(3) $125x^3$
(4) $625x^3$

Part II

Answer four questions from this part. Clearly indicate the necessary steps, including appropriate formula substitutions, diagrams, graphs, charts, etc. Calculations that may be obtained by mental arithmetic or the calculator do not need to be shown.

36 *a* On the same set of coordinate axes, graph the following system of inequalities:

$$y \geq 3x + 6$$
$$x + y < -2$$

b Using the graphs drawn in part *a*, state the coordinates of a point that is in the solution of $y \geq 3x + 6$ but is not in the solution of $x + y < -2$.

37 Angie has one penny, one nickel, one dime, and one quarter in her pocket. Angie picks one coin at random. Without replacing the coin, she picks a second coin.

a Draw a tree diagram or list the sample space showing all possible outcomes.

b Find the probability that the two coins picked have a sum of

(1) *exactly* 15 cents
(2) more than 25 cents
(3) less than 40 cents

38 A movie theater charges $7 for an adult's ticket and $4 for a child's ticket. On a recent night, the sale of child's tickets was three times the sale of adult's tickets. If the total amount collected for ticket sales was not more than $2,000, what is the greatest number of adults who could have purchased tickets? [*Show or explain the procedure used to obtain your answer.*]

39 In a trapezoid, the smaller base is 3 more than the height, the larger base is 5 less than 3 times the height, and the area of the trapezoid is 45 square centimeters. Find, in centimeters, the height of the trapezoid. [*Only an algebraic solution will be accepted.*]

40 Solve the following system of equations algebraically and check:

$$3x + y = 3$$
$$y = 2x - 7$$

41 The graph below shows the distribution of scores of 30 students on a mathematics test.

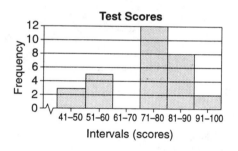

Test Scores

Intervals (scores)

a *On your answer paper,* copy and complete the cumulative frequency table below, using the data in the frequency histogram.

Scores	Cumulative Frequency
41–100	
41–90	
41–80	
41–70	
41–60	
41–50	3

b Construct a cumulative frequency histogram using the table completed in part *a.*

c Which interval contains the median score?

(1) 51–60   (3) 71–80
(2) 61–70   (4) 81–90

d Which interval contains the lower quartile?

(1) 41–50   (3) 71–80
(2) 51–60   (4) 81–90

---

42 a *On your answer sheet,* copy and complete the truth table for the statement $\sim (p \rightarrow q) \leftrightarrow (\sim p \vee q)$.

p	q	$p \rightarrow q$	$\sim (p \rightarrow q)$	$\sim p$	$\sim p \vee q$	$\sim (p \rightarrow q) \leftrightarrow (\sim p \vee q)$
T	T					
T	F					
F	T					
F	F					

b Based on the truth table completed in part *a,* is $\sim (p \rightarrow q) \leftrightarrow (\sim p \vee q)$ a tautology?

c Justify the answer given in part *b.*

REGENTS EXAMINATION

Three-Year Sequence for High School Mathematics-Course I

Part I

Answer 30 questions from this part. Each correct answer will receive 2 credits. No partial credit will be allowed. Write your answers in the spaces provided on the separate answer sheet. Where applicable, answers may be left in terms of π or in radical form.

1 Let p represent "The Yankees won last night" and let q represent "The Yankees stayed in first place." Using p and q, write in symbolic form the statement: "If the Yankees did not win last night, then they did not stay in first place."

2 What is the mode of the set of values 25, 35, 35, 35, 45, 55, 85, and 95?

3 In the accompanying diagram, $ABCD$ is a quadrilateral, $AD = x$, $AB = x + 1$, $BC = 9$, and $DC = 2x + 5$. Find the value of x if the perimeter of quadrilateral $ABCD$ is 35.

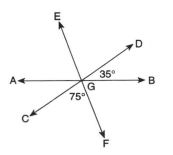

4 In the accompanying diagram, lines $\overleftrightarrow{AB}$, $\overleftrightarrow{CD}$, and $\overleftrightarrow{EF}$ intersect at G. If $m\angle DGB = 35$ and $m\angle CGF = 75$, find $m\angle AGE$.

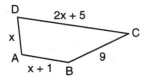

5 In the accompanying diagram, $\triangle ABC$ is similar to $\triangle A'B'C'$, $AB = 24$, $BC = 30$, and $CA = 40$. If the shortest side of $\triangle A'B'C'$ is 6, find the length of the longest side of $\triangle A'B'C'$.

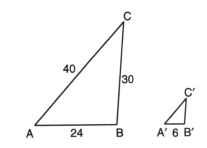

6 In the accompanying diagram, parallel lines $\overleftrightarrow{AB}$ and $\overleftrightarrow{CD}$ are intersected by transversal $\overleftrightarrow{EF}$ at G and H, respectively. If $m\angle AGH = 4x + 30$ and $m\angle GHD = 7x - 9$, what is the value of x?

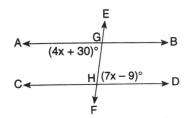

7 Solve for x: $0.4x - 3 = 1.4$

8 Solve for x: $3x + 4 = 5(x - 8)$

9 Express as a trinomial:
$$(2x - 3)(x + 7)$$

10 Write, in symbolic form, the inverse of $\sim p \rightarrow \sim q$.

11 In the accompanying diagram, parallelogram $ABCD$ has vertices $A(2,1)$, $B(8,1)$, $C(11,5)$, and $D(5,5)$. What is the area of parallelogram $ABCD$?

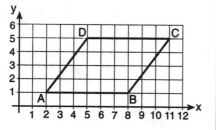

12 What is the slope of the line whose equation is $y = 3x - 5$?

13 If x varies directly as y and $x = 3$ when $y = 8$, find the value of y when $x = 9$.

14 If the legs of a right triangle have measures of 9 and 12, what is the length of the hypotenuse?

15 Express $x^2 - 5x - 24$ as the product of two binomials.

Directions (16–35): For *each* question chosen, write on the separate answer sheet the *numeral* preceding the word or expression that best completes the statement or answers the question.

16 Let p represent "It is cold" and let q represent "It is snowing." Which expression can be used to represent "It is cold and it is not snowing"?

(1) $\sim p \wedge q$ (3) $p \vee \sim q$
(2) $p \wedge \sim q$ (4) $\sim p \vee q$

17 If Manuel has five different shirts and seven different ties, how many different choices of a shirt and a tie does he have?

(1) 5   (3) 12
(2) 7   (4) 35

18 A bag has five green marbles and four blue marbles. If one marble is drawn at random, what is the probability that it is *not* green?

(1) $\frac{1}{9}$ (3) $\frac{5}{9}$
(2) $\frac{4}{9}$ (4) $\frac{5}{20}$

19 If one side of a regular octagon is represented by $2x - 1$, the perimeter of the octagon can be represented by

(1) $12x - 1$ (3) $12x - 6$
(2) $16x - 1$ (4) $16x - 8$

20 The product $\left(\dfrac{2a^3}{5b}\right)\left(\dfrac{3a^2}{7b}\right)$ is

(1) $\dfrac{5a^5}{12b^2}$ (3) $\dfrac{6a^5}{35b^2}$

(2) $\dfrac{5a^6}{12b}$ (4) $\dfrac{6a^6}{35b}$

21 What is the additive inverse of $-4a$?

(1) $\dfrac{a}{4}$ (3) $-\dfrac{4}{a}$

(2) $4a$ (4) $-\dfrac{1}{4a}$

22 Which inequality is represented in the accompanying graph?

<!-- number line from -6 to 6 with closed at -4 open at 2 -->
$$\xleftarrow{\hspace{1cm}}\substack{-6\ -4\ -2\ \ 0\ \ 2\ \ 4\ \ 6}\xrightarrow{\hspace{1cm}}$$

(1) $-4 \le x \le 2$ (3) $-4 < x \le 2$
(2) $-4 < x < 2$ (4) $-4 \le x < 2$

23 The sum of $3\sqrt{5}$ and $6\sqrt{5}$ is

(1) $18\sqrt{5}$ (3) $9\sqrt{10}$

(2) 45 (4) $9\sqrt{5}$

24 For which value of x will the fraction $\dfrac{7}{2x - 6}$ be undefined?

(1) 0   (3) 3
(2) -3 (4) 6

25 Which pair of angles x and y are supplementary?

(1) $m\angle x = 113$ (3) $m\angle x = 140$
    $m\angle y = 67$ $m\angle y = 190$

(2) $m\angle x = 76$ (4) $m\angle x = 180$
    $m\angle y = 14$ $m\angle y = 180$

26 The numerical value of the expression $\frac{5!}{3!}$ is

(1) 9   (3) 20
(2) 10   (4) 40

27 If $x = 0.6$ and $y = 5$, the value of $2(xy)^3$ is

(1) 5.4   (3) 54
(2) 21.6   (4) 216

28 What is the mean (average) of $4y + 3$ and $2y - 1$?

(1) $3y + 1$ (3) $3y + 4$
(2) $3y + 2$ (4) $y + 1$

29 Which letter has *only* horizontal line symmetry?

(1) **A**   (3) **H**
(2) **D**   (4) **F**

30 If $x + ay = b$, then y equals

(1) $\frac{b - x}{a}$ (3) $b - x - a$

(2) $\frac{b}{x + a}$ (4) $\frac{b - a}{x}$

31 If $28x^3 - 36x^2 + 4x$ is divided by $4x$, the quotient will be

(1) $7x^2 - 9x$ (3) $7x^2 - 9x + 1$
(2) $28x^3 - 36x^2$ (4) $28x^3 - 36x^2 + 1$

32 What is the length of the radius of a circle whose area is 100π?

(1) 5   (3) 20
(2) 10   (4) 25

33 Expressed as a fraction, the sum of $\frac{4y}{5}$ and $\frac{3y}{4}$ is equivalent to

(1) $\frac{31y}{20}$ (3) $\frac{7y}{20}$

(2) $\frac{7y}{9}$ (4) $\frac{31y}{9}$

34 What is the *negative* value of x that satisfies the equation $2x^2 + 5x - 3 = 0$?

(1) -1 (3) -3

(2) $-\frac{1}{2}$ (4) $-\frac{2}{3}$

35 In the accompanying diagram, $\triangle R'S'T'$ is the image of $\triangle RST$.

Which type of transformation is shown in this diagram?

(1) dilation   (3) rotation
(2) reflection   (4) translation

Part II

Answer four questions from this part. Clearly indicate the necessary steps, including appropriate formula substitutions, diagrams, graphs, charts, etc. Calculations that may be obtained by mental arithmetic or the calculator do not need to be shown.

36 The frequency table below shows the distribution of time, in minutes, in which 36 students finished the 5K Firecracker Run.

Interval (minutes)	Frequency
14–16	2
17–19	6
20–22	9
23–25	8
26–28	7
29–31	4

a How many students finished the race in less than 23 minutes?

b Based on the frequency table, which interval contains the median?

c *On your answer paper*, copy and complete the cumulative frequency table below.

Interval	Cumulative Frequency
14–16	2
14–19	
14–22	
14–25	
14–28	
14–31	

d *On graph paper*, using the cumulative frequency table completed in part *c*, construct a cumulative frequency histogram.

37 Let *p* represent: "The stove is hot."
Let *q* represent: "The water is boiling."
Let *r* represent: "The food is cooking."

a Write in symbolic form the converse of the statement: "If the stove is not hot, then the water is not boiling." [2]

b Write in sentence form: $\sim r \rightarrow \sim q$

c Write in sentence form: $p \lor \sim r$

d *On your answer paper*, construct a truth table for the statement $p \land \sim q$.

38 In the accompanying diagram, both circles have the same center *O*. The radii of the circles are 3 and 5.

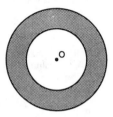

a Find, in terms of π, the area of the shaded region.

b What percent of the diagram is unshaded? [2]

c A dart is thrown and lands on the diagram. Find the probability that the dart will land on the

(1) shaded area
(2) unshaded area

39 Solve algebraically for the positive value of *x*, $x \neq 0$, and check:

$$\frac{2x + 5}{7} = \frac{1}{x}$$

40 The ages of three children in a family can be expressed as consecutive integers. The square of the age of the youngest child is 4 more than 8 times the age of the oldest child. Find the ages of the three children.

41 Solve the following system of equations algebraically or graphically and check:

$$3y = 2x - 6$$
$$x + y = 8$$

42 In $\triangle ABC$, *AB* is $\frac{3}{5}$ of the length of $\overline{BC}$, and *AC* is $\frac{4}{5}$ of the length of $\overline{BC}$. If the perimeter of $\triangle ABC$ is 24, find the lengths of $\overline{AB}$, $\overline{AC}$, and $\overline{BC}$. [*Only an algebraic solution will be accepted.*]

ANSWERS TO REGENTS EXAMINATIONS

JUNE 1995

1. $p \wedge \sim q$
2. 35
3. -1
4. 70
5. 3
6. 34
7. 30
8. $8y - 3$
9. 69
10. 1
11. $\frac{3}{4}$
12. $\frac{3x}{4}$

13. -3
14. 2
15. 10π
16. $q \rightarrow \sim p$
17. 41–60
18. 20
19. 3
20. 2
21. 1
22. 4
23. 4
24. 2

25. 1
26. 3
27. 4
28. 4
29. 3
30. 3
31. 1
32. 2
33. 3
34. 2
35. 1
36. Graph

37. $b(1)\frac{8}{16}$
    $(2)\frac{2}{16}$
    $(3)\frac{13}{16}$
    $(4)\frac{3}{16}$
38. $a\,12$
    $b\,444.5$
39. $c\,25$
40. $x = -7$
    $y = -3$
41. -7 and -6
42. 10

JANUARY 1996

1. 8
2. $\frac{5}{9}$
3. 10
4. 20
5. $a^2 - a - 12$
6. $p \vee \sim q$
7. $q \rightarrow p$
8. 3
9. 9
10. 2
11. 10
12. 8

13. $(2x - 1)(x - 2)$
14. -1
15. 28
16. -5
17. 35
18. 8
19. 2
20. 2
21. 4
22. 4
23. 3
24. 1

25. 3
26. 1
27. 1
28. 4
29. 3
30. 2
31. 1
32. 4
33. 2
34. 1
35. 3
36. Graph

37. $b\,(1)\frac{9}{12}$
    $(2)\frac{6}{12}$
    $(3)\,1$
38. 105
39. Analysis
    5
40. $(2,-3)$
41. $c\,3$
    $d\,2$
42. b No

ANSWERS TO REGENTS EXAMINATIONS

JUNE 1996

1. $\sim p \rightarrow \sim q$
2. 35
3. 5
4. 70
5. 10
6. 13
7. 11
8. 22
9. $2x^2 + 11x - 21$
10. $p \rightarrow q$
11. 24
12. 3

13. 24
14. 15
15. $(x-8)(x+3)$
16. 2
17. 4
18. 2
19. 4
20. 3
21. 2
22. 4
23. 4
24. 3

25. 1
26. 3
27. 3
28. 1
29. 2
30. 1
31. 3
32. 2
33. 1
34. 3
35. 4
36. *a* 17
   b 23–25

37. *a* $\sim q \rightarrow \sim p$
   b If the food is not cooking, then the water is not boiling.
   c The stove is hot or the food is not cooking.
38. *a* 16π
   b 36
   c (1) $\frac{16}{25}$
      (2) $\frac{9}{25}$
39. 1
40. 10, 11, 12
41. (6,2)
42. Analysis
   $AB = 6$
   $AC = 8$
   $BC = 10$

INDEX